ECOLOGY

By

Manju Yadav
Lecturer
Department of Zoology
M.M.H. College
Ghaziabad (U.P.)
(India)

DISCOVERY PUBLISHING HOUSE
NEW DELHI-110002

First Published - 2003

Reprinted - 2017

ISBN: 978-81-7141-716-2

Ecology

Published by:

DISCOVERY PUBLISHING HOUSE PVT. LTD.

4383/4B, Ansari Road, Darya Ganj
New Delhi-110 002 (India)
Phone: +91-11-23279245, 43596064-65
Fax: +91-11-23253475
E-mail: discoverypublishinghouse@gmail.com
sales@discoverypublishinggroup.com
web: www.discoverypublishinggroup.com

Printed at:
Infinity Imaging Systems
Delhi

Preface

Ecology is an inter-disciplinary science and extends to diverse fields such as Zoology, Botany, Earth science and Geography. The present title is an effort to summarize the basic concept and principles of the subject, to present the elementary factual information with which a person to be competent in the field should be familiar, and to show how these principles and facts may be applied in a practical way to the interests and welfare of man. Although the book relates especially to animals enough material is given concerning plants to bring out their essential place in the system of nature, and to emphasize the bioecological point of view.

The present title is designed primarily for undergraduate students though it may also serve an introductory text to those preparing for their master's degree. The title is not intended to be comprehensive, nor could it be at this length, but it concentrates on putting across the basic principles of the subject as briefly and lucidly as possible. It does this with the aid of carefully selected examples, some recent and other classic of the field, and with numerous illustrations.

In the preparation of this book large number of books and research papers have been consulted. So no authenticity is claimed.

Text book can not be written without the support and professional contributions of many people. This book is no exception. The author is grateful to those teachers and colleagues whose stimulating discussions have clarified certain vague points for him, but all errors, omissions, and corrections needed are solely his responsibility.

The author tried hard to be accurate and upto date in statement and realises the impossibility of completely avoiding errors therefore, the author will greatly appreciate having his attention called to any questionable statement.

The author expresses his gratitute to Mr. Vasan and staff of M/s Discovery publishers for their whole hearted co-operation in the publication of this book.

Author

Contents

1
Introduction

The word *ecology,* derived from the Greek words *oikos* meaning habitation, and logos meaning discourse or study, implies a study of the habitations of organisms.

Ecology was first described as a separate field of knowledge in 1866 by the German zoologist Ernst Haeckel, who invented the word *oekologie* for "the relation of the animal to its organic as well as its inorganic environment, particularly its friendly or hostile relations to those animals or plants with which it comes in contact".

Ecology is the study of the interactions between organisms and their environment. There are two distinct components to the 'environment': the *physical environment* (comprising such things as temperature, water availability, wind speed, soil acidity) and *biotic environment,* which comprises any influences on an organism that are exerted by other organisms, including competition, predation, parasitism and cooperation.

OBJECTIVES

Ecology is a distinct science because it is a body of knowledge not similarly organized in any other division of biology; because it uses a special set of techniques and procedures; and because it has a unique point of view. The essence of this science is a comprehensive understanding of the import of these phenomena:

1. The local and geographic distribution and abundance of organisms (habitat, niche, community, biogeography);
2. Temporal changes in the occurrence, abundance, and activities of organisms (seasonal, annual, successional, geological);
3. The interrelations between organisms in populations and communities (population ecology);
4. The structural adaptations and functional adjustments of organisms to their physical environment (physiological ecology);
5. The behaviour of organisms under natural conditions (ethology);
6. The evolutionary development of all these interrelations (evolutionary ecology);
7. The biological productivity of nature and how this may best serve mankind (ecosystem ecology);
8. The development of mathematical models to relate interaction of parameters and predict effects (systems analysis).

A study of organisms in the field may bring to light problems which will be most expediently worked out in the laboratory; but field and laboratory investigations must be integrated. The investigator must often study the morphology of dead organisms in the laboratory, and there perform experiments on living animals and plants held under carefully controlled experimental conditions. But unless such studies are perspective to the normal life of an organism, as it is lived under natural conditions, they are not ecology.

The use of exact quantitative techniques is, of course, a general characteristic of all sciences. But special difficulties arise when such techniques are applied to free-living organisms in natural conditions. For example, size of animal populations has, in the past, often been described in such vague terms as 'rare,' 'common,' or 'abundant.' These are subjective terms, based largely on an impression gained by the observer of the apparent conspicuousness of the species. As James Fisher, an English naturalist, wrote in 1939, a species has usually been indicated as 'rare' when actual numbers expressible in one's and two's could be recorded; 'common' when the observer began to lose count; and 'abundant' when he became bewildered. One of the chief problems of the ecologist is to develop methods by which to measure the absolute size of populations and the productive capacities of different habitats so that the activities of widely varying types of species may be compared. For setting up experiments and organizing and analizing studies under natural conditions, it is becoming more and more essential that the ecologist be familiar with and employ good statistical procedures. An objective of ecological research is the establishment of mathematical models or computer simulation programmes for the various systems involved. Such models give proper weight to all factors so that the effect of variation of any one or combination of factors can be predicted in advance.

As a contribution to human knowledge and understanding, ecology is in the fortunate position of being concerned with the most complicated systems of organization, apart from human societies, with which we have to deal. For this very reason it provides a constant challenge to the imagination as well as to experimental ingenuity. It is more difficult to analyse and isolate the relevant factors in a living community than in a simpler system, but the gain in significant understanding the beauty of its organization is perhaps better in proportion.

Individuals, Populations, Communities and Ecosystems

Ecologyy can be considered on a wide scale, moving from an individual molecule to the entire global eco-system. However, four identifiable subdivisions of scale are of particular interest: (i) *individuals*, (ii) *populations*, (iii) *communities* and (iv) *eco-systems*.

At each scale the subjects of interest to ecologists change. At the individual level the response of *individuals* to their environment (biotic and abiotic) is the key issue, whilst at the level of *populations* of a single species, the determinants of abundance and population fluctuations dominate. Communities are the mixture of populations of different species found in a defined area. Ecologists are interested in the processes determining their composition and structure. *Ecosystem* comprise the biotic community in conjunction with the associated complex of physical factors that characterize the physical environment. Issues of interest at this level include energy flow, food webs and the cycling of nutrients.

It should be noted that the terms 'population', 'communities' and 'ecosystems' are often ill-defined. It is often not possible to clearly delineate where one population stops and another starts, and the same problem occurs with communities and ecosystems. To some degree, these terms simply represent convenient simplification by which we can categorize the natural world.

SUBDIVISIONS OF ECOLOGY

Ecology may be studied with particular reference to animals or to plants, hence *animal ecology* and *plant ecology*. Animal ecology, however, cannot be adequately understood except against a considerable background of plant ecology. When animals and plants are given equal emphasis, the term *bioecology* is often used. Courses in plant ecology usually dismiss animals as but one of many factors in the environment. *Synecology* is the study of communities, and *autecology* the study of species. There is some confusion in these terms since Europeans commonly use 'ecology' in a narrower sense—meaning the environmental relations of organisms or of communities. The broader study of communities, including species interrelations and community structure and function as well as environmental relations (synecology), is generally termed 'biocenology' or 'biosociology' by Europeans.

In this book we shall survey the fundamentals and basic facts of ecology as they relate to animals and have application to man. We will study *community ecology*, the local distribution of animals in various habitats, the recognition and composition of community units, and succession; *eco-system dynamics*, the processes of soil formation, nutrient cycling, energy flow, and productivity; *population ecology*, the manner of population growth, structure and regulation; *evolutionary ecology*, the problems of niche segregation and speciation; and *geographic ecology*, concerned with distribution, paleoecology, and biomes. We will be interested throughout the text with how organisms respond and adjust physiologically to the physical factors of their environment, but a full study of *physiological ecology* must be left to another time and place. We will also be concerned throughout with systems ecology, that is, the possibility of translating ecological concepts into mathematical models, although we will not go deeply into the actual statistical manipulations involved. This new field is becoming very important in ecological philosophy, changing the emphasis of research from the empirical to the theoretical. This has potential value in rendering ecology a more exact science so that future events may be predicted when any of several inherent parameters vary. Finally, in several parts of the book we will deal with *human ecology*, involving the population ecology of man and man's relation to the environment, especially man's effects on the biosphere and the implication of these effects for man.

When special consideration of their ecology is given to one or another taxonomic group, we speak of *mammalian ecology, avian ecology, insect ecology, parasitology, human ecology*. When emphasis is placed on habitat, we speak of *oceanography*, the study of marine ecology; *limnology*, the studyof fresh-water ecology; *terrestrial ecology*; and so on. *Ethology* is the interpretation of animal behaviour under natural conditions; often, detailed life history studies of particular species are amassed. *Sociology* is really the ecology and ethology of mankind.

Ecological concepts, which may be grouped together as *applied ecology*, have many practical applications; notably *wildlife management, range management, forestry, conservation, insect control, epidemiology, animal husbandry*, and even *agriculture*.

This preview of ecology indicates the great breadth and unique character of the subject material, which justifies the view of ecology as one of the three basic divisions of general biological philosophy.

THE SCOPE OF ECOLOGY

Taylor (1936) has said: "Ecology is the science of all the relations of all the organisms to all their environment." Since the plant and animal inhabitants may be very abundant and diverse, and since environmental conditions are extremely variable, the possible scope of ecology becomes very great. The central task of ecology, however, is to delineate the general principles under which the natural community and its component parts operate. These may then be applied to the interpretation of the activities of the particular plants and animals present under the existing specific conditions of a given situation.

Although the fauna and flora of an area must be identified and enumerated, and although the physical forces at work in the area must be recognized, neither an account of the biota nor a description of the habitat constitutes an ecological investigation. Similarly, if a man arises at daybreak and makes a list of the birds he sees without any consideration of the relation of the occurrence of these species to other factors, he is not an ecologist. Modern ecology is concerned with the *functional interdependencies* between living things and their surroundings. Ecology is primarily a field subject. Nevertheless, a knowledge of the principles and problems of ecology should be acquired before attempting to evaluate a natural situation, where the multiplicity of ecological activities may be bewildering. Many ecological relationships can be effectively analyzed under the simplified and controlled conditions of the laboratory.

Because of a lack of understanding of ecological principles the efforts of well-intentioned conservationists and agriculturists are frequently badly misdirected. A story is told of certain sheep ranchers who became convinced that coyotes were robbing them of their young sheep. As a result, the community rose up and by every possible means slaughtered all the coyotes that could be located for miles around. Following the destruction of the coyotes, the rabbits, field mice, and other small rodents of the region increased tremendously and made serious inroads upon the grass of the pastures. When this development was realized, the sheep men executed an about-face, abruptly stopped killing the coyotes, and instituted an elaborate programme for the poisoning of the rodents. The coyotes filtered in from surrounding areas and multiplied, but finding their natural rodent food now scarce they were forced to turn to the young sheep as their only available source of food.

An understanding of ecological principles provides a background for further investigations not only into the fundamental relationships of the natural community but also into sciences dealing with particular environments such as the forest, soil, ocean, or inland waters. Many practical applications of ecology are found in agriculture, biological surveys, game management, pest control, forestry, and fishery biology. Knowledge of ecology is critically important for intelligent conservation whether in relation to soil, forest, wildlife, water supply or fishery resources.

Ecology is significant also in a wider sense for us as citizens. It gives us an insight into how the world works. In addition, man himself is a most important element in the environment. Man almost always has a modifying influence, and, without proper regulation, he often has a destructive effect. Man is himself an organism with an environment, and this fact has been particularly emphasized in the development of human *ecology*. A knowledge of the general principles of ecology thus provides a background for the understanding of human relations just as a study of general zoology is necessary as a groundwork for medicine. Like other animals man is influenced by the physical features of his environment, he is absolutely dependent upon other species, and he must adjust to other individuals of his own species. At the moment man is suffering from lack of these adjustments.

APPROACH TO THE STUDY OF ECOLOGY

The study of ecology is best begun through the analytical approach. This involves the delineation of the individual influence of the environment and the recognition separately of the various activities of the organisms present as steps towards building an understanding of the entire dynamic interaction between the complete environment and its inhabitants.

The fundamental relationships are most readily grasped by analyzing the simplest situations first. Contrast, if you will, the ecological dependencies of an alga living near the surface in the open ocean with those of a tree growing on land. The tissues of the alga receive their energy supply directly from the sun and they carry on their interchange of materials directly with the surrounding water, which is uniform and extremely constant in respect to the ecological factors concerned. The tree, on the other hand, is partly in the light and partly in the dark. Part of the tree is surrounded by the atmosphere with its widely fluctuating temperature and humidity; part is in the soil, where it is subject to a very different temperature and is alternately flooded with air and with water. The part of the tree that is above ground must deal with one set of organisms, and the part of the tree below ground is concerned with an almost entirely different set.

Another reason for adopting the analytical approach is that this procedure is more likely to reveal limiting factors. All animals and plants tend to grow, to reproduce, and to disperse until checked by some influence of the environment. The factor that first stops the growth or spread of the organism is called the *limiting factor*. It is not always easy to single out the limiting factor, and sometimes two or more factors combine to provide the limiting influence. Nevertheless, it is extemely desirable whenever possible to determine what agent or agents control the natural tendency of the plants and animals present to increase in size, numbers, and range. In the investigation of any natural area correlations will be found between features of the environment and the activities of organisms present. Analysis of the action of individual influences at work in the habitat is necessary in determining which of the correlated factors are actually causal factors. Suppose, for example, that we discovered a correlation between the occurrence of the factor *A* and organism *B*. Should we conclude that *A* causes *B*?

$$A \rightarrow B$$

It might very well be that no direct causal relation exists between *A* and *B* whatsoever, but that both are controlled by a third influence, *C*.

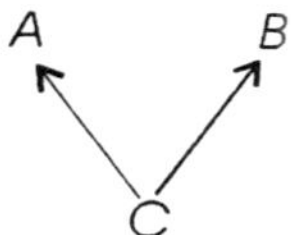

Or factor *A* may influence *C*, which in turn influences *B*, thus:

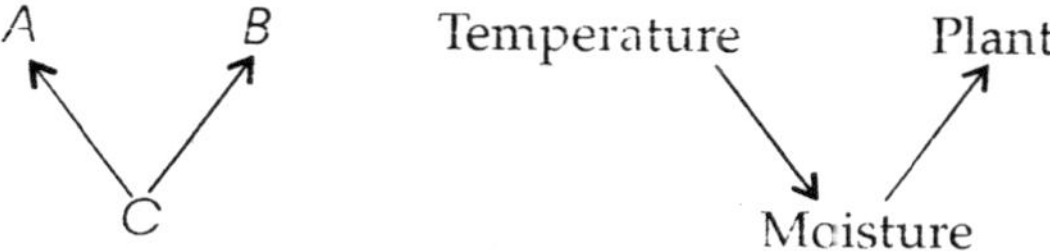

In considering the factors of the environment separately in order to distinguish and to measure the influence of each, we must remain thoroughly aware that in nature the factors are never acting alone. Animals and plants are subject to many influences at the same time, and the effect of one factor is often modified by action of other factors. The 'real life' of the organism, on which its growth, distribution, and multiplication depend, necessarily involves the simultaneous and continuing impact of all existing factors and also influences that occurred at earlier stages in the organism's experience.

An important difference exists in the extent to which factors can be modified by living organisms. Some features of the environment are largely unaffected by the activities of the organisms present; these are *unmodifiable* or *conservative factors*. The salinity of the ocean is an example of a conservative factor. The volume of the ocean is so great that, although animals and plants living in it are continually adding or withdrawing salts, the amounts have an immeasurably small effect upon the total salt content of the water. The *modifiable* or *non-conservative factors* of the environment are susceptible to change caused by the inhabitants of the area. The oxygen in a small pond, for example, may be so depleted by the respiration of a large population of fish that an unfavourable or even a lethal condition for the fish is produced; or the concentration of oxygen in the pond may be increased by the photosynthesis of algae—a modification that will benefit the fish. Heather (*Calluna*) tends to increase greatly the acidity of the soil in which it is growing, and this condition favours the further development of this plant, but it is unfavourable to most other plants. Through modification of its own environment, heather often comes to dominate the vegetation in large areas, as may be seen in Jutland.

No sharp division exists between modifiable and unmodifiable factors. All gradations exist, and a given factor may be modifiable in some situations or for some organisms and quite unmodifiable for others. In the succeeding chapters the more general, often unmodifiable, factors will be scrutinized first, and will be followed by a discussion of the more commonly modifiable factors. This will lay the foundation for a consideration in the later chapters of the composition and functioning of the community as a whole.

ECOLOGY AND DIVERSITY

If theoretical ecology is winning its spurs, one of the toughest rides will be to determine how much biodiversity matters. During the latter half of the 1980s, the reduction of the earth's biological diversity emerged as a critical issue and was perceived as a matter of public policy. A major concern was that the loss of plant and animal resources would impair future development of important products and processes in agriculture, medicine, and industry. For example, *Zea diploperennis*, an ancient wild relative of corn, could be worth billions of dollars to corn growers around the world because of its resistance to seven major diseases that plague domesticated corn. Two species of wild green tomatoes discovered in an isolated area of the Peruvian highlands in the early 1960s have contributed genes for a marked increase in fruit pigmentation and soluble-solids content currently worth $5 million per year to the tomato-processing industry. Loss of tropical forests could mean loss of billions of dollars in potential plant-derived pharmaceutical products. About 25 per cent of the prescription drugs in the United States are derived from plants, and as long ago as 1980 their total market value was $8 billion per year. On a smaller scale, individual species often thought worthless can actually be very valuable for research purposes. Armadillos, for example, are the only known species, other than humans, that can be used in research on leprosy. Desert pupfishes, found in the U.S. Southwest, tolerate salinity twice that of seawater and are valuable models for research on human kidney diseases. The technology does not exist to recreate eco-systems or even individual species. Once a species or a system is gone, it is lost forever.

More than this, humans use not just individual species but whole eco-systems too. Forests soak up carbon dioxide, preserve soil fertility, and retain water, preventing floods. The loss of biodiversity could disrupt an eco-system's ability to carry out these functions. Having convinced governments of the value of biodiversity, we now have to determine just how far eco-systems can be altered before they cease to function in an acceptable way. There are two contradictory theories about this. The 'rivet hypothesis' of Stanford ecologists Paul and Anne Ehrlich likens species to rivets on an airplane. The loss to each rivet weakens the plane a small amount until it eventually crashes. The 'passenger' hypothesis of Australia's Brian Walker asserts that species are like people on the plane, not rivets. Most are superfluous to requirements, and only a few key species (the pilot and crew) keep the plane in the air. Only recently have data become available to indicate that increased diversity increases eco-system performance. Naeem *et al.* (1994) built multitrophic-level communities of 9, 15 and 31 species, with the species-rich communities representing the more diverse communities. The reduction in diversity was cut across each of four trophic levels so that all communities had similar numbers of trophic levels. The experiment ran for six months and all plants grew from seedlings to flowering adults. A variety of eco-system attributes such as retention rates for nitrogen, phosphorus, and other elements were measured, along with plant productivity. Plant productivity increased two to threefold with biodiversity—perhaps because increased diversity increased the number of plant-leaf canopy levels and hence light interception. However, this experiment was done in an indoor laboratory setting. It still remains a challenge for ecologists to show that diversity increases productivity in the field. The temperate forests in the Northern Hemisphere show great differences in species richness (East Asian forests include 876 tree and shrub species, North American forests 158, and European forests 106), but they are virtually identical in productivity. Species diversity may increase productivity up to a certain point, but beyond that not much happens because a lot of the essential 'niches' have been filled.

Finally, good arguments can be made against ecological mismanagement and loss of biotic diversity on moral and ethical grounds. We simply have no right to destroy species and the environment around us. Philosophers such as Tom Regan argue that animals are to be treated with respect because they have a life of their own and therefore have value apart from anyone else's interests (Gunn 1990). Other philosophers such as Christopher Stone, a law professor at the University of Southern California, have argued that entities such as trees, or even natural features such as lakes, could be given legal standing just as corporations, by a legal fiction, are treated as persons for certain purposes. Corporations can sue and be sued; the 'interests' of corporations are represented legally by actual persons.

Can animals (or plants) 'count' in their own right? Consider a tragic accident in 1974 in which a schooner sank off the eastern coast of the United States (Johnson 1990). The captain, his wife, their 80-pound Labrador retriever, and an injured crewman occupied a lifeboat to which two youths, aged 19 and 20, as well as a 47-year-old Navy veteran, tied themselves with ropes and floated in the freezing waters. The captain refused repeated requests by the swimmers that he throw the dog overboard to make room for (some of) those in the water. He later explained that he could not bear to do it. After nine hours, the youths perished from exposure. All the occupants of the lifeboat, dog included, were subsequently rescued. After an initial investigation, the Coast Guard recommended that no criminal action be brought. However, in May 1975, the captain was indicted in a federal court for manslaughter for refusing to eject his dog to make room for some of the swimmers who died. Can a dog count morally, count directly, for its own sake, rather than because of some human's interest in it? Can wildlife count for more than humans in some cases?

A more utilitarian argument is that about 95 million U.S. citizens participate every year in non-consumptive recreational uses of wildlife (observation, photography, and so on). In addition, 54 million people fish and 19 million hunt. These figures may have more impact when it is considered that in 1991 $24 billion was spent on lodging, transportation, licences, stamps, tags, permits, and equipment for fishing, $18 billion for non-consumptive uses, and $12 billion for hunting. Recreational hunters in North America pursue some 90 species (Prescott-Allen and Prescott-Allen 1986). In other countries, money from the observation of wildlife can be extremely important to the economy. In 1985, Kenya netted about $300 million from almost 300,000 visitors, making wildlife tourism the country's biggest earner of foreign exchange (Achiron and Wilkinson 1986). The techniques for maintaining existing ecological habitats and plant animal species Prominent among these are the preservation of natural areas. Perhaps one of the soundest arguments for conservation of biodiversity is that, at the least, it keeps our options open for the future.

Perception of the Environment

The way a person perceives the environment reflects one's previous experience, education, lifestyle, and interests. For example, a special interest in its physical aspects may lead an earth scientist to the study of geology, meteorology, and hydrology. To completely understand these areas, however, scientists must consider how biological events influence their area of primary concern. A chemist might develop an initial interest in chemicals in the environment. This pursuit could lead to a discovery of the fate of pollutants discharged into the

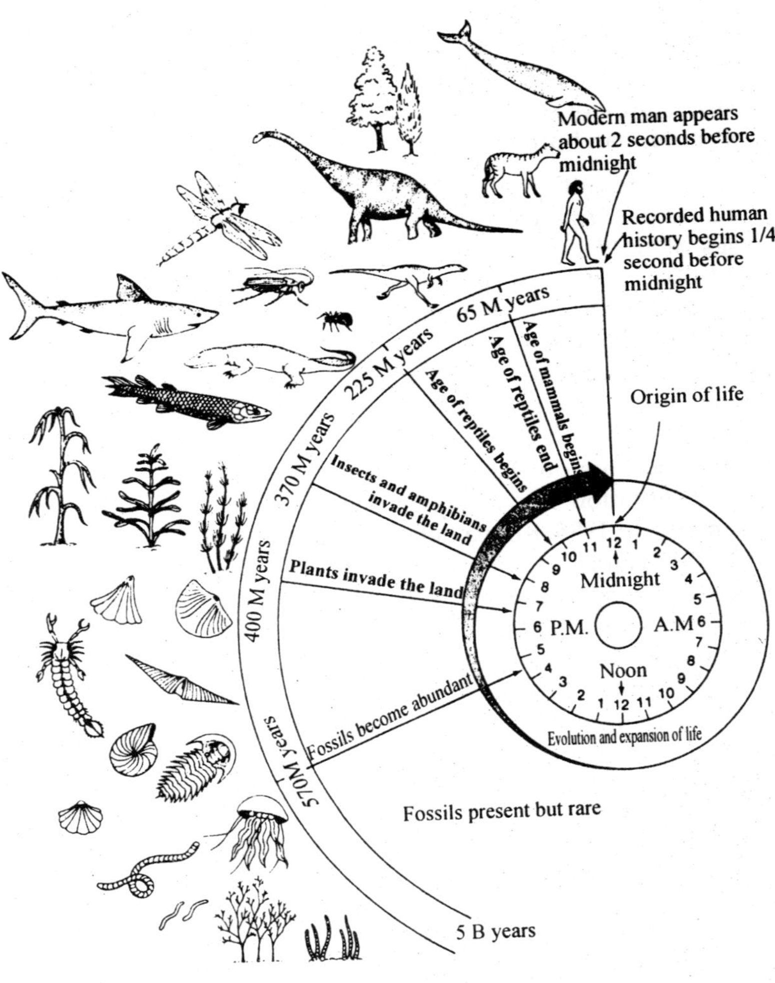

Fig. 1.1 : The history of life condensed to a 24-hour scale.

environment. An ecologist might investigate what happens when organisms consume the pollutants and how, or if, the concentration increases as one species eats another.

Worldwide different cultures look at the environment differently. Some remove 'valuable' components for themselves. In poor countries people take all they can for survival. People in the United States often have difficulty in relating to methods by which other cultures perceive the environment.

Ecology and Environmental Science

Many people view ecology and environmental science as synonymous; thus environmentalists are frequently considered ecologists. Actually, anyone interested in, and concerned about, the environment is an environmentalist. Ecology is one of the disciplines constituting the core of environmental science. Within ecology there are many subdisciplines; thus, an aquatic ecologist has different advanced training than a radiation ecologist.

The word *ecology* is derived from the Greek word *oikos*, meaning a place to live. Most ecologists therefore define ecology as the study of the relationship of an organism or group of organisms to their environment. The ecologist Eugene Odum points out that ecologists are concerned with the biology of groups of organisms and with functional processes. He defines ecology as 'the study of the structure and the function of nature.'

While other disciplines also study the interactions of life and the environment, ecologists are concerned about what limits life, how living things use resources such as minerals and energy, and how living things interact. Ecology is the study of these processes; environmental science is the application of this knowledge to managing the environment.

Divisions within the Environment

The focus of our study is on places where life is found—where all the elements needed to support life are present. This relatively thin layer at the earth's surface is called the biosphere.

By nature the scope of the *biosphere* is restricted. Few living things are found above 3100 meters (10,000 feet) or in the depths of the earth. Although life exists in all parts of the oceans, it is dependent entirely on the food producers within the areas of sunlight penetration.

Within the biosphere are functional units or *ecosystem,* consisting of all living organisms plus non-living components. Deserts, forests, and lakes are examples of these units. For the sake of convenience, we study these areas individually or in small parts. They are not isolated from the surrounding biosphere. Thus, interactions occur at the interface—between pond and shore, forest and field, or ocean and land.

Environmental Modification

Within recent history, people have modified the biosphere to such an extent that major changes have begun. Massive cities, variations in the atmosphere's mineral content, new nutrient balances in water systems, and major alterations in the earth's vegetation patterns have all combined to alter the environment. People have begun to realize that these transitions are causing life itself to change. As the structure of natural communities alters, we can expect these long-term modifications to result in different species of life.

We can already create artificial surroundings that mimic natural environments. These surroundings enable people to live in formerly adverse locations, such as the ocean and outer space. By controlling pressure and providing oxygen, we can descend to the ocean depths in submarines and bathospheres. Someday people may even live in enclosed cities on the ocean floor. In space travel, full pressure suits create an artificial balance of environmental conditions around astronauts so they can carry their life-supporting environment with them.

In recent years there has been an increasing emphasis on monitoring environmental modification on a global level and on recommending strategies for improving the well-being of people and sustaining the eco-systems on which they depend. It is now generally recognized that ecologically sound environmental management and economic development, particularly in developing countries, are closely linked. In 1980, the International Union for the Conservation of Nature and Natural Resources (IUCN), the U.N. Environmental Programme (UNEP) and the World Wildlife Fund (WWF) collaborated to develop a World Conservation Strategy. The Strategy outlines goals and actions to achieve three major objectives:

1. To maintain essential ecological processes and life support systems;
2. To preserve genetic diversity;
3. To ensure that the utilization of living resources, and the eco-systems in which they are found, is sustainable

These objectives are based on the recognition that successful economic development and an improved quality of life take into account the capabilities and limitations of the natural environment, as well as the needs of future generations.

More recently these ideas have evolved into the idea of *sustainable development*. This is development that meets the needs and aspirations of the present without conflicting with natural resource conservation, environmental protection, and sustainable eco-systems. To understand sustainable development, we must examine all the processes that are occurring in our environment.

An Approach to Environmental Science

In this book, we attempt to follow a path of scientific inquiry into how natural systems function, explore the past influences, present problems, and future prospects for people and the environment, and develop rational thought processes for evaluating environmental issues. Our goal is to propose a systematic approach to managing the environment with a view to maintaining our eco-systems.

Environmental science encompasses many disciplines, each with its own principles and concepts. These concepts unify scientific inquiry into a holistic understanding of the environment. The major concepts we use as guidelines are homeostasis, energy, limits, symbiosis, systems and models.

Homeostasis

Because environmental science involves understanding and analysis of many complex interactions, we find a variety of approaches used in its study. *Cybernetics,* the science of controls, is one approach. How is the environment controlled? How are all the interactions within and among the physical and biological systems regulated? The control of such systems requires information in the form of feedback. *Feedback* is the return of output, or part of the output, to a system as input. That is, it gives the system information that will cause it to change so as to maintain a particular state. An example of a cybernetic control by means of feedback is the thermostat in your home. When you set a thermostat to a desired temperature, the information is fed to a temperature-sensing device. If the room temperature is lower than the temperature you indicate, the furnace starts up; if it is higher, the furnace cuts off. Room temperature is continually monitored as signals are sent to the furnace to regulate its operation.

Two forms of feedback occur in this example. When the temperature falls, more heat is supplied. This negative or reverse relationship between input of information and response is called *negative feedback*. When a change in the system in one direction is converted into a command to change the system in the same direction, *positive feedback* occurs. Generally, negative feedback keeps a system in equilibrium and positive feedback disrupts equilibrium and causes the system to become unstable.

Living systems, including groups of organisms living together in the same environment as well as individual organisms, have cybernetic or self-regulating feedback mechanisms that maintain their equilibrium. This tendency for biological systems to resist change and remain in a state of equilibrium is called *homeostasis*. Physiologists study many forms of homeostatic control. Regulation of body temperature and blood chemical content are examples of homeostatic control in individuals. Equilibrium between organisms and the environment can also be maintained by feedback mechanisms. Such processes are important in the balance of nature. Only recently have people begun to analyze the interplay of energy and materials in sustaining life.

The homeostatic mechanisms in living organisms generally operate in a common manner. The initial information input, or *stimulus*, activates a sensing device. The information is transmitted over a sensory pathway to response selector that transmits a signal to effectors which, in turn, initiate a *response*. Thus, an animal that sees danger runs away or prepares to fight.

Response involving whole groups of organisms is much slower. Without predatory organisms, a prey population increases; and predator populations increase in response to a large number of prey animals.

Living systems interact in many ways. We can study the interactions of two individuals or two populations to determine which mechanisms influence their relationship. Likewise, we can examine different levels of biological organization until we encompass the whole universe. Each new bit of information supplies another piece in our puzzle of understanding life and the complicated interactions that take place in the natural system.

Throughout this book, consider how environmental actions influence the established equilibrium. What forms of feedback operate to maintain equilibrium? What are the consequences of disruption?

Energy

Physicists define *energy* as the ability to do work. Work can mean the movement of a car, growth and reproduction of a plant, or explosion of a bomb. Virtually all human actions require energy. Wasteful use of energy, such as needlessly discarding food or excessive use of cars, converts energy into less usable forms.

We introduce energy as a concept because it has an impact on almost all our actions. Since it actually imposes structure on living systems, energy supplies the driving force and determines the limitations of life on earth. For each topic in this book, consider the type of energy involved in the concept and the limits energy imposes.

Limits

The concept of a *limit* is more controversial than the other concepts we have described. Let us consider a small pond on which water lilies grow. In this pond, the lilies double in number each day. The size of the pond is such that this rate of growth will cause the pond to be completely covered in thirty days. When is the last day the pond can be cleared of lilies before it will be completely covered? On the twenty-ninth day the pond will be half covered, and the next day it will be completely covered.

It is because of this concept that some scientists speculate on the capacity of the earth to maintain an ever-increasing human population with its food, energy, and mineral consumption and waste generation. While the increase in human population is not as dramatic as the growth rate of the water lilies, a small increase every day will ultimately reach the capacity of the environment to support consumption or waste assimilation, whatever that capacity is. The area of speculation is not so much whether such a capacity exists, but when it will be reached. How much time do we have for making decisions?

Symbiosis

Symbiosis refers to dissimilar organisms living together. It is an important interaction maintained by homeostatic mechanisms. Various kinds of environmental disruptions result in a breakdown of many symbiotic relationship. We will discuss symbiosis and its ecological importance in Chapter 2.

Systems

The concept of *system* is used to describe many things: a transportation system, the health care system, a school system, a highway system, an eco-system. *Closed systems* have virtually no input from the outside. A space capsule, a submarine, or the earth as a whole are nearly closed systems. In contrast, forests, estuaries, air and river basins, and urban communities are more *open systems*. In general, a number of related interactions are necessary to the functioning of a system. A change in one part has repercussions in the rest. If a mechanic fails to repair a bus properly, it might not run The bus system could be thrown off schedule as a result. If one link in a highway system is missing, there might be traffic jams on the interconnecting roads.

A system's boundaries are selected for convenience. A total transportation system would include not only the bus and highway system, but the materials that supply power as well. Trucks and miscellaneous delivery vehicles might also be essential parts. Gasoline stations for private autos and the electric generating stations that supply a rail or trolley system could be added. However, a line has to be drawn somewhere; so inter-city transportation might be excluded from a consideration of the transportation system of a local community. Even so, the intercity connections influence the inter-city system and have to be considered.

In a living system, interactions are numerous and complex. Species interact not only with each other, but also with the environment. The movement of energy between the sun and groups of organisms and the role of minerals in the inter-relationships between the living and non-living environments are examples of such interactions.

2

Soil

Communities cannot exist without a habitat, nor is a habitat likely to remain long without a community developing in it. The functional inter-relations between community and habitat are many and complex, constituting an *ecosystem*. Most important are soil formation, nutrient cycling, and energy flow. Human interference in these processes often causes pollution, and exploitation often brings exhaustion of natural resources.

We have already considered many of the reactions of plants upon the habitat, such as reduction of light and wind intensities, mitigation of temperature extremes, interception of rainfall, and increase in relative humidity. Plants also exert important effects on the formation, structure, and characteristics of the soil or substratum produced by accumulation of dead plant remains: they further the weathering of rock through acid excretion and the mechanical action of roots; they offer obstruction to wind- and water-borne materials; they help stabilize moving sand and talus slopes and help prevent erosion generally; they variously increase or decrease the water content of soil; they foster decomposition of raw humus into usable nutrients; and so forth. Water plants form marl. It is by these reactions that plants exert dominance in terrestrial communities and establish the physical conditions of the habitat, which must be acceptable to all minor plants and animals that dwell there. Succession of plant stages eventually brings the interactions between habitat and community into equilibrium upon the establishment of the climax. In this chapter we will be primarily concerned with soils as dynamic components of eco-systems, including the cycling of nutrients between the soil and the biota.

The Components of Soil

The vast majority of soils have four recognisable components; inorganic matter, organic matter, soil water and soil atmosphere. These are generally intimately mixed and may be difficult to separate. In poorly drained areas, such as bogs and swamps, the decay of plant material is delayed by the anaerobic (oxygen deficient) conditions that prevail. Under these conditions 80-95% of the soil solids may be organic. Organic soils such as these, while they may be of considerable economic and environmental importance, are not the norm. Most soils are termed inorganic (or mineral); these contain 1-10% organic matter (Fig. 2.1).

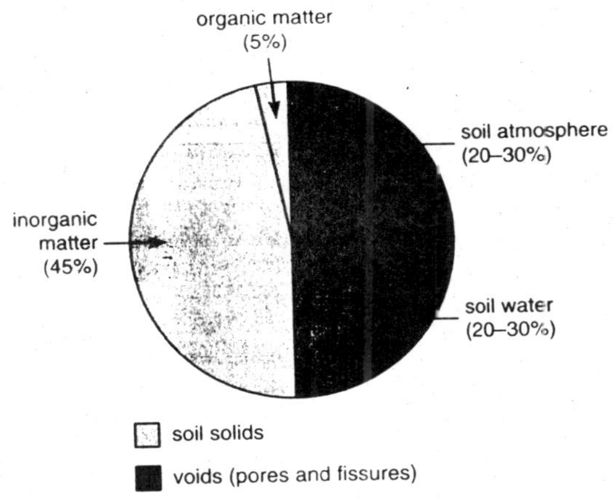

Fig. 2.1 The components of soil. This diagram relates to an inorganic soil; the figures shown in parentheses are typical of those obtained from a slit loam topscil. The proportion of voids occupied by water is extremely variable.

The surface of virtually all rocks is covered with a layer of unconsolidated material called the *regolith*. It is from this that the inorganic fraction of a soil is generally derived. The regolith is either formed by the chemical and physical weathering of the bedrock or has been brought in from outside by the action of wind or moving water or ice.

The particle size distribution of the inorganic fraction of a soil is of considerable importance in determining its ability to support higher plant life and productive agriculture. For example, soils with a high proportion of fine particles (called *fine-textured* soils) have high moisture and nutrient holding capacities but have a tendency to be poorly drained, and are slow to warm in the spring. Conversely, *coarse-textured* soils have a high proportion of largersized particles. Such soils are generally free draining and quick to warm, while often suffering from low nutrient and moisture holding capacities. Soils that contain a fairly even mixture of particles of various sizes are said to be *medium-textured*. These tend to share the advantageous properties of both fine and coarse-textured soils and are therefore ideal for the growth of many plants.

Any one soil will contain inorganic particles with a variety of sizes. These particles may be grouped according to the size range into which they fall. Table 2.1 shows the various classification schemes of particle size in current use. While

Table 2.1 Soil particle size classification schemes.

Size class	*Size range/mm*		
	a	*b*	*c*
Clay	< 0.002	< 0.002	< 0.002
Silt	0.002-0.02	0.002-0.05	0.002-0.06
Sand	0.02-2	0.05-2	0.06-2
Gravel	> 2	> 2	> 2

Size ranges:

a= The international or Atterberg system (this also divides sand into fine sand coarse sand).

b= USDA system (this divides sand into fine sand, medium sand and coarse sand).

c= System adopted by the Soil Survey of England and Wales, British Standards, and the Massachusetts Institute of Technology (this divides sand into fine sand, medium sand and coarse sand).

varying in detail, all of these share the recognition of four gross size ranges, namely, in order of decreasing particle size: gravel, sand silt and clay. The particle size distribution of a particular soil is indicated by allocating it to the appropriate textural class. This is done according to the proportion of sand, slit and clay sized particles that it contains (collectively known as the fine earth), the grit having been removed and ignored (Fig. 2.2). Hence, a soil that contains 20% clay, 40% sand and 40% silt in its fine earth fraction will be a clay loam, a medium-textured soil.

Soil organic matter is a mixture of partially decayed plant and animal remains and material synthesised by micro-

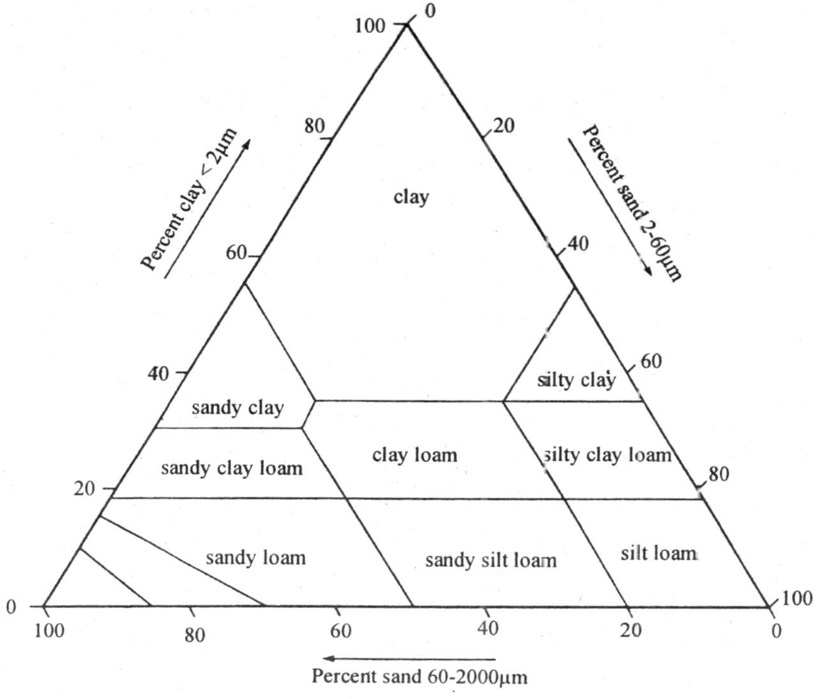

Fig. 2.2 : Soil textural classes (from White, 1979, after SSEW, 1974).

organisms. In many soils, it is a major source of the essential elements sulfur and phosphorus. For plants that are not capable of nitrogen fixation from the atmosphere soil organic matter is essentially the only source of this element.

Humus is the organic matter that is more resistant to microbial decay. The structure of humus is complex and not fully understood. However, certain features have been established. Humus has a very important role in the determination of the fertility of a given inorganic soil. It has a very high nutrient and water holding capacity compared with the inorganic fraction. For example, humus may be responsible for one-third of the moisture retaining ability of a medium-textured soil while representing only 5% of its mass. In addition, it acts as an adhesive, causing the particles of the inorganic fraction of the soil to be bound together into relatively robust pellets called *peds*. The formation of peds is generally accompanied by the appearance of voids. These provide pathways for effective drainage and the ingress of air from above. Thus relatively fine-textured soils can be free draining and well aerated if their humus content is sufficiently high.

Water that is lost from a soil sample on maintaining it at 105°C for at least 24 hours is defined as *soil water*. Water that is bound within the structure of the soil minerals (i.e. water of crystallisation) is not lost under these conditions and is therefore *not* considered to be soil water. Distinction is also made between soil water and *ground water*, the latter being the water that is below the water table. Soil water together with the dissolved salts it contains is called the *soil solution*.

Plants are not necessarily capable of utilising all of the soil water. The water that is lost by drainage from a saturated soil is not available, nor is the water that is held in very small voids or tightly bound to the surface of soil solids. Once all of the available water has been removed the plants lose turgor and wilt. The water content of the soil at this stage is termed the *wilting point* (WP). The water content of a soil after drainage is its *field capacity* (FC). The difference between FC and WP is called the *available water capacity* (AWC). Soils with high values of AWC are less prone to drought than those with low AWC values. As can be seen from Fig. 2.3, medium to fine-textured soils are much less prone to drought than are coarse-textured soils.

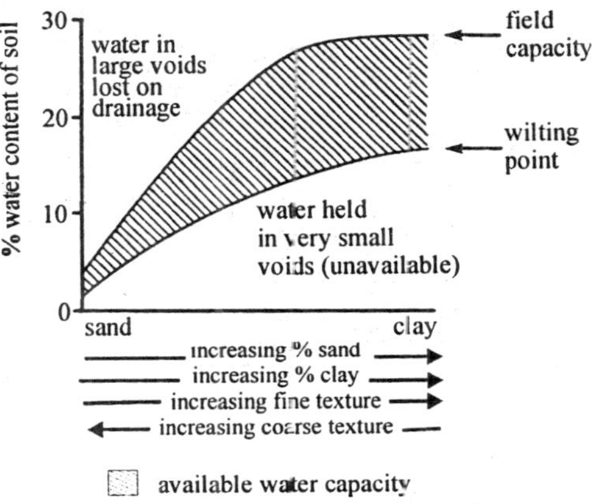

Fig. 2.3 : The generally observed relationship between soil texture and soil moisture characteristics. Note that the available water capacity of a given mineral soil will be markedly increased by a modest rise in humus content.

The major constituents of the atmosphere are nitrogen, N_2 (78%), oxygen, O_2 (21%), argon, Ar (1%) and carbon dioxide, CO_2 (0.03%) (all percentages are on a volume basis and refer to dry air). The air entrained within the soil is referred to as the *soil atmosphere*; it is in slow exchange with the bulk atmosphere above.

The respiratory activity of organisms within the soil, coupled with the low rate of exchange, ensures that the soil atmosphere has a different composition from that of the bulk atmosphere. In particular, in comparison with the bulk atmosphere, its oxygen levels are depleted and carbon dioxide levels are enhanced (Table 2.2). Under waterlogged conditions this alteration can become severe; oxygen concentration can drop to zero, whilst those of carbon dioxide may increase by a factor of greater than 20. Such oxygen deficient conditions (referred to as anaerobic) provoke considerable changes in the soil environment. Micro-organisms capable of utilising respiratory oxidants other than oxygen (micro-organisms called *anaerobes*) multiply in numbers. Their activity, when coupled with the changed redox potential of the soil environment, promotes increasing concentrations of reduced forms of nitrogen (N_2 is produced, as is N_2O), iron (e.g. $Fe^{2+}_{(aq)}$), manganese (e.g.

$Mn^{2+}_{(aq)}$) and sulfur (e.g. H_2S). In addition, volatile organic species that either would not be produced under aerobic conditions, or would be oxidised in the presence of oxygen, accumulate. These include methane, ethene and various volatile fatty acids. The roots of many plants are damaged in waterlogged soils as a consequence of both the low oxygen levels and the presence of highly toxic species such as H_2S.

Table 2.2 The composition of the soil atmosphere in well-aerated soils (Russell, 1973)

Soil and land use	*Usual composition* O_2	CO_2
	(% by volume)	
Arable land		
fallow	20.7	0.1
unmanured	20.4	0.2
manured	20.3	0.4
Sandy soil, manured and cropped with		
potatoes	20.3	0.6
serradella	20.7	0.2
Pasture land	18-20	0.5-1.5

Soil Structure

The arrangement of voids, individual soil particles and aggregates of these particles is referred to as the *soil structure.* This has both macroscopic and microscopic manifestations. The former of these can be seen with the naked eye or with the aid of a hand lens, while the latter only become evident with the application of optical or electron microscopy and will not be discussed here.

The solid material in most soils is aggregated at least to some extent. The greater the degree and stability of the aggregation the more developed the structure is said to be. Soils that do not contain aggregated material have structures that are described as either massive if the material is consolidated or *single-grain* if it is not. Soils with highly developed structures, consisting of small (1-5 mm), stable aggregates and interconnecting voids will be less prone to waterlogging than their massive equivalents.

Aggregates of soil solids may be of several different types. Of these, *peds* are the most significant in natural soils. These are relatively robust aggregates, capable of withstanding several cycles of wetting and drying. They are clearly visible and easily identifiable, being separated from one another by voids or lines of weakness. Peds may be categorised into one of four groups according to their shapes.

Other macroscopic structural units recognised are:

- *fragments,* made by breaking peds *across* natural lines of weakness;
- *clods,* which are lumps of soil formed by tillage;
- *concretions* and *nodules,* both of which are groups of soil particles permanently cemented together by insoluble material;
- *pans,* which are horizontal regions in which the soil particles have become aggregated over considerable areas. They may be formed by mechanical action or by the accumulation of cementing materials washed from higher in the soil;
- *pores,* which are cylindrical voids;
- *fissures,* which are voids that are approximately planar.

The soil profile :

The side of a trench in any well-vegetated soil reveals a succession of layers in the soil. At the surface is a litter of dead or rotting plant parts, with one or more distinctly different layers underneath separating the surface litter from the subsoil a few feet down. These layers have been formed by weathering processes working down from the surface. It is convenient to describe them, and they have come to be called *'horizons'*. Soil horizons form as rotting plant parts mix with the upper layers of the mineral soil and drainage water percolates down through the litter slowly washing the lower horizons. The thickness of earth affected by these processes constitutes the soil. The subsoil underneath is the earth from which the soil was made and is therefore called the *parent material.* This layered appearance is called the soil profile which is the set of *soil horizons* between the undifferentiated parent material and the surface litter. Soil scientists recognize three kinds of soil horizon above the parent material: the A, B and C horizons. It is usually easier to separate a soil into these three horizons just by looking at it. The principles of this simple classification are that:

- *A* horizons have lost material from leaching, although they have gained organic matter as deposits;
- *B* horizons have gained material from leaching and by synthesis *in situ*, particularly of clay minerals;
- *C* horizons are parent material that has been weathered, usually being oxidized or, in dry climates, having deposits of evaporites.

At the top of the soil profile Fig. 2.4, underneath the litter of leaves, the mineral soil is coloured and structured by the organic particles mixed into it by soil animals or roots and by the presence of various organic materials produced by decomposition, called *'humic acids'*. The percolating water dissolves anything soluble in the surface of the mineral soil and carries it down to deeper horizons from the *A* horizon. Below the *A* horizons are a second set of horizons, the *B* horizons, which contain material washed down from the top layers that has been caught and redeposited. *C* horizons contain weathered parent material. Under these three groups of horizons is the unaltered plant material. These are known as *R* horizons. The parent material of an *R* horizon can be any geological formation; bedrock, gravel, sand, clay etc. The thickness and complexity of a soil profile is a function of time and weather, as well as the plants that grow there. In the desert the soil profile may be just a thin surface layer. The soil profile is therefore an instant indicator of important eco-system processes.

The soil profile may be considered to be an open system Fig. 2.6. Material inputs to this system result from aqueous precipitation (rain, snow, hail etc.), natural organic matter input (e.g. litter fall), deposition of debris eroded from elsewhere, lateral inflow from adjacent soils, the input of weathered material from the parent material, and man-made additions including fertilisers and other agrochemicals. Material outputs

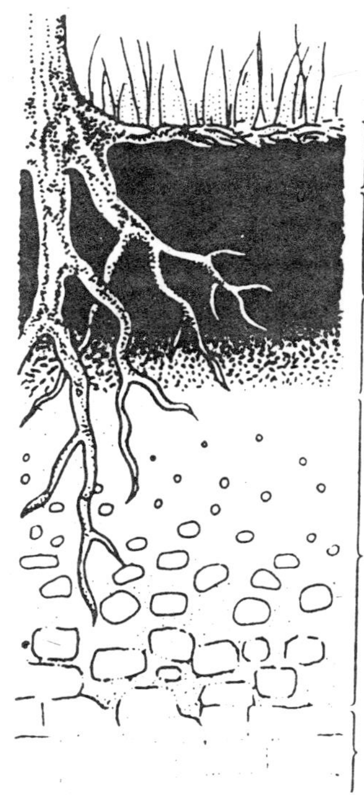

A horizons includes litter,humus and top mineral soil.A horizons are those from which material has been leached

B horizons.Mostly colored mineral soil,but includes humus layers and plant roots.B horizons are horizon of accumulation which receive materials washed out of A horizons.

C horizons Weathered parent material. The base of the C horizon is generaly the time limit to oxidation of the surface crust.

R horizons Parent material.

Fig.2.4 Idealized soil profile. From *Ecology* 2, P.Colinvaux, 1993. Reprinted with permission from John Wiley & Sons, Inc.

result from leaching, lateral outflow, evapo-transpiration, plant uptake, erosion and gaseous losses from the soil atmosphere. Within the system, soil-forming chemical, biological and physical processes occur that result in the alteration and translocation of material. These include biological activity, including humus formation mineral dissolution and secondary mineral formation, *leaching, gleying, lessivage, poazolisation, upward translocation* and *physical mixing*.

Leaching is a process by which percolating water moves matter from one place to another. In its early stages it will move material down the profile, concentrating it in the lower horizons, while at a later stage it may remove the material from

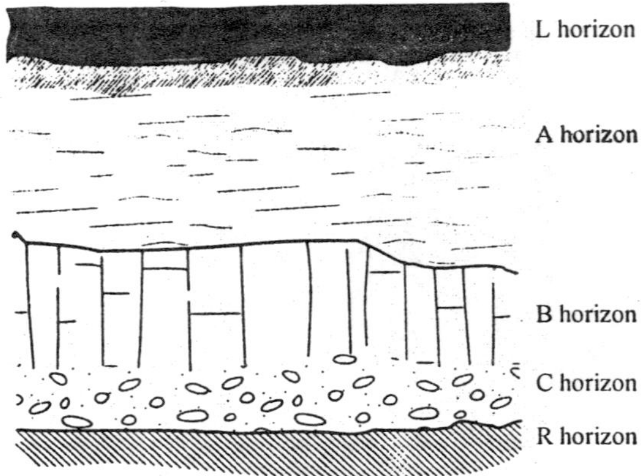

Fig. 2.5 : A typical soil profile.

the soil altogether. There are two principal factors that determine the rate at which a given material is leached, its solubility and rainfall. In general, the more soluble a material is, the greater is its susceptibility to leaching and the more mobile it appears in the soil environment. The highly soluble sodium chloride, for instance, is rarely found in soils except in the most arid areas. In contrast, the sparingly soluble calcium carbonate, present in calcareous parent materials, may remain in the profile for thousands of years, even in humid climates. The behaviour of calcium carbonate also serves to illustrate the rule that higher rainfall leads to more rapid leaching, the trend shown in Figure 2.7 being typical.

Different compounds that contain the same element may have strikingly different solubilities. It follows that the chemical form of a given element will have a considerable influence on its susceptibility to leaching (elements in different chemical forms are said to be in different chemical species). For example, leaching of calcium becomes more rapid as the percolating water becomes more acidic. This can be understood when it is realised that sparingly soluble calcium carbonate ($CaCO_3$) may be converted to the more soluble calcium hydrogen carbonate [$Ca(HCO_3)_2$] and aqueous calcium ions by the following reaction:

$$2CaCO_{3(s)} + 2H^{+}_{(aq)} \rightarrow Ca^{2+}_{(aq)}\ Ca(HCO_3)_{2(aq)}$$

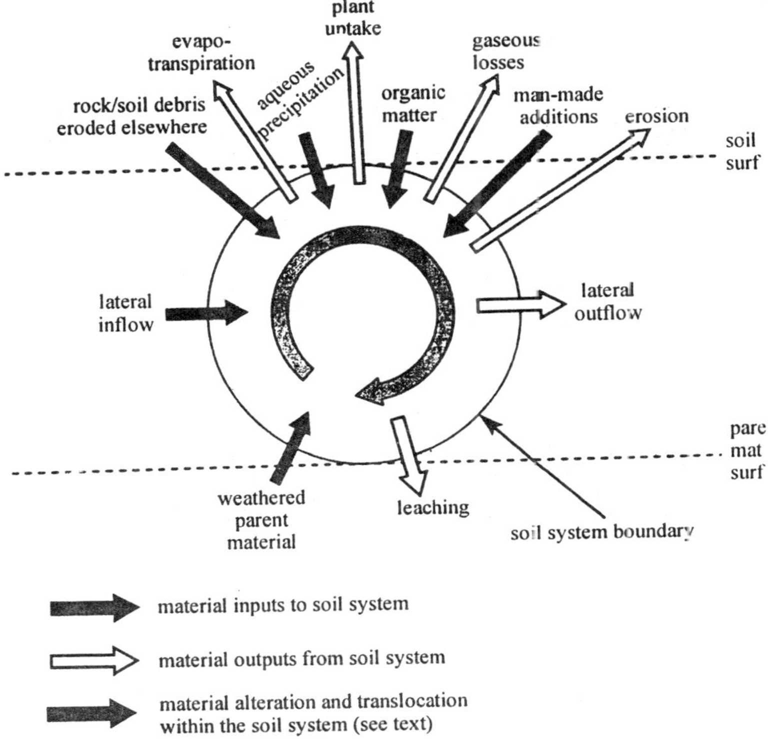

Fig. 2.6 : The systems view of soil.

Gleying is a process that involves changes in chemical specialisation that result in changes in solubility. It is characterised by the reduction and subsequent partial solubilisation and leaching of iron. It occurs in waterlogged mineral soils that nonetheless contain some organic matter. Under the anaerobic conditions that prevail in such soil any iron (III) present may be reduced to iron (II). This occurs by a combination of purely chemical and microbiologically mediated reactions, the latter being dominant. Most of the iron (II) produced is in the form of either soluble complexes or solid blue-green-grey coloured mixed hydroxides of iron (II) and iron (III). A gleyed horizon may therefore be identified by the

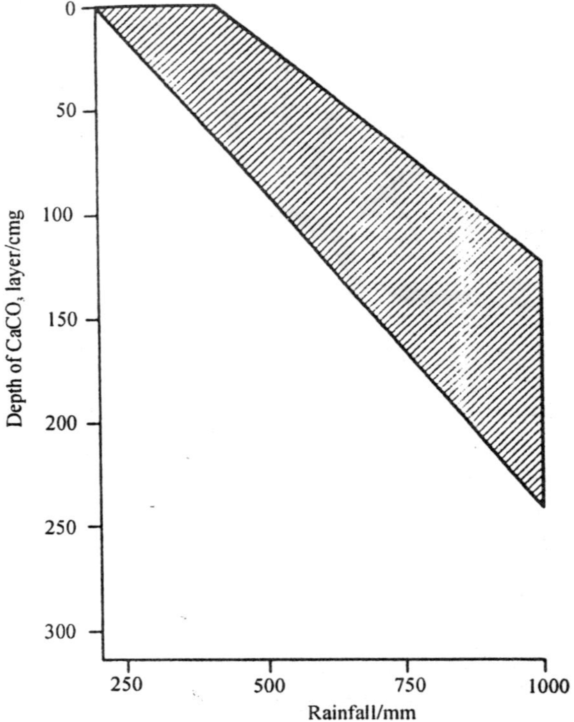

Fig.2.7: The relationship between rainfall and depth of the calcium carbonate layer in soils derived from calcareous loess (a wind-blown parent material) (after White, 1979, after Jenny, 1941)

presence of blue-green to blue-grey colours. These are frequently accompanied in the more aerobic areas, such as root channels, by the orange-brown colours of insoluble iron (III) compounds. These compounds form and precipitate when the soluble iron (II) complexes formed on gleying are transported to the more aerobic areas by leaching.

The solubilisation of iron as iron (II) complexes is not unique to gleying. Similar, while not identical, phenomena occur in the processes of lessivage and podzolisation. Let us consider these in turn.

Lessivage is the eluvation of clay and iron oxides from the *A* horizon. This is essentially a mechanical leaching process. However, it is probable that the destabilisation of soil aggregates within the *A* horizon is a necessary prerequisite of this process. It is likely that this happens by the removal of cementatious iron oxides by the formation of transient, soluble iron(II) complexes with phenolic leaf leachates.

Podzolisation may be considered to be a more extreme form of lessivage. Here, polycarboxylic acid humic compounds and polyphenolic leaf leachates form chelates with aluminium cations and iron (II) respectively. These soluble complexes are leached from the *A* horizon. As a result of the removal of coloured iron and organic matter, this horizon becomes characteristically pale in colour, and is redesignated as an *E* horizon (E horizons are eluvial horizons that are low in organic matter). Once translocated to the *B* horizon the iron complexes break down and the aluminium complexes precipitate. The resulting deposits of iron(III) hydroxides and organic matter cause a darkening of this horizon. The process in the *B* horizon is complex. However, it is primarily a consequence of microbial oxidation of the organic chelating agents and the flocculating effect of relatively high concentrations of polyvalent cations such as Ca^{2+}, Mg^{2+} and Al^{3+} found near the weathering parent material. The end product of this phenomenon is a soil called a *podzol*, the characteristics of which are the presence of mor humus, a leached *E* horizon and an accumulation of iron, aluminium and humic material in the *B* horizon.

Upward translocation is a feature of arid soils where the rate of evapo-transpiration exceeds rainfall. Under these conditions water is moved up through the profile by capillary action. Its evaporation at the surface leads to the precipitation of the dissolved materials it contained. The net result is the upward movement of soluble and sparingly soluble material. When sodium compounds dominate the material translocated (which they frequently do) the process is called salinisation.

Physical mixing of materials within the profile may be medicated by biological, chemical, physical and/or human activity, all of which serve to disrupt horizon development and maintenance. The most significant of these processes are probably those associated with tillage (ploughing, etc.), root growth, burrowing animals and frost action. The last of these is a feature of climates that facilitate the repeated freezing and thawing of water within the soil. During freezing, growing water crystals may disrupt structural features. On melting, the space occupied by any large crystals may be filled with fine-textured material. The net result is frequently one of mixing.

The Control of Soil Development

The major process involved in soil formation are under environmental control. The environmental factors that exert greatest control are parent material, climate, vegetation and relief (topography); the influence of each of these varies with time. The complex nature of the interaction of these controls and processes has led to the generation of a wide variety of soil types. Below are given two examples that illustrate these interactions.

The first example illustrates the influence of time, showing the sequence of development of a soil from its infancy to maturity. It concerns the formation of the brown earth, typical mature soils of the deciduous woodlands of cool humid areas of Europe, North America and Asia.

In the early stages of development (Fig.2.8) the soil consists of little more than a mantle of poorly humified organic matter over a shallow layer of weathered parent material. Typically, such soils support pioneer communities of plants including liverworts, mosses and lichens. During the next two to three hundred years species diversity increases and the vegetation gradually transforms until it is dominated by grasses, sedges and shrubs. The soil is now much deeper (typically 20 to 30cm) and consists of a thick litter layer over a recognisable *A* horizon that contains large amounts of well-humified organic matter with a gradual transition to weathered parent material below. The species diversity continues to increase over the next few centuries, and the vegetation becomes dominated by deciduous woods, the climax community. The numbers of burrowing animals, such as earthworms, increase, resulting in the incorporation of the organic matter into the mineral horizons, producing a very thin litter layer. *A*, *B* and *C* horizons are well developed in the mature soil, although the boundaries between them are gradual.

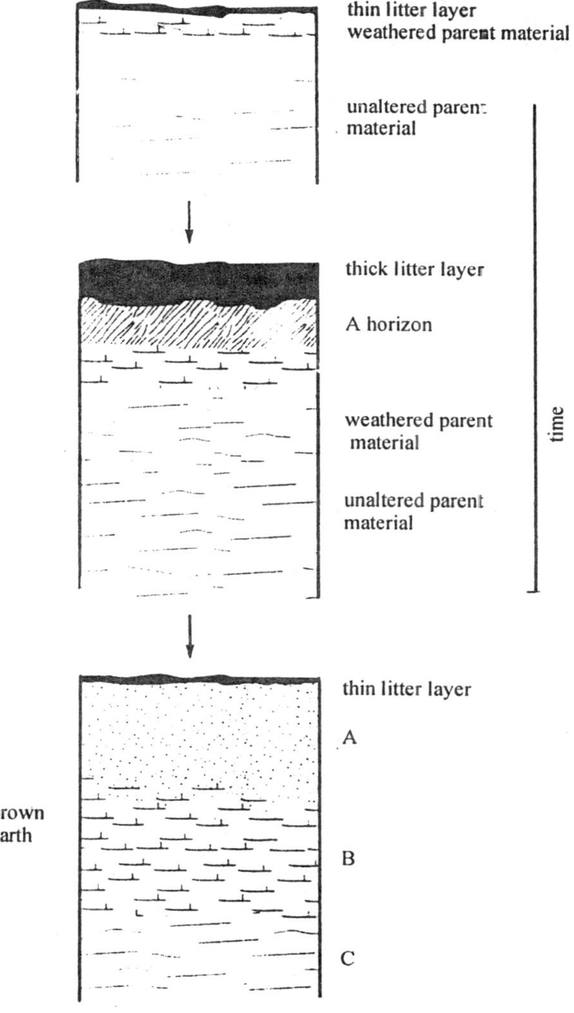

Fig. 2.8 The stages in the development of a brown earth.

The second example concerns a specific location: a valley in south-east Scotland, as described by Simpson (1983). This area has a cool temperate climate. Figure 2.9 shows a section across the valley concerned. This example is useful as it allows the roles of parent material, topography and vegetation to be illustrated, together with those of local changes in climate.

The soils of the hill-tops are typical podzols. These soils have formed under the influence of a combination of moderately high rainfall (1000 mm per year), adequate drainage, and vegetation that produces chelating polyphenolic leachates.

The impervious fine-textured glacial till that dominates the parent material in the valley bottom has created an area of poor drainage. This has facilitated the formation of gleys and peats.

Interestingly, there is an asymmetry evident across this transact. This is attributable to local differences in climate, the south-facing slope being warmer and drier than the north-facing one. There are two distinct consequences of this difference. Firstly, podzolisation extends further down the north-facing slope as there is more water here to facilitate leaching. Secondly, peat formation is less advanced on the south-facing slope as there is less moisture available for waterlogging.

It is noteworthy that in this example there is a strong relationship between natural vegetation cover and soil type. While these relationships are common, particularly in mature soils, caution is required in their interpretation. It is not always possible to ascribe a direct causal relationship between soil type and vegetation. It is true that the soil type can influence the plants that are able to thrive in a particular locality. It is also true that the plants growing in a particular place will determine some of the characteristics of the soil in which they grow. However, the role of climate, topography, parent material and time must not be overlooked. A second and more pragmatic need for caution follows from the fact that a given plant species will generally thrive on more than one type of soil. Therefore, the boundary between soil type does not always coincide with a change in the vegetation.

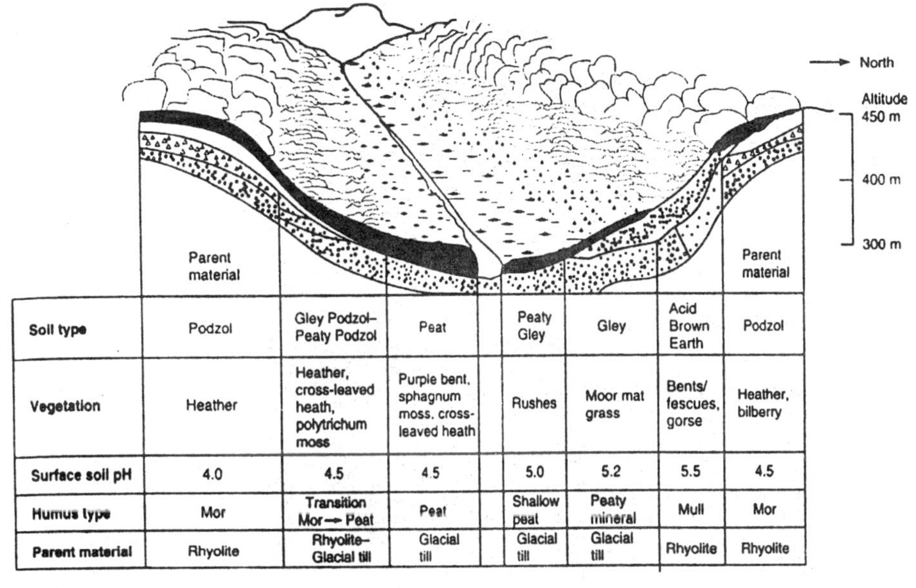

Soil type	Podzol	Gley Podzol–Peaty Podzol	Peat	Peaty Gley	Gley	Acid Brown Earth	Podzol
Vegetation	Heather	Heather, cross-leaved heath, polytrichum moss	Purple bent, sphagnum moss, cross-leaved heath	Rushes	Moor mat grass	Bents/ fescues, gorse	Heather, bilberry
Surface soil pH	4.0	4.5	4.5	5.0	5.2	5.5	4.5
Humus type	Mor	Transition Mor → Peat	Peat	Shallow peat	Peaty mineral	Mull	Mor
Parent material	Rhyolite	Rhyolite–Glacial till	Glacial till	Glacial till	Glacial till	Rhyolite	Rhyolite

Fig. 2.9 : The soils of a valley in south-east Scotland (from Simpson, 1983).

While the examples chosen above were taken from temperate zones, the processes of soil formation are essentially the same the world over. However, the rate and extent of soil formation vary considerably. Higher temperature and higher rainfall favour rapid development, while stable conditions favour maturity. Full soil development may take only tens to hundreds of years in highly permeable deposits in the humid tropics, while requiring several thousand years in cool, dry climates. In either extreme, soil maturity will not be achieved under unstable conditions engendered by, for example, high rates of erosion or rapidly changing vegetation.

Attempts have been made to develop predictive models of soil development. One approach is to identify the combination of environmental conditions that optimises each of the individual soil-forming processes. Once this has been done it is possible to predict the combination of soil-forming processes that are dominant under a given set of environmental conditions. On this basis it is possible to predict the soil type that should be formed.

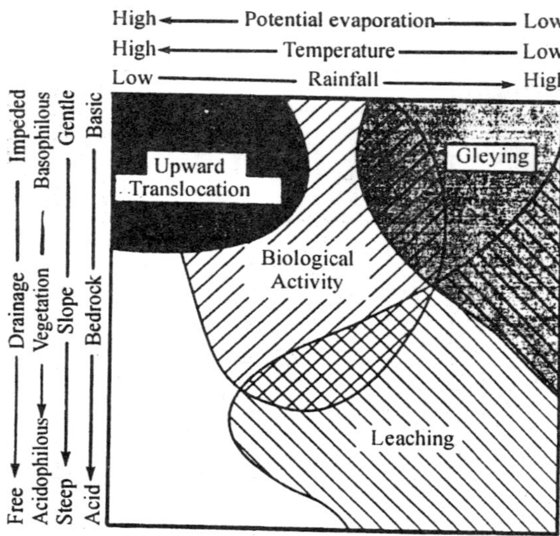

Fig. 2.10: The relationship between environmental factors and the soil-forming processes of leaching, biological activity, gleying and upward translocation. The zones show the optimal conditions for each of these processes (from Briggs and Smithson, 1985).

By way of illustration of this approach, Figure 2.10 shows the environmental conditions that favour maximal upward translocation, biological activity, gleying and leaching, four of the most important soil-forming processes. The reader may find it useful to evaluate the success of this model using the second example discussed above and illustrated earlier in Figure 2.9.

Soil-Group Divisions

Soils are classified into groups, and in the United States soil groups fall into two great divisions: the pedocals and the pedalfers, in the terminology used by Wolfanger (1950) (Fig. 2.11) the aridic and the humid soils, in the somewhat different classification used by Lyon, Buckman, and Brady (1952). The *pedocal* division is composed of incompletely leached soils found characteristically in the arid Great Plains of the West. The slight rainfall of these regions does not saturate the soil to a depth sufficient to reach the water table deep in the ground,

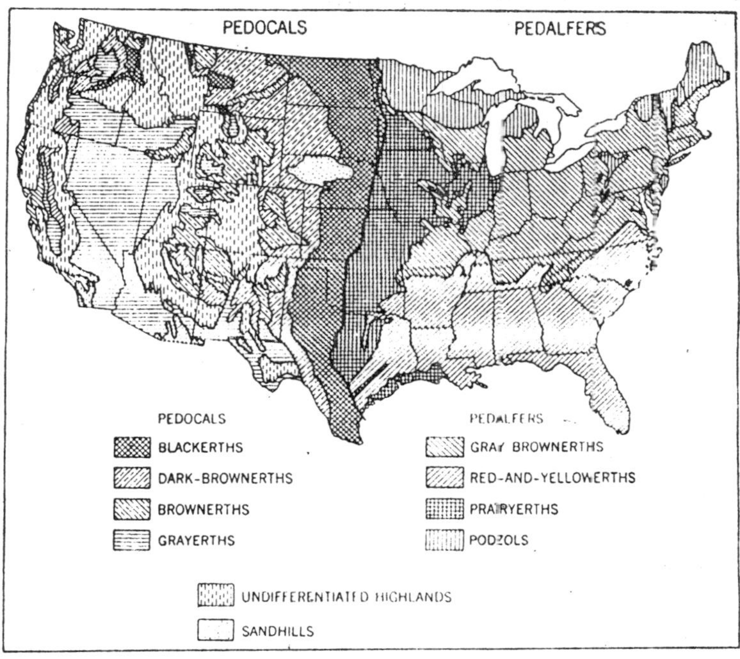

Fig. 2.11 The soil groups of the United States and their division into Pedocals and Pedalfers.

and lime tends to be deposited in the *B* horizon. When evaporation from the soil begins, the ground water is drawn upward again toward the surface, carrying some of the solubles with it. The water in this kind of soil has been said to be "hung from the top, like Monday's wash!" The roots of the grasses and shrubs that are characteristic of the pedocal soils absorb the water and the contained salts and also carry them toward the surface. Grasses particularly tend to absorb a considerable amount of calcium and to restore this material to the upper layers. This process is aided by the low growth habit of these plants and the great development of rhizomes and other structures near the surface of the ground. For these various reasons calcium tends to accumulate in the upper layers of the soil, and from this fact the name "ped-o-cal" is derived. The retention of calcium carbonate and magnesium carbonate helps to prevent pedocals from becoming acid.

An outstanding example of the interaction of climatic, edaphic, and biological agents in the development of a pedocal soil is furnished by the *chernozems,* or blackerths, that occur from the Bakotas Southward. The climate is less arid than it is farther west, and a good grass cover is typically present. The action of the ground water and of the vegetation in retaining salts and other nutrient materials in the upper layers and in maintaining a neutral or slightly alkaline condition (Fig. 2.12). Organic matter tends to accumulate in the *A* horizon, resulting in the development of the rich, dark soil condition familiar in the region. Chernozems are consequently especially valuable for grazing or for farming because when left undisturbed, or properly managed, the ecological processes present tend to perpetuate soil fertility.

The *pedalfer* division consists of soil groups found principally in the more humid regions of the eastern half of the United States. Here the rainfall is heavier with usually more than 75 cm falling per year. The solubles of the soil tend to be carried beyond the reach of the roots by the large amount of water percolating through the upper horizons. Much of this water filtering through the soil reaches the water table and drains off valuable nutrients. The roots of the typical forest vegetation growing on such soils extract salts but do not remove relatively as much calcium as do the grasses. The annual leaf-fall is on the surface, and some of the organic matter resulting from decay is carried away by the run-off, particularly in hilly

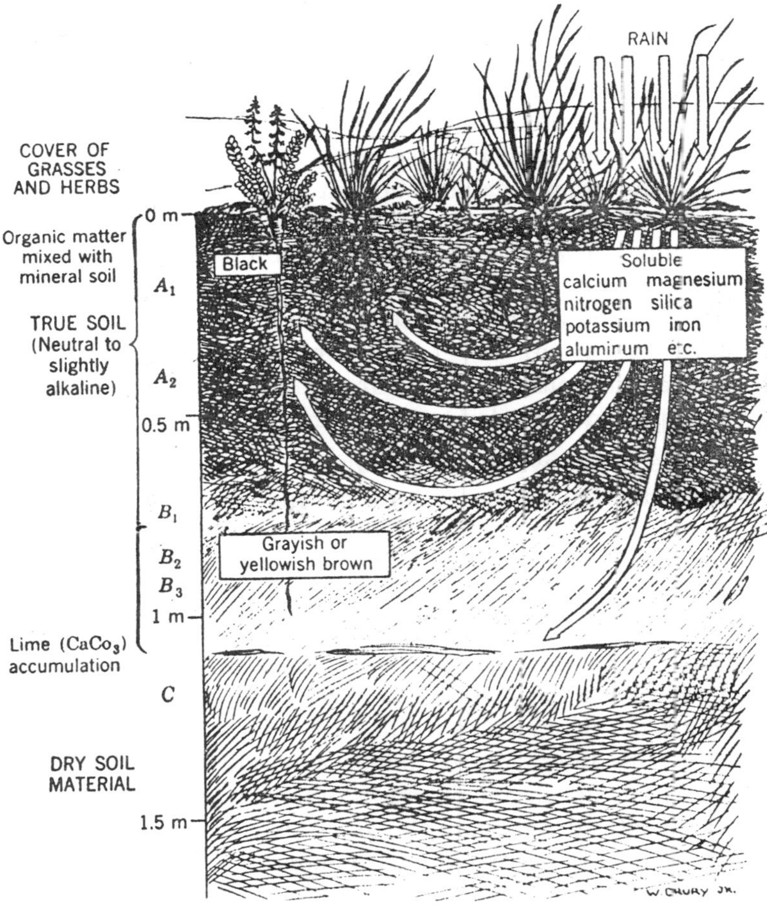

Fig. 2.12 Schematic vertical section through a forest soil of the chernozem group as representative of the pedocal soil-group division. The subdivisions of the soil profile are indicated in relation to the vegetation and the movement of soil water. The depth scale is representative, but varies greatly.

regions. This loss of nutrient materials is in marked contrast to the situation with the pedocals and the grass vegetation of the prairies.

In the north-central and north-eastern parts of the United States the increased rainfall and lower temperatures tend to accentuate the processes just described. Soils of the *podzol* group, which occur in these regions, represent an extreme development of the conditions found in the pedalfer division (Fig. 2.13). In the podzols the solubles are rapidly lost by leaching. The *A* horizon becomes acid because of the accumulation of leaves, needles, and other organic debris on and in the surface starta. Calcium is dissolved from the upper layers and is often almost completely extracted from the soil. Aluminum and iron are similarly carried down but tend to be precipitated again as silicates in the less acid *B* horizon, and may cause the formation of hard-pan at this level. The tendency toward relative accumulation of aluminum and iron generally in this soil-group division provides the derivation for the term 'ped-al-fer.'

In the region where the pedalfers exist there is generally sufficient rainfall for agriculture, but the soil tends to lose its nutrients. In the extreme podzol type the interaction of the climate and vegetation is such that solubles are rapidly carried away and the soil becomes progressively poorer and more acid, whereas in the chernozem soil fertility tends to be perpetuated. The contrast between the chernozem soil as a representative of the pedocal and the podzol as a representative of the pedalfer illustrates the effectiveness with which the activity of organisms coupled with differences in climate can modify the substratum. The soil in its development and its organization is thus seen to be an outstanding example of a system formed by the interaction of organism and environment.

The review of the substratum as an ecological factor given in this chapter has revealed the great variety of surfaces and solid materials on and in which animals and plants live. The manner in which the nature of the substratum controls the distribution and the growth of different species has been indicated. We have also stressed the fact that the activities of

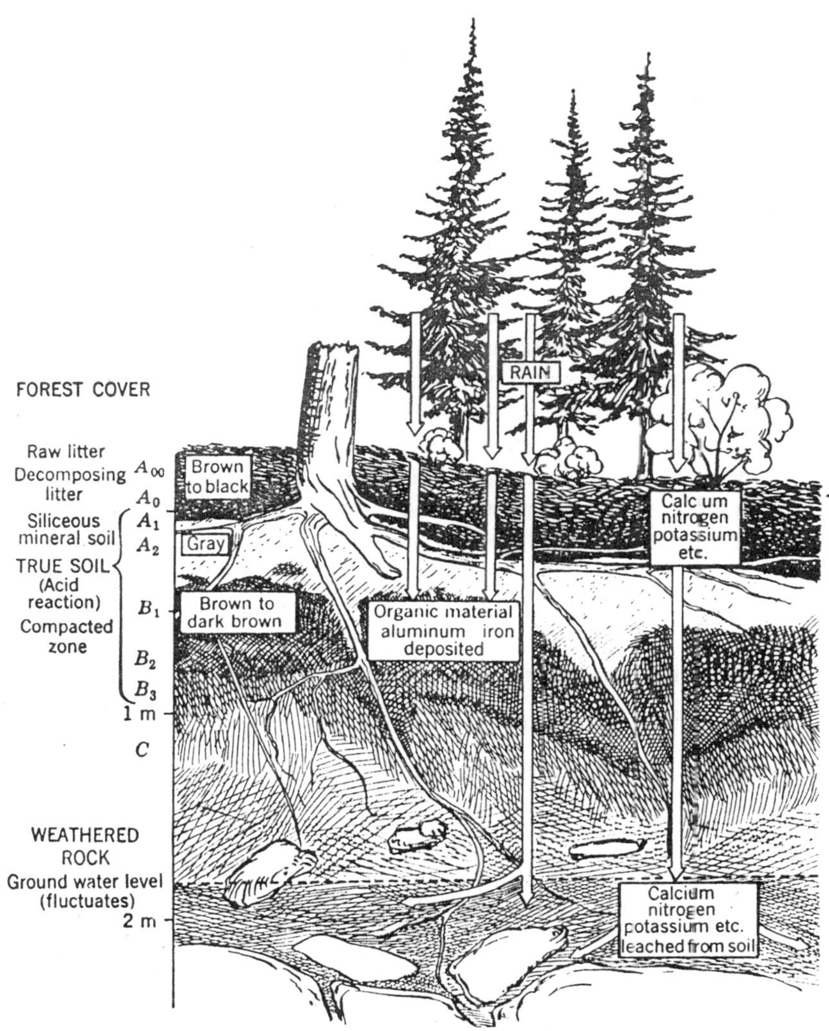

Fig. 2.13 Schematic vertical section through a forest soil of the podzol group as representative of the pedalfer soil-group division. The subdivisions of the soil profile are indicated in relation to the vegetation and the movement of soil water. The depth scale is representative, but varies greatly.

living organisms—and the material of their bodies after death—may profoundly alter the substratum in both the aquatic and the terrestrial environments. Sometimes this process itself brings about the further progressive changes in the fauna and flora known as ecological succession. Most of all, our present discussion focuses attention upon the reciprocally dependent relation existing between the living community and the substratum. The nature of the mud on the bottom of a pond or of the soil on the surface of the land is partly the result of the biological influences and is at the same time partly responsible for their existence.

The physical conditions of the water or the land originally allow certain plants and animals to exist in a given area. These living agents may then modify the substratum, whereupon the substratum may be further affected by the 'climatic' factors, perhaps resulting in additional changes in the fauna and flora. When we visit the area after these activities have gone on for a long time, and are still going on, it is not easy to distinguish cause and effect. Nevertheless, with a good grasp of the relations summarized in this chapter we are better equipped to determine the specific action of climatic, physiographic, and biological influences in bringing about the existing conditions.

Land Surfaces and Animals

Physical differences in the land surfaces are correlated with special adaptations of animals inhabiting them. On rocky terrains and in regions with hard, open ground the running speed of animals is improved by the possession of small resistant feet usually with a reduced number of toes, as in the deer, antelope and ostrich. Animals living in areas of soft sand, marsh, or snow are characterized by spread-out feet, like those of the camel, or of toes that present a large surface, like those of wading birds and the snow-shoe rabbit. Sand-dwelling lizards and insects similarly are enabled to move over loose sand by toes or legs widened by lateral scales or hairs. Other animals have feet especially adapted for climbing trees (squirrels), or for clinging to branches or leaves (tree frogs), or for dealing with other special substrata.

In contrast to the species requiring rapid locomotion over the land surface is a large group of animals that burrow into the substratum. Many rodents, some birds and reptiles, and a great many insects, as well as other types of invertebrates, are built for effective digging in the ground. Certain species such as the mole, the earthworm, and many insects spend most of their lives underground. All these burrowing forms are limited to regions in which suitable soil conditions exist. For example a tongue of soil running across the Florida panhandle that is too dry for the burrowing of crayfish acts as an ecological barrier separating certain west Florida species of crayfish from species limited to areas farther to the east.

The chemical composition of the soil substratum affects animals both directly and indirectly through their food. Land snails with calcareous shells are especially abundant on soils rich in lime, but the abundance of the shell-less slugs is not affected in this way. The shell of one species of *Helix* was found to weigh 35 per cent of the total weight of the snail in limestone regions but only about 20 per cent of the total weight in areas with soils poor in lime. The bones of mammals likewise are heavier on limestone soils; this is especially true of deer that annually must grow new antlers sometimes weighing as much as 7 kg. It is no accident that strongly built race horses are raised in the bluegrass pastures on the limestone soils of Kentucky. Many herbivorous mammals require an abundant supply of salt (NaCl) because sufficient sodium must be taken in to maintain a proper ionic balance with the large amount of potassium contained in their plant food. If an adequate amount of salt is not available in the halophytes of salt meadows, ruminants are forced to travel to 'salt licks,' or are excluded from the region entirely.

The amount of organic matter in the soil is of vital concern to earthworms and other small invertebrates that use this material as source of food. A great number of variety of small insects and spiders, of still smaller nematodes, and of microscopic Protozoa, which live permanently in the soil, also depend upon this organic matter for their nutrition. Thus the microfauna also is controlled by the chemical qualities of the soil, as well as by its compactness, dryness, and other physical characteristics.

Action of Organisms on Soil

Having seen the many ways in which the nature of the land subtratum may influence the lives of organisms, we now may inquire to what extent the action is reversed. The fact is soon revealed that animals and plants play a very important part in modifying their substratum on land just as they do in water. This activity on the part of terrestrial organisms is particularly striking in relation to the formation and development of soil. It has been truly said that if it were not for organisms there would be no soil—at least none of biological importance. The soil is an outstanding example of the result of the organism and the environment acting as a reciprocating system.

Abundance of Organisms in Soil: The great abundance of organisms which live wholly within the soil and the far-reaching extent of the underground parts of organisms are not always appreciated. Anyone who spades up a garden or transplants a shrub should be impressed with the number of roots and the bulk of the root systems of even small plants. The roots of the typical plant are so finely divided and subdivided into rootlets and root hairs that a tremendous surface is provided for the exchange between the organism and its surroundings. Roots are frequently sufficiently abundant to produce a continuous mat or network extending several feet into the soil. The root system of a maize plant may extend over 1 m laterally and 2½ m deep. The roots of 17-year-old apple trees were found to have grown to a depth of 10½ m.

In the animal kingdom the number of species that burrow through the soil and thus influence it is also very large. Burrowing rodents and moles of one kind or another exist almost everywhere, and they are much more abundant than is generally realized. Although in some instances the actual number of the larger forms may not be impressive, their digging activities may be remarkably extensive. Prairie-dog burrows more than 4 m deep have been reported. In certain parts of

California systematic trapping has shown that as many as 50 mice inhabit each hectare (2½ acres) under normal conditions. Periodically, as we shall see later, the rodent population tends to increase greatly in numbers. Even when the population of mice and other burrowing forms is at low ebb, a considerable influence on the soil may be produced in the course of a year. In addition to a variety of mammals, many kinds of reptiles and amphibians as well as a few species of birds spend at least a part of their lives burrowing in the soil.

Of smaller animal forms, the numbers present in the soil are much greater. Earthworms have been estimated at hundreds of thousands per hectare, and their burrows may extend to depths greater than 2 m. Insects, especially in the larval stages, are very numerous in the upper centimeters of the soil (Salt *et al.*, 1948). In some regions population densities of several million soil insects per hectare have been found. Spiders, tardigrades, millipedes, and isopods are also abundant inhabitants of the land substratum. An extensive study in Illinois showed that the invertebrates of the soil reached an average summer maximum of 3300 per sq m—or roughly one animal under every 3 sq cm of surface. Since many of these forms are short-lived and populations succeed one another in the soil, the study indicated that at least one or two invertebrate animals had existed during the year under every square centimeter of soil surface. In mineral soils of Jutland nematodes have been found to range in number from 175,000 to 20,000,000 per square meter.

The numbers of microorganisms in the soil are, of course, much greater and produce a profound effect on the substratum. Protozoa may exist in concentrations of hundreds of thousands per gram of soil. The mycelia of fungi penetrate the soil wherever suitable conditons are found, and becteria are extemely abundant almost everywhere. In raw humus as many as 20,000 becteria per gram are a common occurrence. In rich loam the bacteria population may rise to 50 or even 100 million cells per gram of soil. All these denizens of the soil from sizable burrowing animals to the smallest microorganisms add their influence to that of the underground parts of plants in modifying the substratum.

Soil and its Action on Plants: An adequate discussion of the nature of the soil, its changes in time and space, and its influence on plants, and indirectly on animals through its effect on vegetation, would require a whole book in itself. Nothing more than an introduction to the subject can be given here. For a more extensive treatment of soil itself the reader is referred to Lyon, Buckman, and Brady (1952), Kellogg (1941), and the Yearbooks of Agriculture issued by the U.S. Department of Agriculture. Excellent chapters on the ecological relations of soils in relation to plants are to be found in Oosting (1948) and Daubenmire (1947).

Besides its ecological importance as a substratum, soil has immeasurably great economic importance. Fortunately we are rapidly becoming aware of the critical value of the productive capacity of soil as a support for civilization. Wolfanger (1950) has stated that "the soil of a nation is its most valuable material heritage." The crucial need for immediate soil conservation in almost all countries of the world has been ably pointed out by Osborn (1948), Vogt (1948), and others. A thorough understanding of the ecological relationships involved is essential for the intelligent use of our existing soils, for the prevention of further soil loss and degradation, and for the restoration of the fertility of worn-out soils.

Soil represents an extremely complex matrix consisting of minerals derived from the parent rock of the area, organic matter of local origin, and substances carried in by various agents. The physical nature of a soil depends first of all upon its texture and structure. *Texture* is determined by the size of the constituent particles, and *structure* is dependent upon the aggregation of these particles in the undisturbed soil into grains, clumps, and flakes. Soil particles are classified according to size in the accompanying table.

Sand	1.00-0.05	mm in diameter
Silt	0.05-0.002	
Clay	<0.002	

In a good loam all three of the categories are well represented. The type of structure into which the particles are arranged affects profoundly the porosity of the soil. It also controls the amount of surface which is presented on the one hand to the air and water moving through the soil, and on the other to the hairs of roots growing in the soil. The relative proportions of the soil constituents and of the air and water present are extremely variable. In an average good soil about half the volume is commonly represented by pore space of which half may be occupied by air and half by water. The solid material of such a soil may consist of 95 per cent mineral particles and 5 per cent organic matter. In tropical soils, however, organic matter may be less than 1 per cent, and in peaty soil it may approach 100 per cent of the dry material. In addition to differences in texture and structure, soils vary physically in the type of layering that they develop as they mature under biological and climatic influences as will be discussed in the next section.

The chemical nature of soils is even more diverse and variable. Upon the disintegration of the parent rocks the whole spectrum of minerals present becomes available for incorporation into the soil. Added to these are a wide variety of organic substances derived from animals and plants and other materials introduced from the air and ground water. Further chemical changes take place within the soil as climatic and biological agents work on it. As a consequence, soils and soil water differ widely in chemical composition, organic content, and total salinity, as well as in degree of acidity, oxidation-reduction potential, and other physiocochemical characteristics. Some of these features of the soil are intimately interrelated with the physical characteristics. For example, the smallest particles involved in the texture of a soil are colloids, and their behaviour and reactions are also involved in the chemistry and physical chemistry of the soil. The abundance and type of the colloids present affect the amount of water retained by the soil and its availability to plants. At the same time the colloids influence the chemical composition of the soil water.

In the present section we are concerned with soil primarily in its physical nature as a substratum for land organisms. The foregoing brief sketch of soil has indicated the extremely complicated nature of this substratum and the degree to which its physical characteristics are bound up with its chemical and its physicochemical feature. Certain of the latter will be further discussed in relation to other ecological factors considered in later chapters.

When gross differences in the land substrata exist, factors limiting plant growth can frequently be distinguished. On a solid rock substratum lichens and certain mosses are characteristically the only plants that can survive. In situations with a coarse, shifting substratum, like a sand dune or a gravel slide, the vegetation is limited to specially adapted forms. Dune grasses with their network of horizontal rhizomes, and certain other plants, such as *Paronychia,* with strong and extensive root systems are among the few plants that can maintain a foothold. When the soil is very hard owing to a high silt and clay content or to the development of a hardpan, the roots of many species cannot penetrate. At the other extreme the presence of very soft soils prevents the establishment of plants that require firm anchorage. Many evidences of this relation were seen in New England after the hurricane of 1938 when whole groves of trees on loose soil were uprooted but neighbouring groups of the same species on hard ground remained standing after the storm.

In other situations a gross difference in such factors as the chemical composition, moisture, or temperature of the soil may be distinguishable as the prime influence controlling the vegetation, and examples of these will be considered later in the appropriate chapters. Too often for the peace of mind of the ecologist, however, very complicated or subtle differences occur in the soil, and frequently two or more interdependent factors appear to act mutually in limiting plant growth. More detailed treatments of the ecology of soils, such as those referred to earlier in this section, should be consulted for a further discussion and examples of situations of this type. In some habitats investigators disagree not only as to which of the soil influences is critical in determining the composition of the vegetation but also even as to whether the climatic factors are not more important than the edaphic (soil) factors.

The ecological relations in the soil also frequently illustrate the *principle of partial equivalence:* an increase in one factor may sometimes partially make up for a deficiency in another factor. The lack of moisture in a sandy soil may be compensated for to some extent by a greater rainfall in certain localities, or, conversely, plant species for which a given region is generally too humid may find the effective moisture conditions sufficiently reduced in a local area with a sandy substratum.

Biota

Plants and animals have a highly important role in the formation of soil, both as they affect its structure and as they aid in the production of humus. Plants contribute to the mechanical and chemical weathering of rock. Plant roots, especially those of trees, can split large rocks. Lichens, mosses, and even bacteria and fungi excrete acids in the course of metabolism, which dissolve the substances that cement rock granules together. When plant roots die, fungi convert them to dry, soft, spongy material (punk), used as food by saprophytic microarthropods. Usually the bark of the root remains intact the longest. Hollow tubes are thus formed that permit water and air to penetrate considerable depths into the soil. These channels gradually become filled with silt and animal excreta.

The addition of plant and animal organic matter to heavy compact soils or clay tends to open the soil, making it more porous. Addition of organic matter to sandy soils binds the particles closer together, making the soil less porous.

Earthworms may be divided into deep- and shallow-working species. Deep-working species dig narrow tube-like channels which may reach 2-3 m down through overlying soil to parent rock. Earthworms ingest soil while burrowing, digest and absorb organic matter from it, and egest the residue in a semi-liquid form which is used to cement the walls of the burrow or else is deposited at the surface as castings.

Earthworms prefer easily digested succulent vegetation and dung for the purpose, but in the autumn may pull the freshly fallen leaves down into their burrows to use as food or nest linings. Ejected petioles may form midden piles around burrow entrances. In an undisturbed virgin prairie in Texas, earthworm casts made a layer 2-3 mm thick over the entire ground surface and when air-dried weighed about 2400 g/m^2 (10.7 tons/acre). Earthworms are not, however, important soil builders in disturbed grassland; they may be absent altogether in arid regions. In other studies, the dry weight of casts brought to the surface annually by earthworms varied from 475 g/m^2 (2.1 tones/acre) in a moderately hot dry climate, to 24,000 g/m^2 (107 tons/acre) in the White River valley of the Sudan, during the rainy season. Earthworm casts compared with the surrounding soil show higher total nitrogen, organic carbon, exchangeable calcium, exchangeable magnesium, available phosphorus, exchangeable potassium, organic matter, base capacity, pH, and moisture equivalent. Only certain species make these surface castings; other species void the ingested soil into subterranean spaces.

The ant *Lasius niger neoniger* spends most of its time in its underground burrows and deposits excavated soil upon the ground surface around burrow entrances. In an oldfield community in Michigan such deposits amounted, at one sampling, to 85.5 g/m^2 (750 lb/acre). However, entrances are abandoned and new ones made, so that in the course of a few weeks a much larger quantity of soil is brought up.

In the semi-arid Great Plains of western North America there is at least one species of ant that excavates extensively underground and builds a conical mound of this excavated material above the surface. A single such mound weighs approximately 77 kg (170 lb); there are as many as 50 such mounds per hectare (20 per acre) in some localities. Plainly, these little excavators move prodigious amounts of soil. The relatively sterile subsoil is gradually mixed with organic material and spread over the surface of the ground, thus increasing the depth of the fertile topsoil. Scarabeid beetles, bees, wasps, and in tropical regions mound-building termites also move considerable subsoil to the surface.

The crayfish *Cambarus diogenes* often occurs in poorly drained fields; it burrows down to the water table, sometimes a depth of 3 m. Excavated material is brought to the surface and built into chimney-like affairs which may be 20 cm high and almost that much in diameter. Where crayfish are abundant, as much as 600 to 2000 g/m2 (2.7 to 8.9 tons/acre) of soil per year may thus be moved.

The burrows of prairie dogs and badgers may extend 2 to 3 m below the surface, and a single mound of excavated dirt weigh from 100 to 10,000 kg. Mounds made by pocket gophers and ground squirrels weigh from 7 to 180 kg each; it is not unusual to find 42 such mounds per hectare (17 per acre). These animals thus move from 7 to 9 kg of subsoil for each m^2 of surface (30-40 tons/acre) in a period of several months.

Large terrestrial animals trample the soil into greater compaction and destroy vegetation at sites where numerous individuals foregather; around water holes in grassland where bison and antelope come to drink, for instance, or winter yards of deer and moose, trails on hillsides, wallowing places, and so on. These reactions are usually very local, however.

As animals burrow and bring large quantities of loose soil to ground surface exposure, the likelihood of water and wind erosion destruction is greatly increased, especially true if the burrowing is done on hillsides where the flow of water is faster and where the animals always tend to deposit the soil on the downslope side of burrow entrances. On the other hand, the very same activities may decrease erosion where a soil is, in consequence, made more porous so that there is less water runoff.

Humus. In soil, organic matter that is partly or entirely decomposed is called *humus*. The amount of humus varies from less than 1 per cent to as much as 20 per cent of the soil; peat soil may be largely organic material, but much of it resists decomposition, and hence is not true humus. Decomposition breaks down complex organic compounds into simpler ones that are washed back into the soil, thus becoming available again as nutrients.

On virgin prairie in Texas the ground litter of dead grasses and herbs amounted to over 300 g/m² when measured in April. The annual dry weight of leaves that fall to the ground in deciduous and coniferous forests varies from site to site and with the density of the trees but is commonly in the range of 50 to 400 g/m² (Olson 1963). In mature climax forests the rate of decomposition of the litter and re-absorption by plants of the nutrients thus yielded keeps pace with the annual accumulation so that an equilibrium is established. In seral stages, decomposition and utilization do not keep up with the annual accumulation, so that the organic content of the forest floor increases with time (Fig. 2.14). Litter production is lower in arctic than in tropical regions.

	Metric Tons/Hectare-Year		
Region	*Leaves*	*Other*	*Total*
Arctic-alpine	0.7	0.4	1.0
Cool temperate	2.5	0.9	3.5
Warm temperate	3.6	1.9	5.5
Equatorial	6.8	3.5	10.9

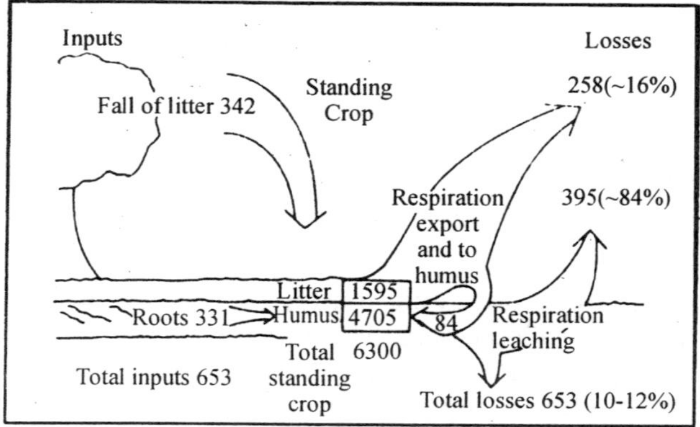

Fig. 2.14 Annual budget of organic matter in soil of an oak-pine forest (g dry weight/m²).

However, in warm tropical regions the amount of humus that accumulates on the forest floor is low because of the high rate of decomposition, runoff, and leaching (Fig. 2.15).

The thick organic layer on the ground moderates extremes in the daily and seasonal rhythms of soil temperature, retards freezing of the ground in the autumn and thawing in the spring, and retains soil moisture. Because of humus formation (involving oxidation) and the respiration of plant parts and animals underground, soil air contains little oxygen but much carbon dioxide, and it possesses a higher moisture content than does the general atmosphere above ground. This is especially marked in warm summer months when these processes go on more rapidly. The decay of organic matter usually makes the top soil somewhat acid (most commonly pH 5 to 7), but in the mineral subsoil, the acids are often neutralized by the basic salts commonly present.

The mineral content of leaf fall varies according to the species of tree, but in the northern United States it averages about as follows (in g/m^2):

Element	*Hardwood forest*	*Coniferous forests*
Calcium	7.3	3.0
Nitrogen	1.8	2.6
Potassium	1.5	0.7
Magnesium	1.0	0.5
Phosphorus	0.4	0.2

Silicon, copper, manganese, carbon, and zinc are also present in the leaves of hardwood trees. Carbon is relatively more abundant and nitrogen less abundant in coniferous than in deciduous leaves, but commonly the carbon/nitrogen ratio is 55 : 20.

Both plants and animals are important agents effecting the decomposition of organic matter and the formation of humus. An animal digests and metabolizes plant foods, much of which is returned to the soil, in part as the excreta of the living animal, in part as the body of the dead animal, in part as gas. Fully

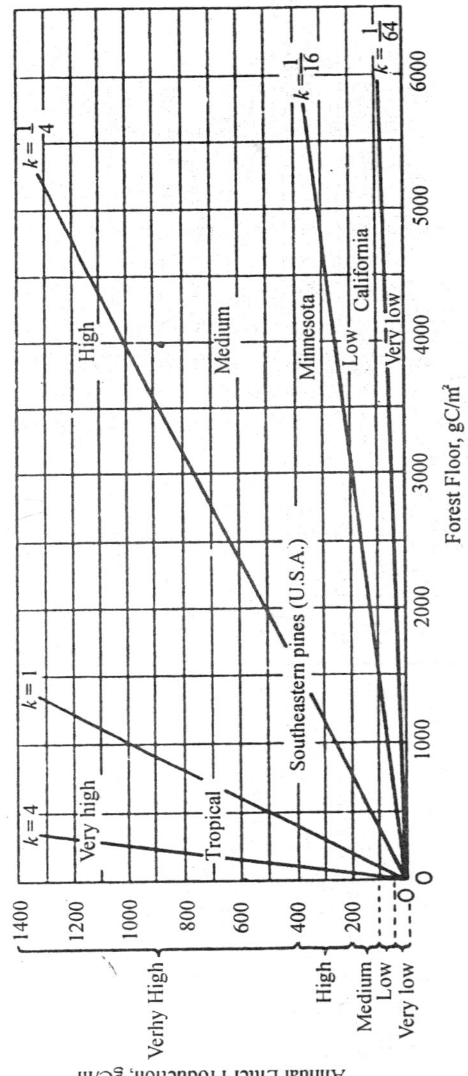

Fig. 2.15: Relation between annual production of litter (g C/m²) and amount stored on forest floor (k = fraction of stored carbon lost annually).

formed humus is, in fact, derived mostly of fecal material. The larger herbivorous and carnivorous animals pass urine and feces containing simple nitrogenous compounds and compounds of phosphorus, potassium, and traces of calcium, magnesium, sulfur, and other elements. Humus is but one point in a continuous cycle of decomposition of plant and animal organic matter, absorption of decomposition products by plants, ingestion and metabolization of plant matter by animals, decomposition of plant and animal organic matter—ad infinitum, the consumption by saprovores and herbivores of living and dead plant matter the consumption of herbivores by carnivores neither add nor subtract from the total nutrient supply of an eco-system. The chemical elements available in the air, water, and soil of an eco-system pass, in one compound or another, from one organism to another, and from one stage in the cycle to another. They continue thus to circulate within the eco-system unless and until they are physically withdrawn from it. To remove plant and animal crops from an eco-system is to withdraw nutrients from it, and thus to reduce the fertility of the system. Fertility can then be maintained only if the nutrient supply is kept replenished by artificial fertilization.

Kangaroo rats defecate promiscuously throughout their underground burrow systems. The soluble nitrate content of the soil in the region of one burrow system averaged 221 ppm and in another one 570 ppm, compared with a maximum of 15 ppm in the surrounding desert soil generally. It is a reasonable estimate that the total bird population in a deciduous forest would deposit 0.1 g dry weight of organic excrement per m^2 in a year's time; the mammal population, perhaps 0.5 g; and the total invertebrate fauna, possibly 2-3 g. The accumulation of excrement under the roosts of birds is sometimes enough to kill the ground vegetation and even the trees. The guano deposits on the coast of and islands off Peru and elsewhere in the world were originally several meters thick, as the result of centuries of occupancy by nesting colonies of marine birds, but have now been largely depleted by man for use as crop fertilizer. Bat excrement, deposited in caves, was exploited in years past as a source of saltpeter for gunpowder.

The conversion of raw organic matter into materials suitable for re-absorption and utilization by plants is a complicated process and depends almost entirely on the reactions of plants and animals. Especially important soil animals are the mites, earthworms and enchytraeids, woodlice, millipedes, springtails, and other saprophagous insects. Decomposition of litter proceeds more rapidly with these soil animals present than in their absence (Fig. 2.16). The role of beetles and flies in the decomposition of trees and of carrion on the ground surface have been mentioned in connection with the microseres that are formed. The digestion of animals produces both mechanical changes in raw humus that can be measured quantitatively. The non-nitrogenous substances in fresh litter are sugars, starches, pectins, pentosans, celluloses, cutins, tannins, lignins, oils, fats, waxes, and resins. Most of these substances are readily broken down in the soil by fungi, actinomycetes, bacteria, and protozoans, but tannins, lignins waxes, and resins decompose very slowly. The end products of complete decomposition are H_2O and CO_2, but sometimes decomposition is incomplete and organic acids are formed instead.

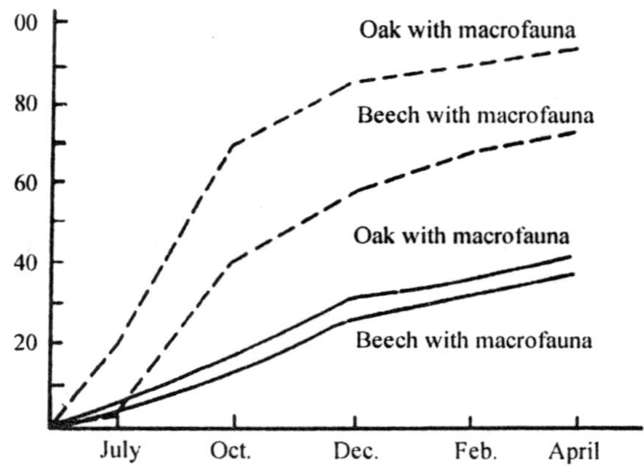

Fig. 2.16 Rate of litter decomposition with and without larger invertebrates present.

The most important soil organisms concerned in the decomposition of the litter are the bacteria, both aerobic and anaerobic forms. They are commonly divided into two types.

Heterotrophic bacteria obtain their energy from the oxidation of the carbohydrates and fatty substances as described above. They use this energy for the synthesis of cell substances and the production of enzymes that break down complex compounds in the litter into simpler compounds, including proteins into ammonia compounds. They then use part of the ammonia compounds in synthesizing the amino acids they need in building their own proteins. *Autotrophic* bacteria, in turn, are of two types: *chemosynthetic* species, which obtain their energy from the oxidation of inorganic compounds (hydrogen, sulfur, hydrogen sulfide, iron, ammonia) and *photosynthetic* species, which include purple and green sulfur bacteria, possess a form of chlorophyll, and utilize the energy of sunlight. Chemosynthetic bacteria convert ammonia compounds into nitrites and nitrates, part of which they use in their own anabolism, the rest becoming available for plants to absorb. Photosynthetic bacteria use the ammonia compounds in their own anabolism and do not render them directly available to plants. Chemosynthetic bacteria are more abundant than photosynthetic bacteria in soil; photosynthetic bacteria are the more abundant in water.

The decomposition products available to plats are reabsorbed by the roots and built up into plant tissue. Plants are eaten by animals, and as plants and animals die the minerals are returned to the soil. All essential minerals cycle repeatedly through the eco-system, the cycle of each mineral differing in various ways from the cycle of every other mineral.

Nitrogen Cycle. In the nitrogen cycle proteins are broken down, yielding ammonia (NH_3) compounds in the course of the metabolic processes of all animals and by the activities of heterotrophic bacteria, filamentous fungi, and actinomycetes (Fig. 2.17). The process is called *ammonification*. Some of the ammonia is oxidized to form nitrites (NO_2) and nitrates (NO_3) through the action of autotrophic bacteria; the process is called *nitrification*. Other types of bacteria act on ammonia in the process of *denitrification*, by which nitrogen (N_2) is liberated into the atmosphere. Nitrogen is removed from the air by the *nitrogen-fixing* bacteria which live either freely in the soil or as

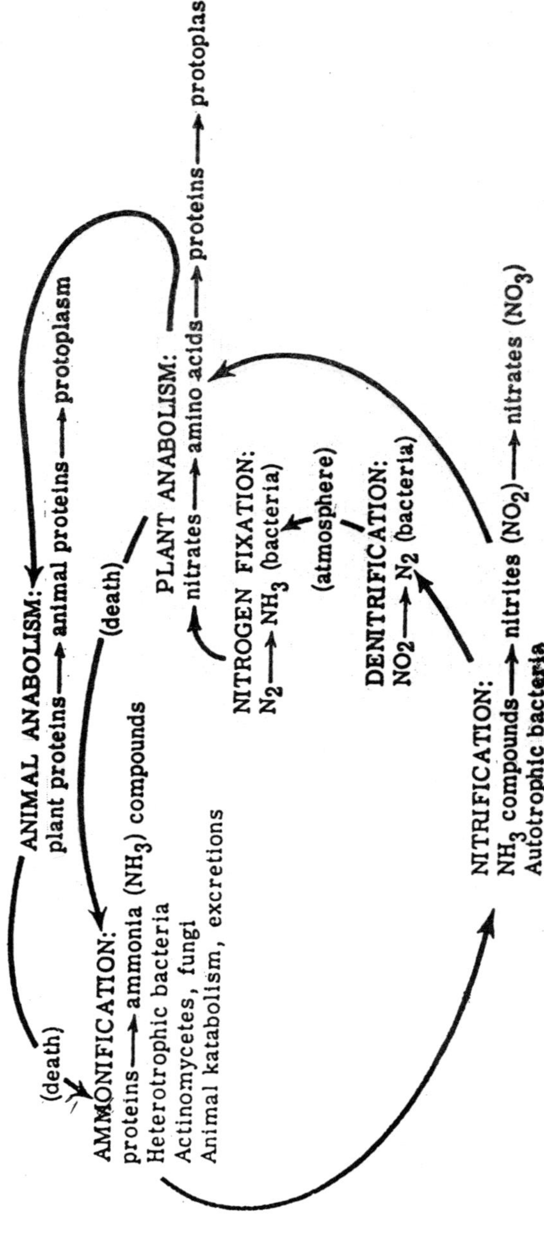

Fig. 2.17 Steps and processes in the nitrogen cycle.

symbionts in the root and leaf nodules of many legumes and some non-legumes, *Ceanothus, Elaeagnus, Alnus,* and *Myrica,* among others. Blue-green algae also fix nitrogen in moist soil and are more important than bacteria in this regard in aquatic habitats. Ammonia compounds, nitrates, and other substances are added to the soil in small amounts with rainfall; sources of these nitrogen compounds are volcanic eruptions, terrestrial decomposition, and atmosphere nitrogen fixed by lightning. An attempt to estimate the quantities of nitrogen involved in the different parts of the cycle has been made by Hutchinson (1944). Man has developed an important chemical fertilizer by artificially fixing nitrogen from the air, and much research is underway for the better understanding of the enzymatic mechanisms of biological fixation.

Hardpan

In arid regions, evaporation may be in excess of rainfall. Moisture in the soil has no opportunity to percolate downward; rather, it rises to the surface of the ground and is lost. Where rainfall is inadequate for efficient leaching, *hardpan* may form in the B horizon as the result of deposition here of ferric oxide, alumina, colloidal clay, or calcium salts (Fig. 2.18). This layer becomes so compact and hard that it is impervious to root penetration and the burrowing of animals, although during periods of wet weather it disappears temporarily. In arid regions, the hardpan may be at or close to the surface; but in more humid climates it occurs at progressively greater depths until it disappears altogether.

Mull and Mor Humus

Distinction between mull and mor humus is made primarily for forest soils. Both humus types of soil may be subdivided, but these subdivisions need not concern us here.

Mull (endorganic layer) is a porous, friable humus layer of crumbly or granular structure and only slightly matted, if at all. The A_1 horizon is well developed; bacteria are abundant, annelids numerous, and nitrification occurs.

Mor (ectorganic layer) is a strongly matted or compacted humus layer. There is no A_1 horizon; the transition from humus

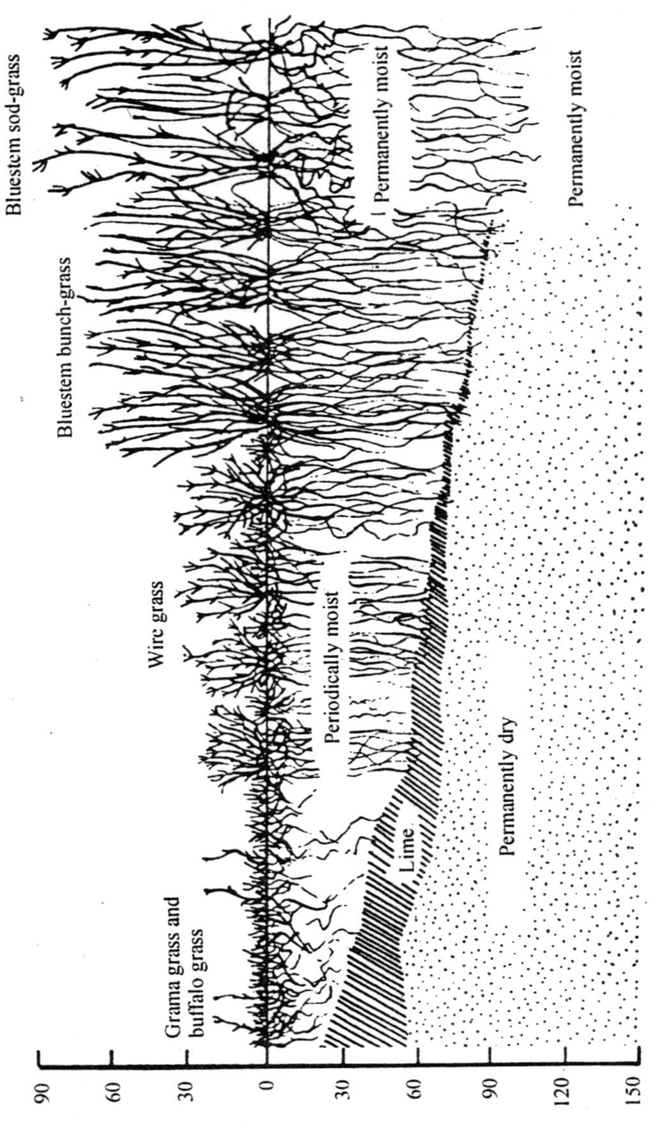

Fig. 2.18 Relation of lime hardpan to types of prairie vegetation extending from west to east in central North America.

layer to mineral soil is abrupt. The underlying soil is more acid, bacteria are much less numerous, and nitrification is usually reduced if not absent. Abundant fungi reduce the raw humus to punky material; thereafter it is worked on by the small arthropods, particularly springtails, mites, flies, and ants. Decomposition of mor is much less rapid than of mull. Annelids are few or absent, and moles are less common than in mull soils. Snails are scarce in acid soils. In general, the biomass of organisms inhabiting a mor soil is smaller, species are less diverse, and individuals are smaller than in mull soils.

Mor humus is common in cold regions and at high elevations, but is not limited to such climatic zones. It occurs especially under coniferous forests, even in warm climates, and under ericaceous vegetation. It is also found in very wet or very dry habitats, where there is accumulation of poorly decomposed or sharply delimited humus layers on top of the mineral soil, sand, or rock underlying. Mull humus commonly develops in warm, humid climates and is found especially under hardwood or deciduous forests. Patterns of animal and plant distribution correlate closely with these two humus types; indeed, the formation of each is a result, in the main, of unique combinations of biota reactions, climate, and mineral soil characteristics.

DEPTH DISTRIBUTION OF ORGANISMS

The small animals in the soil, including protozoans and nematodes, are most abundant in the *L*, *F*, and *H* horizons, becoming rapidly less abundant in the mineral soil see (Table 2.3). Bacteria, actinomycetes, and fungi are also most abundant in these top layers, especially in *F* and *H*, although they occur well down into the *B* and *C* horizons.

The depth distribution of soil animals depends on temperature and varies with season. When the top layers freeze during the winter months, much of the fauna keeps well below the frost line, although many species are tolerant of freezing. During the cold months the depth at which most soil insects, mollusks, and annelids occur varies from about 9 cm in silty clay-loam to 38 cm in gravelly clay-soil. With the return of warm weather, the fauna ascends to the top horizons.

Table 2.3 : Depth of distribution of soil arthropods in the Adirondack Mountains. The figures indicate approximate number of individuals per square meter in the total thickness of each layer.

Soil horizon	*Depth (cm)*	*Mites*	*Springtails*	*All others*
	Mor humus under red spruce and balsam fir			
L and F	0-5	150,000	19,800	1,400
H	5-25	62,000	19,200	400
A_2	25-33	1,500	0	0
B	33-58	1,900	100	0
	Mull humus under beech, sugar maple, and yellow birch			
L and F	0-5	62,000	17,800	2,800
A_1	5-15	15,000	6,200	600
B	15-55	6,200	5,800	200

Soil Erosion

As rain water drains over the surface of the ground, it picks up and transports particles of soil more or less proportional to the rate of flow. This is a continuous process that has gone on throughout geological time, resulting in base-leveling of mountains, the filling in of valleys, and the formation of river deltas. Normally this is a slow process, with changes becoming apparent only after thousands of years. With the ground covered with vegetation, the flow of water is retarded and more of it percolates into the ground or is evaporated. In fertile areas, new soil is formed faster than it is eroded or at least there is stabilization.

With removal of the native vegetation and cultivation of the soil in agriculture, surface particles are more easily moved and surface water drainage is more rapid. As a result, topsoil is being lost at an alarming rate. Hugh H. Bennett, former chief of the U.S. Soil Conservation Service, estimates that of the original 23 cm, the average depth of the topsoil over the country, one-third has already been washed away. Much land has been so badly damaged that there is little likelihood that it will ever again be productive of crops. Conservation of the fertile topsoil is one of the major problems facing modern agriculture. Putting in underground drainage tiles in level farmlands helps to draw rainwater into the soil and to decrease surface runoff and evaporation. Contour plowing of slopes decreases the rate of surface runoff and erosion. *Soil conservation* procedures should be strictly followed.

MAJOR SOIL GROUPS

The interrelations between the basic mineral content of the parent substrate, biotic reactions, and climate can be seen through an analysis of the development of the great soil groups of the world. We will give only a simple classification of them. A detailed classification would include many subdivisions and intermediate categories.

Podzolic soils are formed in humid temperate climates, under forest vegetation. The A_2 horizon is moderately well developed, for there is sustained leaching. Soils are more or less acid and only moderately fertile. *Podzols* develop under coniferous forest and have a mor type of humus. *Gray-brown* and *brown* podzolic soils are found under hardwood forests and have a mull humus.

Latosolic soils develop in humid tropical or semi-tropical forest regions. Humus is quickly oxidized by action of microorganisms and hence does not accumulate. The soil fauna is depauperate. Chemical weathering of the parental material is intense. Water drainage through the porous soil is rapid, so leaching is extensive. In early stages of its formation, the soil is neutral or slightly alkaline, but as leaching continues, it becomes acidic. The soil has a thin organic layer (A_0 and A_1 horizons) on a reddish, leached soil (A_2 horizon) that extends to great depths below the surface.

Chernozemic soils occur in humid to semiarid temperate climates under grass vegetation. The grasses on dying return considerable organic matter to the soil. The A_1 horizon is consequently dark in color and of great thickness. The soil contains more bases and hence is less acid than in the two types above. The $_B$ horizon in humid regions is indistinct, but where there is less rainfall, calcium salt may accumulate to form a hardpan. Prairie soils in temperate climates are among the most fertile soils of the world, but fertility decreases in the tropical and desert climates.

Desertic soils are characteristic of arid climates and contain very little organic matter. A profile is poorly developed. The surface soil is brownish gray, and grades quickly into the calcium carbonate horizon which usually forms a hardpan just below the surface. Wind erosion removes the finer soil particles, leaving the coarser material to form a hard pavement. The soils are but slightly weathered and leached; lacking nitrogen, they are infertile.

Mountain and *mountain valley* soils vary from shallow layers on eroding rocks to deep organic soils of valleys and swampy areas.

Tundra soils occur in cold northern areas where the substratum remains continuously frozen and the vegetation of lichens, mosses, herbs, and shrubs makes a peaty surface layer. The region is poorly drained and characterized by many scattered shallow ponds.

Alluvial soils may be important locally. These soils are mostly without a developed profile and are the result of deposition by streams. They are usually very fertile and support luxuriant vegetation.

Saline soils are found in dry climates where rapid evaporation of water results in surface deposition and accumulation of salts leached from surrounding upland areas.

This classification shows clearly the correlation that exists among soil groups, climate, and vegetation. The composition and particularly the abundance of animals is also affected. The average biomass of soil animals under different categories of vegetaion-soil-climate systems is as follows:

Moss tundra:	3 g/m^2
Coniferous forest:	20 g/m^2
Broad-leaved deciduous forest:	100 g/m^2
Short Grass plains:	25 g/m^2
Semi-desert:	1 g/m^2

REACTIONS IN WATER

Considerable attention has already been paid to the reactions of animals and plants in streams, lakes, and ponds. These involve changes both in the chemical and physical characteristics of the habitat and are fundamentally the same as occur on land.

Water plants, especially those, such as water lilies and water hyacinths, that float, and surface concentrations of both zoo- and phytoplankton reduce light intensities like forest canopies. There is accumulation of plant and animal remains and fecal material on the bottom of the water bodies just as on land, and this material is worked over by bacteria and a large variety of microorganisms which differ only in taxonomic composition, not in activities, from those on land. Water plants obstruct the flow of water and cause deposition of suspended materials, simulating the reduction of wind velocities inside forests. An important physical reaction is the damming of streams by beavers so that ponds are formed. Such beaver activity can sometimes be usefully co-ordinated with waterflow management. These ponds eventually fill with silt and organic matter, succession occurs, and so-called beaver meadows are produced.

The nitrogen cycle and carbon cycle in aquatic eco-systems are essentially the same as on land. The absorption of oxygen by organisms and by the decomposition of organic matter in some lakes and ponds causes in the habitat a seasonal change of profound importance.

3

Environmental Factors

WATER

Water plays several different roles in the ecological relations of plants and animals. As a material entering the organism water is important as a necessary and abundant constituent of protoplasm, and the plant or animal body as a whole generally contains a large percentage of water—sometimes 90 per cent or more. Water is essential also as one material taking part in the photosynthetic reaction, through which energy becomes available either directly or indirectly to all living beings. Water is necessary as a solvent for food and as an agent for the chemical transformation of the materials within the body. Plants obtain mineral nutrients from the soil after they are in aqueous solution. Although most animals take in their food in solid form, this material must be dissolved before it can be absorbed by the blood and tissues. The intestines could be stuffed full of solid food and yet the animal would starve if no water were available for its digestion. Furthermore, water serves as a vehicle for transport or circulation within the bodies of organisms. Minerals are carried up the stem of the plant by the transpiration stream. Water is the principal constituent of the circulatory, excretory, and reproductive fluids of animals, and it is necessary as a transfer agent at respiratory and olfactory surfaces. Water also acts as a regulator of temperature for plants and animals.

For these essential purposes the proper concentration of water must be maintained inside the organism, and, at the same time, the transfer of water must be suitably regulated. If the organism could live hermetically sealed in a capsule, the retention of the necessary amount of water would be easy, but exchange of this material between the organism and the outside world is necessary. The crucial need for a proper water balance is demonstrated by the consequences of even small water losses in some instances. During starvation, for example, man may lose as much as 40 per cent of his body weight including half of the proteins and nearly all the glycogen and fat without serious danger, but if 10 per cent of the water content of the human body is lost, serious disorders result. If as much as 20 per cent of the water is lost, death follows.

The concentration of water divides the environment into aquatic and terrestrial habitats. At first sight, one might say that more than enough water exists in the aquatic environment and that water is a problem only on land. Closer scrutiny shows, however, that water is tending to enter or to leave the organism too fast in almost every situation. In certain terrestrial habitats the water supply may be excessive for many organisms, as in some topical rain forests where the air is often 100 per cent saturated and the ground is completely permeated. A visitor to such a habitat sees moisture condensing on every surface and hears the steady dripping of water from the vegetation.

In the ocean in spite of the thousands of cubic miles of sea water a scarcity of water would nevertheless exist for many plants and animals! For these organisms the concentration of water is too low relative to the abundance of salt for proper osmotic equilibrium. Fresh-water environments, viewed similarly in relation to osmotic balance, tend to contain too much water. This contrast between marine and inland aquatic situations is, in a sense, analogous to the difference between dry and humid climates on land. The maintenance of the proper water balance is thus a problem that must be considered in all habitats for all types of organisms.

THE PROPERTIES OF WATER

Chemical Properties of Water

Water is the universal internal medium of all organisms. Living matter is made up of more than 90% water. Water is also the external medium of all aquatic lifeforms and can function as a resource, condition and a habitat. The unique relationship between water and living organisms stems from the fact that water is a universal solvent: almost anything will dissolve in water to some degree. The shape of the water molecule, with an HOH angle of 105°, results in the side with the hydrogen being positively charged, *electropositive*, whereas the other negatively charged, *electronegative*. This explains many of water's properties relative to physical and chemical reactions. It also explains why water is attracted to charged ions. Cations such as Na^+, K^+ and Ca^{2+} become hydrated because of their attraction to the negatively charged oxygen end of the water molecule. The polar nature of water also explains hydrogen bond formation by water. The bonding of each water molecule to other water molecules and to other biological components explains the solution properties.

Penetration of Light through Water

The penetration of visible light through water bodies follows a negative exponential *Beer's law* relationship:

$$R_s = R_{sc}e^{-ax}$$

Where R_{sc} is the solar constant and R_s is the solar radiation after passage through a depth *x* of water of extinction coefficient *a*. Since water strongly absorbs infrared radiation, that portion of the solar spectrum will be sharply depleted on penetration through water bodies. Furthermore, turbidity due to suspended soilds (silt) will strongly deplete visible light. Although the prediction of visible light attenuation in pure water is relatively simple, in natural waters it is much more comple x due to the variability in the nature and content of suspended materials, and therefore the simple Beer's law relationship may not be easily applied. This means that photosynthesis can occur in water, at least to certain depths. Consequently, primary producers such as plants and cyanobacteria, the basis of any food chain, can grow.

Water and Temperature

Water surfaces are poor reflectors of solar radiation and therefore serve as a good sink for solar energy. However, water has a high *heat capacity*. The heat capacity of a substance is the amount of heat required to raise 1 cm^3 by 1°C. A high heat capacity means that it can absorb heat energy with only a small increase in temperature when compared with other substances. Because of this high heat capacity and the ability of water to mix and therefore to dissipate heat quickly, aquatic life forms do not have to be adaptable to a wide range of temperatures or to rapidly changing temperatures. Water becomes increasingly dense and therefore heavier at lower temperatures down to 4°C. This is why deeper water feels cooler in a lake. Because the maximum density of water occurs at 4°C, water becomes increasingly lighter at 3°C, 2°C, 1°C and 0°C (freezing). The density of liquid water at 0°C is greater than the density of frozen water at the same temperature. Thus, water is heavier as a liquid than as a solid, so that solid water and warm water can float over cold water (Fig.3.1). Another important consequence of these properties is that water freezes from the top down. This

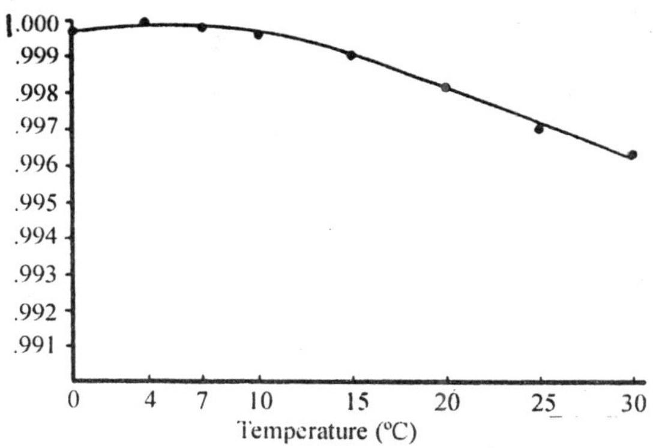

Fig. 3.1: Density of water as a function of temperature.

protects aquatic organisms because ice acts as an insulator to prevent further decreases in temperature in the remaining water and decreases the chances that ponds and lakes will freeze solid.

Energy Transfer and Water Phases

To the meteorologist, water vapor is the single most important constituent of the atmosphere. This is because at terrestrial temperatures, water passes easily from vapour to liquid and solid phases with a large release or absorption of heat. For example, the evaporation of 1 g of water requires

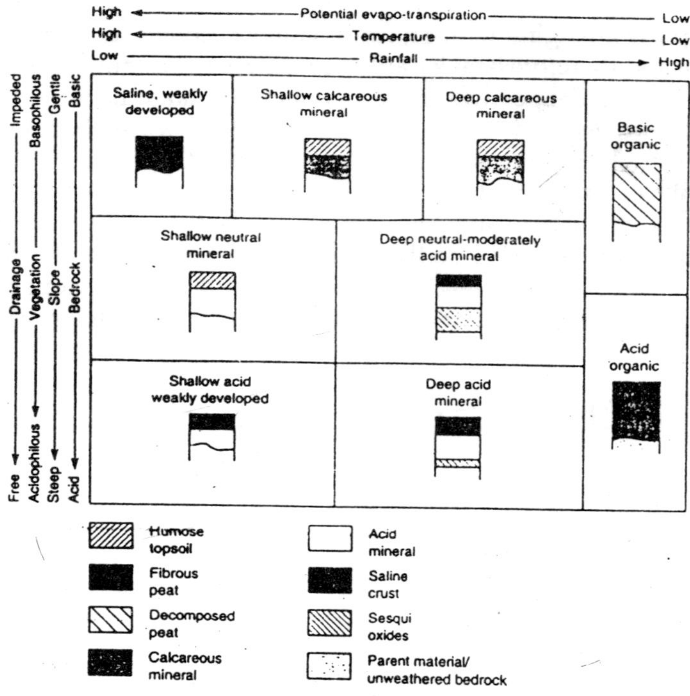

Fig. 3.2 : The relationship between environmental factors and soil type.

about 2430 J of heat, the *latent heat of vapourization*. The vapor is carried in air which rises until a level is attained at which condensation occurs with its concomitant release of the 2430 J g^{-1}. When water freezes, about 335 g^{-1} is released as the *heat of fusion* and the same amount of energy is required to melt the snow. *Sublimation,* the direct transition from solid to vapor phase (and *vice versa*) without passing through an intermediate liquid phase, involves the consumption or release of 335 J g^{-1}. Thus these energy-consuming and releasing processes in water phase changes provide the mechanisms for the transportation

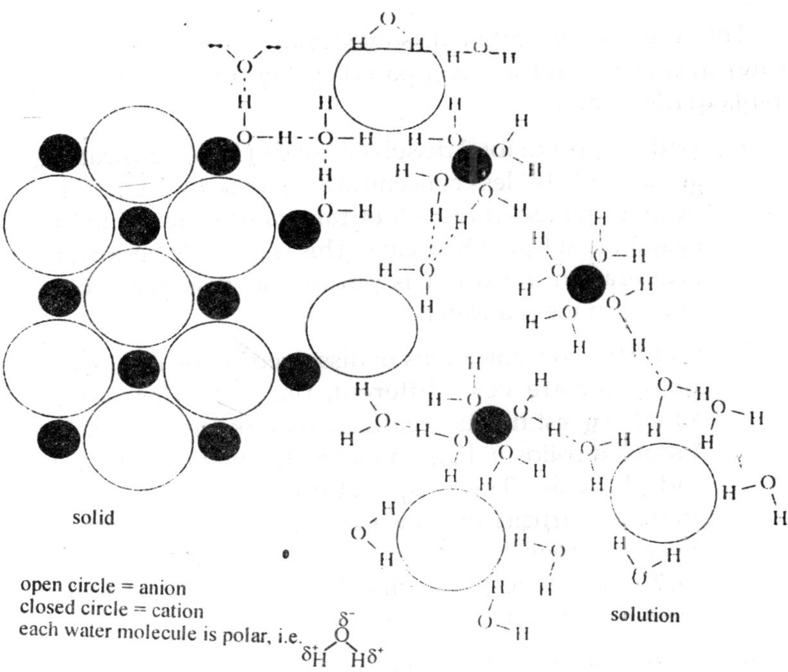

Fig. 3.3 : The dissolution of an ionic solid in water. Note how the polar nature of the water molecule allows both the dissolved cations and anions to be surrounded by solvent molecules (a process called solvation). It is the/water molecules involved in solvation that keep the ions apart and stop them from reforming the solid.

of large quantities of heat to and from the surface of the Earth. These processes play an important role in energy transfer and availability of energy to biological systems.

Many of these properties have their origins in the polar nature of the water molecule and/or the three-dimensional hydrogen bonding network that is found in bulk water. For instance, it is hydrogen bonding that is responsible for the high melting and boiling points of water, while its ability to dissolve ionic compounds can be related to the polarity of its molecules.

Rainwater

The compositions of the dissolved solids in rainwater, river water and sea water are compared in Figure 3.4. The main points to notice are:

1. With respect to total dissolved solids (TDS), rainwater is about 4.8×10^3 less concentrated than sea water. This is despite the fact that 83% of rainwater is generated by evaporation from the oceans. This is possible because on evaporation the water is purified as dissolved solids remain in the sea water.
2. While the concentrations of dissolved solids in rain and sea water are very different, the *proportions* of the different constituents in each are remarkably similar. The dissolved solids in both cases are dominated by sodium and chloride. The evaporation of sea water *spray* produces particles of sea water solids in the atmosphere; these are then dissolved by falling raindrops. This produces, with respect to dissolved solids, rainwater that appears to be very dilute sea water.

In addition to dissolved solids, rainwater contains dust particles that have been washed out of the atmosphere together with dissolved gases. The solubility of the various atmospheric gases in water is related to their atmospheric concentration and to their Henry's law constants by the following expression:

$$P_{x(g)}K_H = [X_{(aq)}] \quad (3.1)$$

where X is a given atmospheric gas, e.g. O_2, Ar, etc.; K_H is the

Henry's law constant for X; $Px_{(g)}$ represents the partial pressure (in atmospheres) of $X_{(g)}$, and $[X_{(aq)}]$ is the molarity of $X_{(aq)}$.

Table 3.1 gives the Henry's law constants for the major atmospheric gases. Of the gases listed in this table, one of them, carbon dioxide, reacts with water to give an acidic solution (i.e. pH < 7). This is important because it means that even

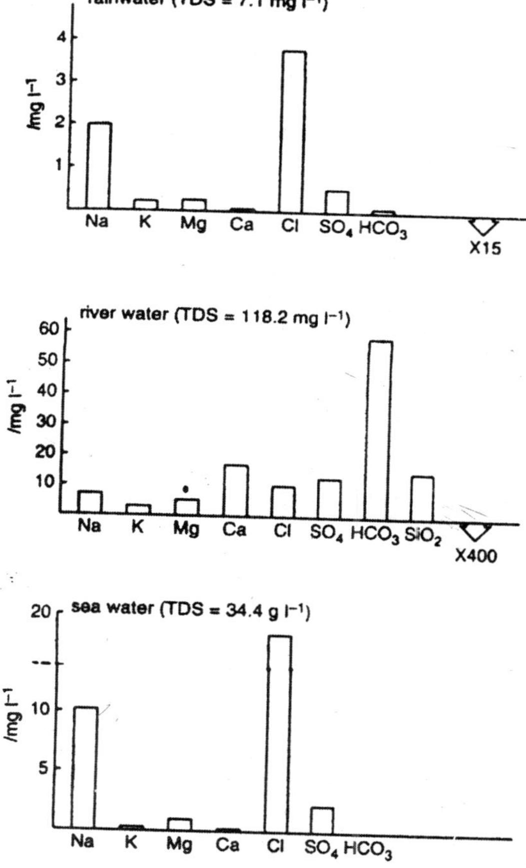

Fig. 3.4 The dissolved solids content of rain, river and sea water (from Raiswell *et al.*, 1980)

unpolluted rainwater is more potent weathering agent than pure water (which has pH = 7).

Table 3.1 Henry's Law constants (K_H) for the major gases of the atmosphere in pure water at 20°C.

Gas	K_H/*moldm*$^{-3}$ *atm*$^{-1}$
Nitrogen (N_2)	6.79×10^{-4}
Oxygen (O_2)	1.34×10^{-3}
Argon (Ar)	1.43×10^{-3}
Carbon dioxide (CO_2)	3.79×10^{-2}

It is possible to predict the pH of unpolluted rainwater via the following argument.

The reaction between carbon dioxide and water may be represented as follows:

$$CO_{2(aq)} + H_2O_{(1)} \rightleftharpoons H_2CO_{3(aq)}$$

While this reaction lies to the left a significant amount of H_2CO_3 (carbonic acid) is present, and *aqueous carbon dioxide may be treated as if it were carbonic acid.* This undergoes reactions of the type:

$$H_3CO_{2(aq)} + H_2O \rightleftharpoons HCO_3^-{}_{(aq)} + H_3O^+{}_{(aq)} \quad (3.2)$$

for which

$$K_1 = 4.5 \times 10^{-7} \simeq \frac{[HCO_3^-][H_3O^+]}{[H_2CO_3]} \quad (3.3)$$

and

$$HCO_3^-{}_{(aq)} + H_2O \rightleftharpoons CO_3^{2-}{}_{(aq)} + H_3O^+{}_{(aq)}$$

for which

$$K_2 = 4.7 \times 10^{-11} \simeq \frac{[CO_3^{2-}][H_3O^+]}{[HCO_3^-]} \quad (3.4)$$

As a result of the involvement of $H_3O^+_{(aq)}$ in these reactions, the relative proportions of the various carbonate species (i.e. carbonic H_2CO_3, hydrogen carbonate, HCO_3^-, and carbonate,

CO_3^{2-}) will be pH dependent. This dependence (Fig. 3.5) can be seen when the contribution from the carbonate ion only becomes significant at pH values > 8.5. For most natural water bodies it can therefore be ignored, as can Equation 3.4, and the total dissolved carbon dioxide concentration may be treated as the sum of the molar concentrations of $H_2CO_{3(aq)}$ and $HCO_3^-{}_{(aq)}$.

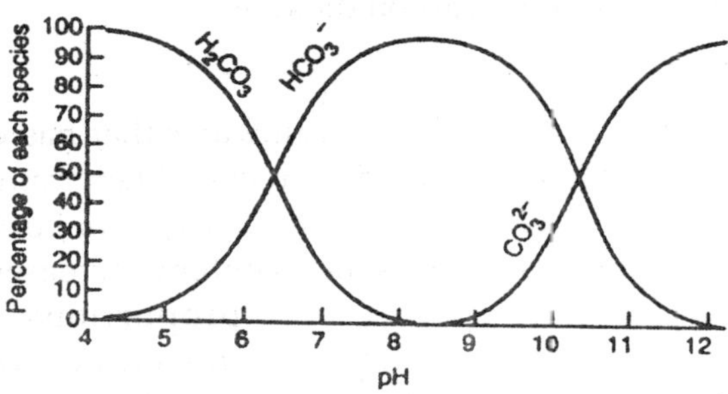

Fig. 3.5 The relationship between carbonate specification and pH.

From Equation 3.1 and the statement given in italics earlier it follows that:

$$P_{CO_2}(g)^{K}{}_{H_2CO_3} = [H_2CO_{3(aq)}] \qquad (3.5)$$

while from Equation 3.3:

$$[HCO_3^-][H_3O^+] \simeq K_1H_2CO_3) \qquad (3.6)$$

Combining Equations 3.5 and 3.6 gives:

$$[HCO_3^-][H_3O^+] \simeq K_1 K_{H_2CO_2} P_{CO_{2(g)}}$$

Noting that if the $[H_3O^+]$ from the auto-ionisation of water is ignored, Equation 3.2 implies that [H3O+] =, HCO_3-, therefore Equation 3.7 becomes:

$$[H_3O^+]^2 \simeq K_1 K_{H_2CO_2} P_{CO_{2(g)}}.$$

Taking square roots,

$$[H_3O^+] \simeq \sqrt{K_1 K_{H_2CO_2} P_{CO_{2(g)}}}$$

and, as $P_{CO_2}(g) = 3.5 \times 10^{-4}$ atmospheres,

$$[H_3O^+] \simeq \sqrt{4.5 \times 10^{-7} \times 3.79 \times 10^{-2} \times 3.5 \times 10^{-4}}$$

Therefore, as $pH = \log_{10}\{H_3O^+\} \simeq \log_{10}[H_3O^+]$,

$$pH \simeq 5.61$$

This is in close agreement with measured values of the pH of unpolluted rainwater, suggesting that its pH is indeed controlled by dissolved carbon dioxide.

River Water

River water is derived from rainwater that has either run off or run through its surrounding rocks. When rainwater hits the ground and percolates through rock its composition changes. Its total dissolved solids content increases and the proportions of its individual dissolved constituents alter. As a result, river water tends to be markedly more concentrated than rainwater and to have calcium rather than sodium as its dominant cation.

However, the situation is more complicated than this simplistic picture would suggest. When the weight ratio of $Na^+/(Na^++Ca^{2+})$ of the world's rivers is plotted against the logarithm of their total dissolved salts content, a distinct pattern emerges (Fig. 3.6).

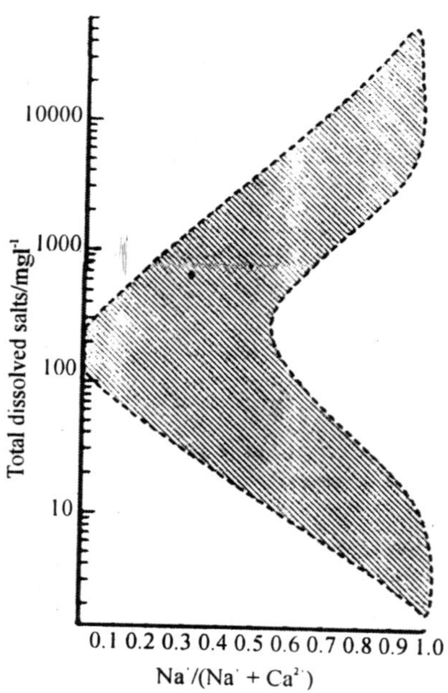

Fig. 3.6 A plot of the weight ration $Na+/(Na^++Ca2^+)$ against total dissolved salts for waters of the hydrosphere. The world's rivers, lakes and oceans are nearly all found within the shaded zone.

The water in rivers with catchment areas that are highly weathered tends to be only slightly different from rainwater. This is because the surrounding rocks contain little material that is easily leached. These rivers are said to be under *precipitation dominance* (Fig. 3.7) and have characteristically low dissolved salts contents and $Na^+/(Na^++Ca^{2+})$ weight ratios of about 0.55 and below.

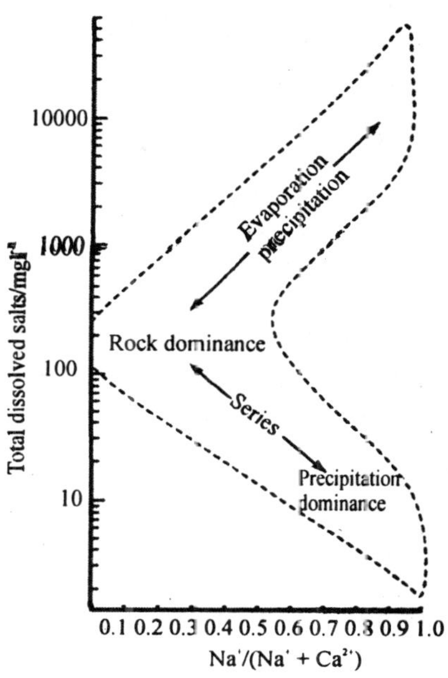

Fig. 3.7 The location of rivers under precipitation dominance, rock dominance and the influence of evaporation precipitation on a plot of the weight ratio $Na^+/(Na^++Ca^{2+})$ against total dissolved salts for waters of the hydrosphere.

Some rivers have dissolved salts contents higher than those under rock dominance. This can arise only in very arid areas, where the rate of evaporation considerably exceeds that of precipitation. Under these conditions the dissolved salts content increases. Eventually, the solubility of calcium carbonate is exceeded and this salt precipitates, leaving a solution with a high dissolved salts content and $Na^+/(Na^++Ca^{2+})$ weight ratio. Such rivers are under the control of *evaporation precipitation.*

Lake Water

Here we consider fresh water lakes, the dissolved solids content of which is largely determined by the river water that flows into them. However, compared to rivers, lakes are calm. This means, as we will see, that unlike rivers their behaviour is determined to a very great extent by the fact that pure water has a density maximum at 4°C.

If sufficiently deep, lakes in temperate zones develop a distinctly layered or *stratified* pattern during the summer months (Figure 3.8). An upper layer of warmer, less dense water, called the *epilimnion,* floats on a cooler, denser layer, called the *hypolimnion;* these are separated by a zone of rapid temperature change called the *thermocline* or *metalimnion*. Photosynthetic biological activity tends to deplete the epilimnion of nutrients; while breakdown of detritus in the hypolimnion depletes it of oxygen. On cooloing in the autumn, the temperature of the epilimnion falls to that of the hypolimnion. At this point the water is the same density throughout and wind-induced vertical mixing is possible, a process called *turnover*. This facilitates the oxygenation of the deeper parts of the lake, while bringing nutrients to the upper waters. In the winter stratification may also occur, with water at less than 4°C floating on water at 4°C in the depths. If the

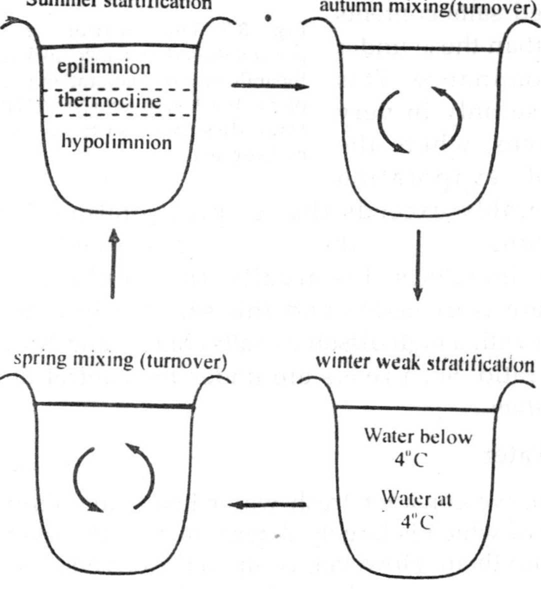

Fig. 3.8 The stratification and mixing of temperate lakes.

water freezes, the lower density of ice ensures that it does so from the top down. The insulating properties of the ice layer retard cooling so inhibiting the complete solidification of the lake. In the spring, the upper layers warm until the waters of the lake are all at the same temperature and density, therefore turnover can once again occur. Continued warming of the upper layers leads to the reintroduction of the summer stratification.

In contrast, lakes in the polar and subpolar regions that are frozen for most of the year do not show stratification in the summer, but exhibit turnover for the duration of this season.

Stratification occurs in tropical and subtropical lakes. In comparison with temperate lakes the temperature of the waters of the epilimnion and hypolimnion is higher, whilst the temperature difference between these strata is less marked. Turnover on cooling is relatively easily induced in these warm water lakes. As a result those in subtropical zones tend to have a winter circulation period, while those in tropical climates with little seasonality may either trun over frequently or have rare or irregular periods of turnover.

Sea Water

Sea water contains both dissolved and suspended solids. The concentration of total dissolved solids is, on average, 35 gl^{-1}; however, this varies between about 33 and 37 gl^{-1} depending on climate, location and season. As shown in Figure 3.4 earlier the dissolved solids composition of sea water is dominated by chloride anions and sodium cations.

Water, dissolved solids and particulate matter are perpetually added to the oceans by the action of rivers, rain, marine volcanoes, and ice and underground water flow (Table 3.2). This material is lost from the oceans, in the case of water by evaporation, and in the case of solids by precipitation and loss to the atmosphere of fine particles formed on the evaporation of sea spray.

Geological evidence suggests that the composition of sea water has changed little over the past 10^8 to 10^9 years. This

Table 3.2 Estimated rates of inflow of suspended and dissolved solids into the oceans.

Source	*Suspended solids/kg a^{-1}*	*Dissolved solids/kg a^{-1}*
Atmosphere	6×10^{11}	2.5×10^{11}
Ice	20×10^{11}	$<7 \times 10^{11}$
Marine erosion	2.5×10^{11}	
Rivers	183×10^{11}	39×10^{11}
Subsurface waters	4.8×10^{11}	4.7×10^{11}
Volcanic activity	1.5×10^{11}	

implies either that change is very slow or that the inputs of water and solids must be matched by their outputs; in general the latter proposition is accepted as true. The fate of suspended solids is, in the main, straightforward as these merely sink to be deposited as sediments. The mechanisms by which dissolved solids are removed from sea water are not all fully understood. However, physical, biological and chemical reactions all play a role. Table 3.3 itemises the main categories of reaction by which dissolved solids are believed to be removed from sea water, together with a numerical indication of their relative importance.

Methods of Meeting Osmotic Problem

If the osmotic pressure inside the organism is the same as that outside, relatively little difficulty is experienced in adjusting water equilibrium. If the osmotic pressure of the fluids of an animal or plant departs greatly from that of the outside medium, a problem results in maintaining the proper water balance. If an animal moves into an area of very different salinity, a further adjustment will be required.

Sample values of the depression of the freezing point reveal the fact that because of the higher salinity the osmotic pressure of Mediterranean water is higher than that of Atlantic water. The latter is very much higher than a typical value for fresh water. The osmotic pressure of the internal fluids of plants and of invertebrates is generally equal to, or higher than, the medium in which they are living. Among elasmobranch fishes the osmotic pressure is maintained at different levels but is

Table 3.3 A budget showing proposed mechanisms of removal of dissolved solids that enter the oceans by river discharge. Note carbonate and silicate formations are largely biologically mediated, while the other processes are either chemically mediated, or in the case of evaporative processes, physically mediated.

	Balance of constituents/10^{10} mol							
	$SO4^{2-}$	*Cl^-*	*Ca^{2+}*	*Mg^{2-}*	*Na^+*	*K^+*	*SiO_2*	*HCO_3^-*
Annual addition by rivers	382	715	1220	554	900	189	710	3118
Removal processes:								
Evaporative processes								
$715Na^+ + 715Cl^- \rightarrow 715NaCl$								
$382Ca^{2+} + 382SO_4^{2-} \rightarrow 382CaSO_4$			838	554	185	189	710	3118
Carbonate formation								
$838Ca^{2+} + 1676HCO_3^- \rightarrow 838CaCO_3 + 838CO_2 + 838H_2O$								
$44Mg^{2+} + 88HCO^{3-} \rightarrow 44M_gCO_3 + 44H_2O + 44CO_2$				510	185	189	710	1354
Silica formation								
$360H_4SiO_4 \rightarrow 360SiO_2 + 360H_2O$				510	185	189	350	1354
Interactions between clays and sea water								
Ca-clay + $189K^+ \rightarrow 95K_{2-}$-clay+$95Ca^{2+}$				500			350	958
Ca-clay + $189Na^+ \rightarrow 93Na_{2-}$clay+$93Ca^{2+}$								
$188Ca^{2+} + 376HCO_3^- \rightarrow 188CaCO_3 + 10CO_2 + 10H_2O$								
$10Mg^{2+} + 20HCO_3^- \rightarrow 10MgCO_3 + 10CO_2 + 10H_2O$								
Reverse weathering								
$100Al_2Si_2O_5(OH)_4 + 500Mg^{2+} + 100SiO_2 + 1000HCO_3^-$								
$\rightarrow 100Mg_5\,Al_2Si_3O_{10}(OH)_8 + 1000CO_2 + 300H_2O$							250	–42

always higher than the surrounding water due to the retention of urea in the blood.

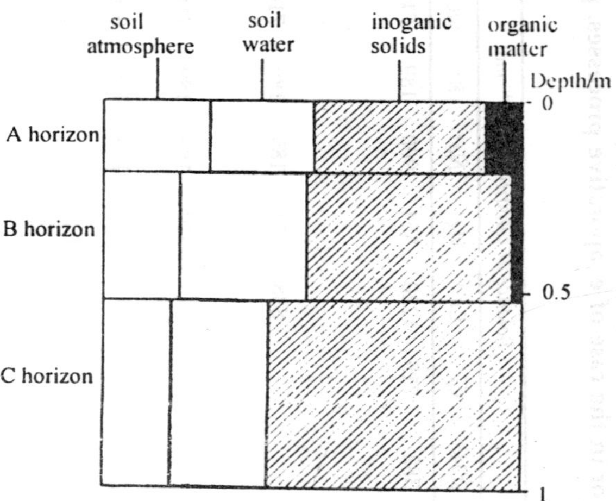

Fig. 3.9 The proportion of soil atmosphere, water, organic and inorganic solids down a typical profile

The osmotic pressure of teleost fishes tends to remain within a relatively narrow range for both fresh-water and marine species. In fresh water the teleost is osmotically superior to its medium, whereas it is osmotically inferior in salt water (See Fig. 3.10). In the sea a strong tendency exists for the fish to lose water from its tissues to the surroundings. A brief calculation will indicate the magnitude of the osmotic force with which such an animal must contend in order to maintain its water balance. If the osmotic pressure of the fish is represented by $\Delta = 0.7$ and that of the surrounding medium by $\Delta = 1.8$, the magnitude of the pressure corresponding to this difference of $\Delta = 1.1$ is:

$$\frac{1.1}{1.84} \times 1700 = 1015 cm\ Hg$$

An osmotic pressure of 1015 cm Hg, or 13 atmospheres, is tending to extract water from the fish's tissues! For the teleost, therefore, and for other animals and plants with low internal osmotic pressures sea water has the effect of being physiologically dry. Fresh-water animals and plants that migrate, or are carried, into the sea are indeed entering an arid climate.

The easiest way for an organism to deal with the osmotic pressure of its environment is to establish an internal osmotic concentration of the same magnitude. Even better, if the tissue fluids can be maintained at a slightly higher pressure, then there will be a tendency for the needed water to enter the organism from the surrounding medium. This is the general method followed by plants and by invertebrate animals.

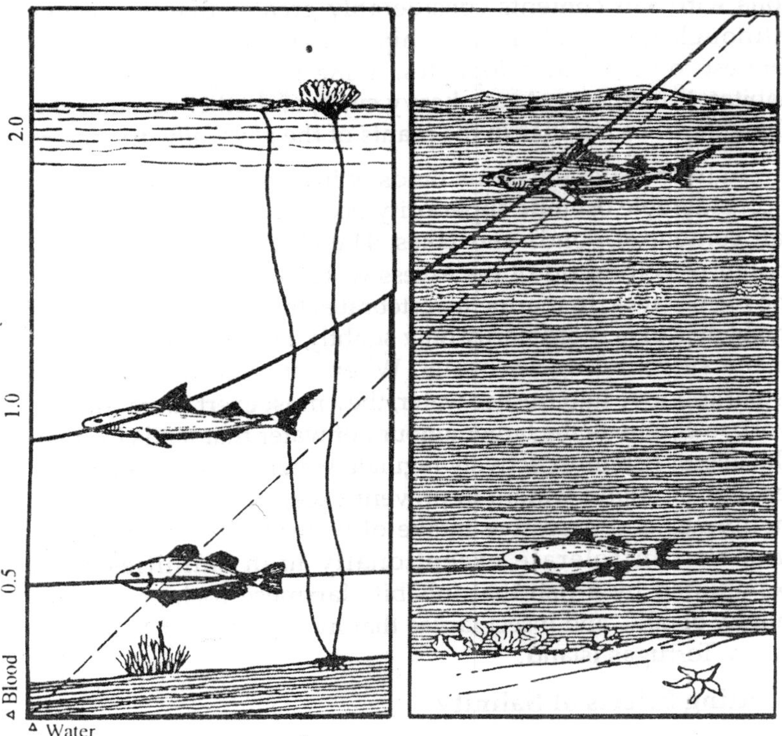

Fig. 3.10 Schematic diagram of osmotic pressure differences between elasmobranch and teleost fishes and their fresh-water (*left*) and marine (*right*) media. The degree of shading indicates the relative values of osmotic pressure.

Most fresh-water and terrestrial plants can maintain an internal osmotic pressure ranging up to 2 atmospheres but they find saline habitats with a higher osmotic pressure physiologically too dry and uninhabitable. However, specially adapted plants, the *halophytes,* are able to tolerate higher salinities and to absorb water by virtue of the fact that the osmotic pressure of their tissue fluids is exceptionally high. Halophytes sometimes exhibit internal osmotic pressures as great as 35 or even 40 atmospheres. Such plants inhabit the margins of the ocean and of salt lakes and can grow in saline soils with salt contents considerably greater than sea water. Although the exact relationship is not understood, it is of interest to note that halophytes, living in a *physiologically* dry habitat, exhibit succulence, heavy cutin, and other xeromorphic characteristics familiar in the plants of the *physically* dry desert.

In fresh water the tissue fluids of invertebrate animals have a considerable osmotic superiority because of the typically low salt content of ponds and lakes. The chief problem here is a matter of getting rid of the excess water which enters through the membranes. In marine invertebrates the osmotic pressure of the body fluids is ordinarily only slightly, if any higher than that of the surrounding sea water. When invertebrates living in estuaries move into fresher water, they must exert more osmotic work to eliminate the larger amount of water that tends to enter. The teleost fish must actively regulate water transfer both in the sea and in inland waters to prevent excessive loss of water in the former and excessive intake of water in the latter. A few animals can tolerate exceptionally high salinities as is exemplified by a fish that inhabits Japanese rock pools with salinity of 60‰ and insect larvae that live in water of 42 to 62‰ at Dry Tortugas Island.

Limiting Effects of Salinity

The extent to which the various species can endure changes in osmotic pressure influences their activity and range. Organisms with a low tolerance for differences in salinity are known as *stenohaline* forms. Species that are limited exclusively to the open ocean or to fresh water are thus classed as stenohaline. Others may be limited to a narrow range of salinity of an intermediate value, but this is relatively rare. Plants and animals that are able to tolerate a wide range of salinities are termed *euryhaline*. Thus organisms inhabiting an estuary and fish that migrate back and forth from fresh to salt water are euryhaline. Such fish as the salmon and the eel are able to regulate their water balance in either a hypertonic or hypotonic medium.

It should be noted that no sharp line or numerical limit distinguishes stenohaline and euryhaline forms—these terms are entirely relative. The terms also do not bring out the important difference in the effect of time of adaptation. Some species can withstand wide differences in salinity if they become adapted slowly, but are unable to tolerate rapidly changing salinity. Relatively few species can endure both a wide range and a rapid change in salinity. Interestingly enough, some euryhaline organisms may grow best in intermediate salinities, or even require them for a part of their life cycles, although our knowledge is scanty in this area. The division rate of certain species of phytoplankton has been shown in laboratory cultures to be twice as great at a salinity of about 20‰ than at salinities to about 10‰ or 30‰. We also know that the American oyster can live in full sea water and can tolerate fresh water for short periods, but observations indicate that the larvae of this species will not settle at salinities above 32‰ nor below 5.6‰. The optimum salinity for settlement in this estuarine animal is 16 to 18.6‰. Changing salinities in estuaries may control the growth, reproduction, and distribution of more of the inhabitants than we now realize.

Limits of distribution are sometimes sharply determined by salinity toleration. The vegetation around the margins of salt lakes and in the spray zone near the ocean shore exhibits a characteristic gradation of species from those most tolerant of salt to those least tolerant. In San Francisco Bay the slight difference in tolerance between two species of pile-boring mollusks was of great economic importance. In the upper regions of the Bay wharves and other wooden structures had been built without any special protection against shipworms because the local species of the genus *Bankia* was unable to grow in salinities below 10‰ which frequently occurred in the area. In 1913 another species of wood-boring mollusk, belonging to the genus *Teredo*, was introduced into San Francisco Bay, and this species could tolerate salinities as low as 6‰. Spreading rapidly, *Teredo* attacked the wooden structures of the area and within a few years had caused destruction amounting to more than $25,000,000.

Many interesting problems in relation to salinity tolerance remain unanswered. In most cases we do not know why one species can tolerate a rapid or extensive change in the salt content of its medium and another cannot. Why the euryhaline species have not come to dominate the open ocean as well as the coastal regions is unknown. Another question of interest is how oceanic birds and mammals can balance their water budgets without a supply of fresh water. Many of these forms eat marine invertebrates whose salt content is nearly as great as an equal volume of sea water. It may be that the water contained in the tissues of the food and the metabolic water are sufficient for the needs of marine birds and mammals, or these animals may have some special adaptation for excreting excess salts.

AMPHIBIOUS SITUATIONS

It is difficult enough for animals and plants to move from the sea to fresh water or vice versa, but, when aquatic organisms attempt to invade dry land, they are flopping out of the frying pan into the fire as far as the water problem is concerned. Nevertheless life on land does have certain advantages. Plants usually find more light, better anchorage, less abrasive action, and more concentrated nutrients. Animals may make use of the more abundant vegetation for food and shelter, the greater oxygen supply, and the possibility of more rapid movement. Against these advantages there is one signal disadvantage—the scarcity of water. Success in colonizing dry land has been dependent on securing and retaining sufficient water. This could be easily arranged, perhaps, if the organism could seal itself up, but, as already stressed, surfaces must be left open for exchange with the external world. Of immediate importance is the problem of how to feed and to respire in air without drying up. Representatives of relatively few phyla have succeeded in becoming wholly independent of the aquatic habitat. However, some plants and animals have come part way out of the water, or come out for short periods, and we shall consider these first.

Swamps and Temporary Pools

Certain organisms have hit upon methods of using the advantages of both the air and the water environment. The emersed vegetation, for example, which grows in swamps or on the margins of lakes has its roots in the water and its upper portions in the air. In this position the roots find plenty of water available and extract nutrients from the mud, while the leaves are in the best position for receiving light from the sun and carbon dioxide from the atmosphere. However, many plants are not able to survive in a partially submerged condition because of inadequate direct supply of oxygen for the roots. Plants that are adapted to existence entirely under water, or with their roots in water or in saturated soil, are known as *hydrophytes*.

Amphibians are similarly able to utilize both air and water environments. Many species can leave the water to forage on land, and, if their pools dry up, they can migrate to other bodies of water. Such animals are not limited by land barriers as most fish are. Yet amphibians must re-enter the water or visit damp places at intervals to keep their skins moist, and most species need water for reproduction.

Organisms living in swamps and pools must be able to deal with periods of drought when they occur. Physiological adaptations making this possible have been evolved sometimes without any accompanying special morphological structures. Certain species of fish in India can live in wet grass for as long as 60 hours, and other animals burrow in the mud where they remain in a dormant condition during the dry period. Water mites have been shown to be able to live under debris for intervals up to 6 months after their pools have dried. The African lung fish, on the other hand, constructs a special mud cocoon and curls up inside for the duration of the dry spell. The fish secretes about its body an impervious sheath which prevents the loss of water from its tissues, and is thus enabled to survive drought for more than two years. Other specialized structures for tiding over dry periods are the spores and seeds of water plants, and the 'resistant' eggs and cysts of various aquatic animals ranging from the Protozoa to the Crustacea.

Another adaptation for meeting the water problem in temporary pools is that of speeding up development and taking quick advantage of water when it is available. In certain animals living in this type of habitat the larval stage is accelerated; in others parthenogenesis has been adopted. If a single female cladoceran such as the water flea, *Daphnia,* finds itself in a temporary pool, it does not have to hunt around for a male. It can go ahead and reproduce rapidly by parthenogenesis. The young animals mature and continue reproducing parthenogenetically until in a relatively short period a large population has built up. Later, however, if the pool tends to dry up, the adverse conditions cause the appearance of males in the population. The females now produce a different type of egg, one which requires fertilization. Only one of these 'winter' eggs is formed by each female and the brood pouch is modified into an ephippium. After the death and disintegration of the female, the ephippium remains as a resistant egg case within which the egg can survive prolonged periods of drought and even freezing. In this condition the egg may be blown into another pond or carried in dried mud by animals to another region. When water is again supplied to it, the egg hatches out and is able to start the cycle over again.

Tidal Zone

Another situation that might be classified as amphibious is the *tidal zone*. Plants and animals living on the seashore between tide marks are subject to alternating droughts and floods twice each day. Organisms living in the center of the tidal zone must be able to be aquatic for about 6¼ hours when the tide is in, terrestrial for about 6¼ hours when the tide has ebbed, and so on, day in and day out. The tidal zone is thus a very difficult habitat, and also a very complicated one because of differences in the timing and height of tide. Since low tide and high tide occur about 1 hour later each day, the intertidal area comes to be exposed at all hours of the day and night. During the first and last quarters of the moon *neap tides* occur, i.e., tides of low amplitude, but during periods of new moon and full moon the tides rise higher and fall lower and are known as *spring tides*. In some locaties, such as the west coast of the United States, one set of high and low tides each day has a much greater amplitude than the other set. The height of the tide differs greatly from locality to locality, and the duration and depth of inundation varies according to the level within the tidal zone at which a given organism exists. For example, an animal or a plant attached near the upper extreme limit of the tide will be covered by sea water for only a brief period about every two weeks, whereas an individual living near the lower extreme tidal limit will receive only a correspondingly short exposure to the air. The tidal zone comprises the upper portion of the *littoral zone*.

Despite these severe conditions a dense population of a very wide variety of plants and animals characteristically inhabits the tidal zone. The fauna and flora consist chiefly of sessile forms, such as the common seaweeds, barnacles, mussels, and clams. Those plants and animals that cannot move out of the tidal area must endure the full effect of the differences in the two media. In addition, certain active forms forage here under favourable conditions and then retreat. At low tide animals come from the land and feed on the material that has been exposed. Shore birds are familiar in this area, but less observed are rats, mice, mink, skunks, and other small mammals that feed chiefly at night. During high tide the reverse situation obtains. Fishes, crabs, and other active animals from deeper water move in for a few hours and then retire. The alternative visitation of these forms remind one of Box and Cox in the famous operetta.

It is of interest to inquire why the tidal zone should be so densely populated since the habitat presents such great difficulties for existence. In the first place all species tend to spread; marine animals and plants are pressing on their boundaries, and terrestrial forms are likewise attempting to invade regions at the margin of their habitats. These species extend their ranges just as far as they can endure the conditions. Furthermore, the tidal zone has certain advantages over conditions in deep water. Shallow water has more light, oxygen, and circulation; and certain predators are excluded. A great amount of organic matter occurs in this zone, and much of this forms a valuable food for the inhabitants. A large part of the organic material exists as detritus resulting from the breakdown of seaweeds and the disintegration products of dead animals. In addition many kinds of smaller planktonic organisms abound in the shallow water and add to the food supply. The adaptations of the intertidal organisms and the control of their activities by the environment will be discussed briefly here; more comprehensive treatment of this special habitat will be found in such books as Yonge (1949) and Wilson (1951).

Practically all the permanent inhabitants of the tidal zone are aquatic organisms, and accordingly their chief problem is dealing with the adverse conditions of the air medium when the tide is out. Many physiological and morphological adaptations are displayed for withstanding the recurring periods of water shortage. Many of the simpler organisms can endure a considerable drying of their tissues without permanent injury. Sea anemones, for example, have been kept out of water 18 days; at the end of this period the animals looked like dried raisins but became active again when replaced in sea water. Attached forms with shells like the barnacles, mussels, and oysters simply close up shop during the period of low tide, and many can remain sealed up in this way for unbelievably long periods without injury. If kept at a low temperature oysters may be stored out of water for 3 months or more and will resume normal activity when replaced in the sea. Snails and starfish move into sheltered crevices when the tide is out; worms, clams, and other burrowing forms dig deeper in the mud and wait for the tide to return.

Certain intertidal forms seem to benefit by periods out of water. A practical application of this fact is made use of by many oyster growers who prefer to raise oysters in situations where they are exposed to the air for at least a portion of each day. The oystermen believe that as a result the muscles which close the shells are strengthened. When the oysters are shipped to market, the shells are held more tightly together thus retaining the fluids within the mantle cavity. These oysters are said to be better able to "hold their liquor!"

Actively swimming or crawling animals may be able to go without food for the period when the tide is out, but they cannot remain active and do without a supply of oxygen for this length of time. There is no lack of oxygen in the air, of course—the problem is to get it without allowing the respiratory membranes to dry out. The gills of many of these animals are placed in partially enclosed cavities of the body where oxygen may reach them but the loss of water from the respiratory surfaces is retarded. A striking gradation in the anatomical adaptations of crabs in the tidal zone has been described by Pearse (1950) for the North Carolina shore. He called attention to the fact that species living higher up on the shoreline show a reduction in the number of their gills and particularly in the ratio of gill surface to the size of the whole animal (Fig. 3.11).

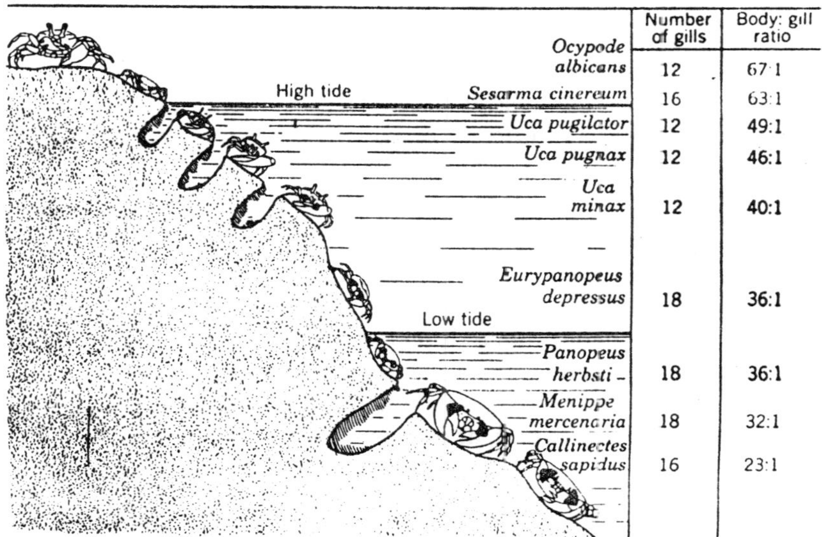

Fig. 3.11 Reduction in number of gills and increase in ratio of body to gill volume in species of crabs living at progressively higher levels in the littoral zone.

Differences in ability to withstand the progressively more difficult conditions from the low-tide mark to the high-tide mark have resulted in a zonation of the plant and animal species. On rocky shores this zonation is often so pronounced that it can be seen at a distance when the tide is out. In the situation, the sharp upper limit for the growth of the rock weeds is easily apparent. At a higher level the longer period of air exposure and of lack of sea water imposed a clearly defined upper boundary upon the barnacle population. Stephenson and Stephenson (1949) have found that on rocky shores in many parts of the world three fundamental zones may be recognized: the Littorina Zone, characterized by snails; the Balanoid Zone, characterized by barnacles; and the Laminarian Zone, characterized by laminarian seaweeds. The relation of these zones to the tide marks is shown in the diagram.

Littorina Zone	
Balanoid Zone	High Spring Tide TIDAL ZONE Low Spring Tide
Laminarian Zone	

The upper part of the Littorina Zone extends into the spray zone, or *supralittoral zone,* where the salt-resistant outposts of the land fauna and flora are to be found as well as those marine organisms most capable of enduring desiccation. The Laminarian Zone extends for varying distances below low-tide mark.

Frequently various subdivisions are superimposed on the main zonation of rocky shores. The vertical limits of the various subzones of fucoid algae and of associated invertebrates are determined not only by relative tolerance in respect to the water factor but also by type of rock, wave action, and relative success in invasion, competition, and resistance to predation.

On muddy and sandy shores zonation of plants and animals between tide marks similarly occurs, but the observer will have to dig to ascertain the full ramifications of the tidal relationships since many of the inhabitants are burrowers. In addition, microscopic examination of samples of bottom material reveal a zonation of the microorganisms in the tidal zone. The distribution of copepodes dwelling in a sandy beach of Cape Cod will serve as an illustration (Fig. 3.12). The different species present were so nearly alike in general appearance that they could be distinguished only by careful study. Nevertheless, one species was found to live only below low tide where it is never subjected to the drying action of air. Another form was most abundant midway between the tide marks. A third species was most numerous just below the high-tide line. Furthermore, when samples of sand from different levels beneath the surface were studied, the numerical abundance of the copepods was found to vary with depth in the sand as well as with position across the tidal zone. At the low-tide mark the largest number of animals was found in the upper 4 cm, whereas at the high-tide mark the largest population occurred at a depth of 12 to 16 cm. In the tidal zone, as in other amphibious situations, the water factor thus exerts a major controlling influence on the minute organism as well as on the more conspicuous species.

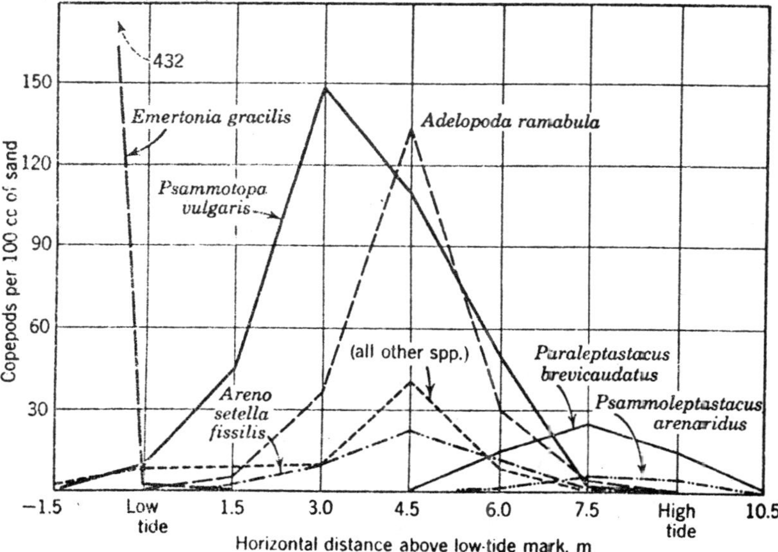

Fig. 3.12 Abundance and distribution of sand-dwelling copepods across the tidal zone at Nobska Beach, Woods Hole.

WATER PROBLEM IN THE TERRESTRIAL ENVIRONMENT

When we turn to dry land as an environment, the water problem becomes extremely acute. Everyone is familiar with the quick death which awaits aquatic organisms brought out on land. Many terrestrial forms can endure the land environment only if they remain in damp places. This is true for a great many species of plants and also for animals without special protection. A salamander that has escaped from its terrarium and run across the dry floor will soon be dead if it is not rescued and put back in a humid atmosphere. Earthworms frequently crawl out on a concrete sidewalk during a spring rain. If the sun comes out and dries the sidewalk, many of the animals will be killed before they can find their way back to the moist ground.

The successful invasion of the dry land is dependent upon the possession of methods for securing and retaining sufficient water and at the same time allowing adequate exchange with the environment. Terrestrial plants and animals as well as aquatic forms must maintain a proper *water balance* between water income and loss. Great variation exists in the moisture content of land organisms and also in the amount of water intake and output that is necessary to provide for metabolic processes. Insects subsisting on dry food probably hold the record for the minimum water exchange. At the other end of the scale are many types of higher plants that are extravagant water users. Some species of plants absorb and transpire more than 2000 grams of water for every gram of dry matter produced by assimilation. An oak tree may transpire as much as 570 liters of water in a single day.

How many taxonomic groups have succeeded in colonizing dry land? Actually, only a very few of the prominent kinds of plants and animals have accomplished this. In the plant kingdom it is chiefly the vascular forms that are able to live in really dry places, but some low-growing bryophytes, algae, and fungi—particularly the rock lichens—form noteworthy additions to the list. Among the vertebrates, only the reptiles, birds, and mammals can live freely exposed to dry air. Although several phyla of invertebrates are reported in the land fauna, only the insects, spider, and snails occur in important numbers in dry habitats. By contrast, there are a great many plant and animal groups that are exclusively aquatic.

Occurrence of Water in Land Environment

The only ultimate source of water for the terrestrial environment is condensation—chiefly in the form of rain. If precipitation were evenly distributed, it would cover the earth to a uniform depth of 1 m each year. Quite the contrary is the case. A glance at Fig. 3.13 will reveal the irregularity in the geographical distribution of rainfall. In the desert areas the rainfall is less than 25 cm per year and is inadequate for most organisms. In the black areas of the chart an excessive rainfall of 200 cm or more is reported each year. The mean annual rainfall on the south side of the Himalayas is recorded as 1232 cm. At the other extreme is Iquique, Chile, with an average precipitation of 0.125 cm per year. Among the Olympic Mountains in the state of Washington the annual rainfall totals as much as 380 cm on the windward side of some ridges and as little as 25 cm on the leeward side.

Seasonal variations in the distribution of rain may be of even greater importance than the geographical aspect. In many regions the division of the year into a rainy season and a dry season is of more ecological significance than the change in temperature between summer and winter. In the climatographs shown in Fig. 3.14 two situations are contrasted. In the Chicago region representing a temperate climate, an average of 5 to 9 cm of rain falls in every month of the year but the average monthly temperature varies from about—4°C in winter to about 22°C in the summer. In the tropical climate of Barro Colorado Island the average monthly temperature changes by not more than 1 or 2°C throughout the year, whereas the precipitation drops to less than 3 cm per month in the dry season and rises to more than 40 cm per month in the wet season.

Wherever the amount of precipitation varies greatly during the course of the year, the time of occurrence of rain in relation to the temperature cycle has great effect on the vegetation. Only

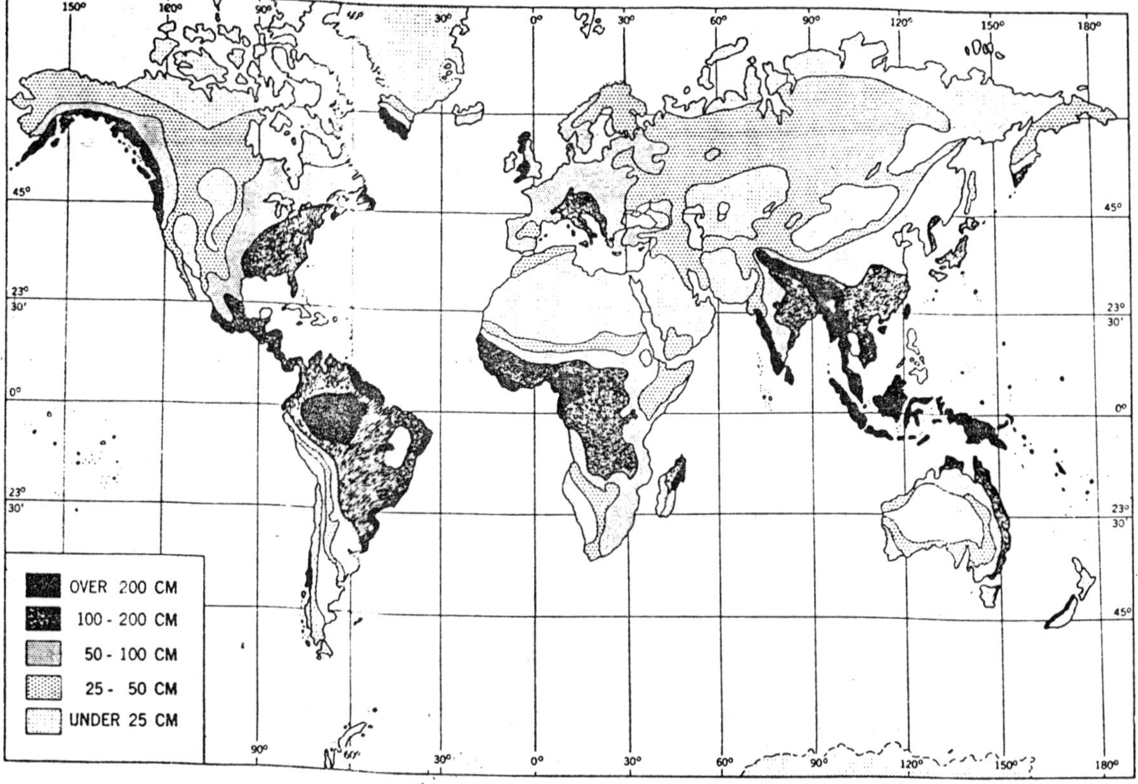

Fig. 3.13: Geographical distribution of mean annual rainfall.

those species will thrive whose varying needs for water in the different life stages are satisfied by the seasonal distribution of available moisture. In many of the grass-covered or forested areas of the temperate zone the major portion of rain occurs in the summer, but in other regions, of which southern California is an example, most of the precipitation occurs in the winter. In the latter situation we find a special type of vegetation with broad evergreen leaves known as the sclerophyllous forest. In

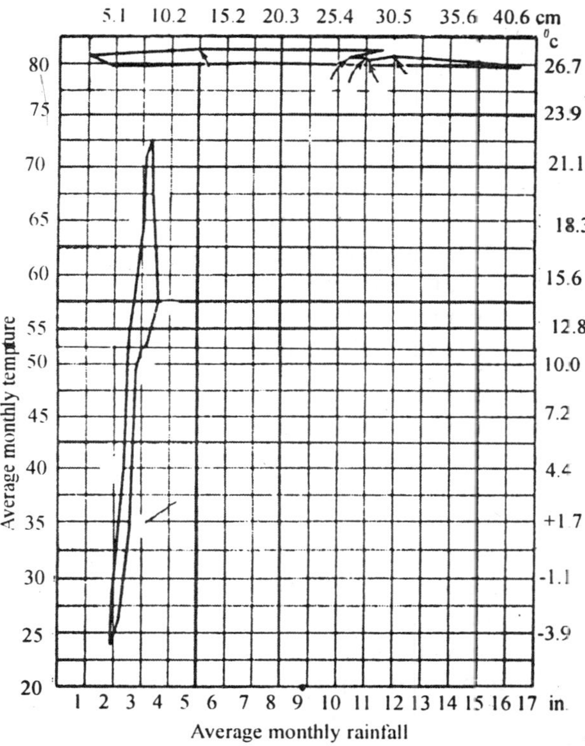

Fig. 3.14 : Average rainfall and temperature for each month of the year as indicated by the numerals.

Lower figure : temperate climate—data for Chicago, Illinois.

Upper figure : tropical climate—data for Barro Colorado Island, Panama Canal Zone.

the prairie provinces of Canada the average annual precipitation of 50 cm would not be sufficient to support agriculture if the rain were evenly distributed through the year. However, the major portion of the rainfall occurs during the spring growing season, and also at a time before the rate of evaporation has become excessive, with the result that the available precipitation is used to best advantage. The existing combination of ecological conditions thus permits western Canada to be an important wheat-growing region.

The foregoing discussion brings into relief the fact that in considering moisture conditions on land the actual amount of the precipitation is only part of the story. The other part, and often the more important part, is the loss of water. Of the rain that falls upon the surface of the soil, part runs off immediately, part sinks in, and another portion is lost by evaporation. The amount of moisture at any one time and place depends upon the relative rates of the supply of water to the soil and evaporation from it—another example of an ecological factor whose value depends upon an equilibrium. In evaluating the water factor in the terrestrial environment both supply and loss processes must be taken into account.

Moisture in the Soil: Since the terrestrial environment is so varied, it is not surprising that the water factor is very complex, and the moisture in the soil and in the air will first be considered separately. When rain water enters the soil, it fills the spaces between the particles. The volume that can be filled is known as the *pore space* and commonly varies between 60 per cent for heavy soils and 40 per cent for light soils. When the rain ceases, a certain portion of the water in the soil soon drains out, and this is known as the *gravitational water*. The portion of the soil water that is held by capillary forces around and between the particles is the *capillary water*. That part of the pore space which is not filled by gravitational or capillary water may be occupied by water vapour. Another form of soil water, termed the *hygroscopic water*, occurs as an extremely thin film on the soil grains but this cannot move as a liquid. A small portion of the soil water is chemically bound with soil materials and is known as *combined water*. The total amount of capillary, hygroscopic, and combined water plus the water vapour constitutes the *field capacity* and is the maximum amount of water that the soil can hold after the gravitational water has drained away Fig.3.15.

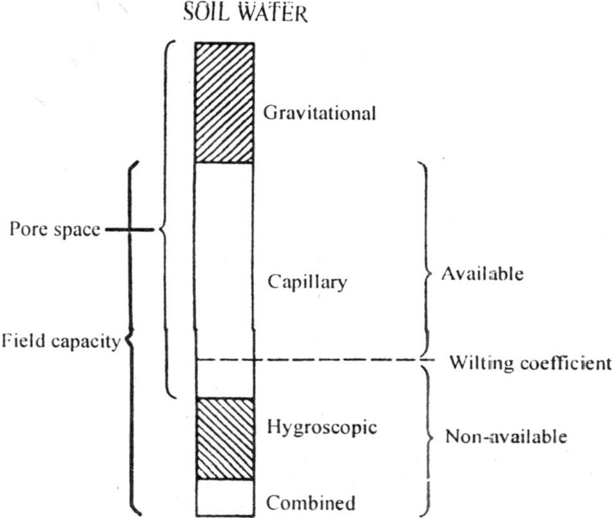

Fig. 3.15 : Generalized diagram of the forms of soil water. The relative proportions and the value of the wilting coefficient differ according to the nature of the soil.

The maximum potential supply of water available for the vegetation is represented by the full field capacity of the soil. As plants draw on the soil water, they gradually reduce the amount held in capillary spaces. Below a certain moisture content plants tend to wilt in the hottest part of the day. When the capillary water has been still further depleted, a point is reached at which the plants will not recover from wilting until more water is added to the soil, regardless of other environmental conditions. At this point absorption of water by the plants has become too slow to replace the water lost by transpiration. The moisture then remaining in the soil is designated as the *permanent wilting percentage*, or the *wilting coefficient*. Since little difference has been found in the abundance of soil water when wilting occurs for plants of various species growing in the same oil, the permanent wilting percentage is primarily a characteristic of the soil, and as such has great ecological significance. The amount of water between the full field capacity, as a maximum, and the wilting coefficient, as a minimum, represents the *available water*. As indicated in Fig. 3.17, a considerable amount of water may still remain in the soil after the wilting coefficient has been reached. A fraction of the capillary water, the hygroscropic water, and the combined water, as well as the water vapour, cannot be obtained by the plant, and these constitute the *non-available water.*

Soils vary considerably in the relative proportions of the different categories of soil water. In some instances the amount of non-available water may actually be larger than the available water. The fact that a portion of the water in the soil is unavailable to land organisms and that water in the sea is unavailable to aquatic organisms not adapted to its osmotic pressure contributes further to the general concept of the universality of the water problem. From the ecological point of view the actual amount of water present in any habitat is not as important as its availability.

Moisture in the Air: For many terrestrial organisms the moisture in the soil is chiefly important as constituting the principal source of water, whereas the moisture in the air is mainly significant as controlling the loss of water. For this reason the *absolute humidity*, or total amount of water in the air, is generally of far less ecological consequence, than the *relative humidity*. The ecologist therefore focuses his attention on the relative humidity or the amount of moisture in the air as a percentage of the amount which the air could hold at saturation at the existing temperature. Since the capacity of air for water vapour increases with temperature, the relative humidity of the atmosphere is reduced in any situation in which an increase of temperature occurs without an accompanying increase in the total moisture content of the air.

In the terrestrial environment the ecological effect of the water factor is consequently strongly influenced by the temperature factor Of two regions having the same rainfall the warmer is the drier in the ecological sense (Fig.3.16). The climate of a locality with a mean annual precipitation of 50 cm would be characterized as humid if the mean annual temperature were —7°C or less. On the other hand, another locality with the same rainfall would be regarded as having a semiarid climate if the annual temperature were 21°C or more. The contrast in climate between the Canadian prairie and the Mexican desert, both with about 50 cm of rain but with very different temperature conditions, is an excellent illustration of this principle.

The geographical variation in relative humidity is very great. Relative humidities of 80 to 100 per cent characterize the tropical rain forest. Regions reporting values of less than 50 per

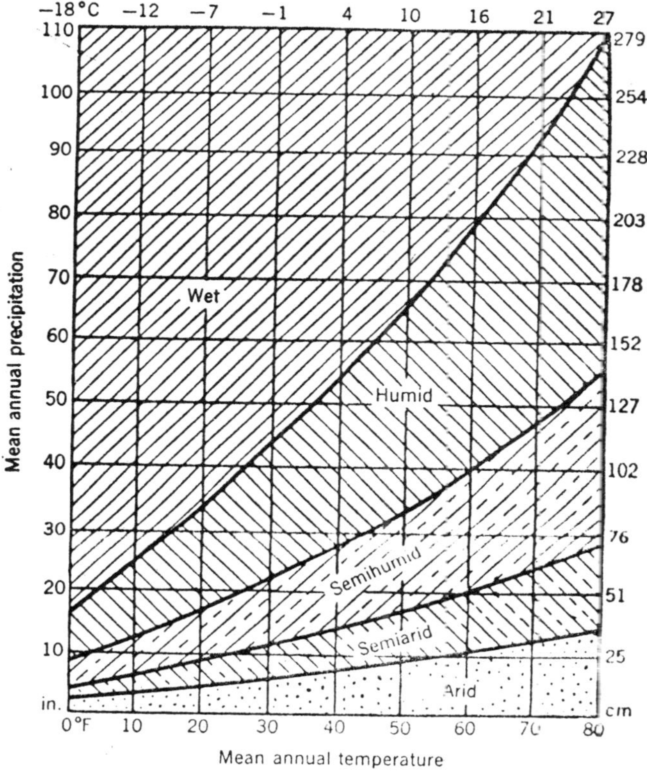

Fig. 3.16 Schematic representation of the influence of rainfall and temperature on climate.

cent are regarded as having dry climates, and those with values of less than 20 per cent are extremely arid. It is of interest to note in passing that in cold winter weather the relative humidity inside our houses is rarely higher than 35 per cent.

At any one locality the relative humidity may remain relatively constant for long periods of time or may vary widely. On many oceanic islands the humidity is very nearly the same throughout the year. In other localities, characterized by wet and dry seasons, the humidity fluctuates widely from one part of the year to another. In certain situations as on the plains and in desert regions considerable changes in moisture content may occur during the course of each day. Records made in a short-grass prairie in the United States showed a variation from a relative humidity of less than 30 per cent in the early afternoon to more than 95 per cent in the middle of the night (Fig. 3.17). Obviously animals and plants living in such habitats must be equipped to withstand these rapid and extensive changes in the water factor, and their lives must be attuned to them. In deserts

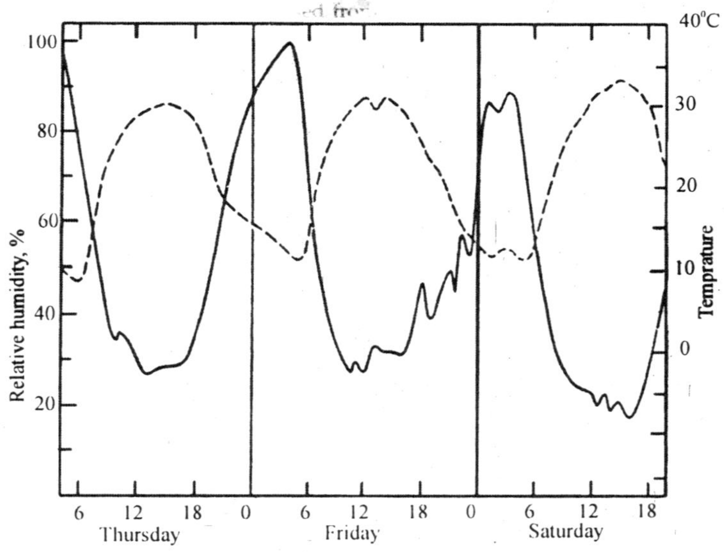

Fig. 3.17 Hygrothermograph record of temperature (*broken line*) and relative humidity (*solid line*) in the short-grass plains of central United States during days in early July.

where daytime humidities are extremely low and evaporation is excessive, the greater relative humidity at night helps to relieve the critical moisture condition. Many desert animals take advantage of this situation by going abroad only during hours of darkness; in some desert plants the stomata open only at night when transpiration loss is at a minimum.

The foregoing has been a brief description of the amount of moisture in the air in terms of its relative humidity. As already implied, the chief ecological significance of the relative humidity is its effect on the rate of water loss. Terrestrial animals and plants lose water directly to the air by evaporation and transpiration. The water supply in their substratum is also reduced by direct evaporation from the soil and indirectly by the transpiration of the vegetation. Evaporation takes place very rapidly from soil because of the great surface area presented by the fine particles and causes the drying of the upper layers. Water is lost from the deeper layers chiefly by abscrption by roots and attendant transpiration.

The magnitude of water loss by transpiration is often not appreciated. Perhaps some idea can be obtained from the fact that a hectare (2½ acres) of mature oak trees will transpire as much as 25,000 liters of water per day. This volume would be equivalent to a layer of water 0.25 cm deep over the whole area. In other words, if transpiration continued at this rate for 10 days, a rainfall of at least 2.5 cm would be required to restore the water. One might think that the maximum rate of water loss would occur from saturated soil or from a pond. Actually the rate of water loss per unit area from the ground due to the transpiration of the vegetation may be nearly twice as great as that from a free water surface due to evaporation.

Three atmospheric conditions that greatly modify the rates of evaporation and transpiration are the saturation deficit, the temperature, and the wind velocity. The *saturation deficit* is the difference between the actual vapour pressure and the maximum possible vapour pressure at the existing temperature. The saturation deficit thus gives more information of ecological significance than the relative humidity alone since the saturation deficit provides a measure of the capacity of the air to take up additional moisture. An increase in the saturation deficit produces a rise in evaporation rate.

An increase in temperature similarly speeds up the evaporation process. A good indication of the magnitude of this influence is obtained from the amounts by which reservoirs are lowered in regions of different temperatures. During one year the water level in a reservoir in Ontario dropped 38 cm, a reservoir in California lost 2.4 m, and one in Egypt was lowered 3.6 m through evaporation alone. The distribution of plants often reveals the influence of temperature on direct and indirect water loss. On mountain slopes vegetation zones are found at different altitudes according to the exposure to the heat of the sun. Moisture-demanding species are restricted to higher levels on the side of the mountain toward the equator than they are on the cooler opposite side. The variation in the amount of rain required for a good growth of short grass furnishes another illustration of the effect of temperature. In Montana a precipitation of about 35 cm is sufficient, but in northwestern Texas a rainfall of 53 cm is needed to produce the same amount of grass. The explanation of this fact is easily apparent when it is realized that evaporation in the 6 summer months in Montana amounts to 84 cm, whereas in Taxas it averages 137 cm.

The wind velocity also exerts a major effect on the loss of water. With a gentle zephyr of only 8 km per hr the transpiration of plants is increased 20 per cent over that which they exhibited in still air. A wind velocity of 16 km per hr increases the transpiration by 35 per cent and one of 24 km per hr increases it by 50 per cent. The desiccating action of warm dry winds sometimes prevents the invasion of windward mountain slopes by vegetation that grows perfectly well on the leeward sides of the same slopes where water loss is much reduced. The harmful effect of excessive evaporation caused by the wind may also be seen in one's own garden. Here winter killing is occasionally due not to very low temperatures, but to a combination of wind and high temperatures! If a warm, dry wind blows for too long a period when the ground is frozen, the plants may lose water faster than their roots can obtain it and they will succumb as a result.

Our discussion of moisture in the soil and in the air has indicated that living organisms—particularly the higher plants—may exert a profound reciprocal influence on the water conditions in the terrestrial environment. Water is a highly modifiable factor in land habitats, and living elements may act either to augment or to diminish the amount of moisture present. Vegetation tends to catch the rain, so that, if the precipitation comes in short showers, the rain may never reach the soil. The plant cover also tends to reduce the water content of the deeper layers of the soil as a consequence of its transpiration, but at the same time it adds moisture to the air. On the other hand, the presence of forest vegetation favours the reduction of evaporation at levels near the ground because it lowers the temperature, slows down the wind, and sometimes causes increased condensation.

The amount of water present at any point in a land habitat and the rates of gain and loss are thus seen to be the result of the equilibrium of processes both climatic and biological. The interrelations of the water factor and the vegetation are frequently very complicated and are the subject of elaborate studies beyond the scope of this book. In addition to presenting the general picture, the present discussion reveals the interdependency of vegetation and moisture as another outstanding example of the organism and the environment acting as a reciprocating system.

4
Temperature

No environmental factor seems so easily measurable, and there is no factor of which we are so readily aware, as temperature. Because of its accessibility a chaotic mass of temperature data has accumulated, most of which is unrelated to the ecological context within which it was published. In fact, it is difficult to measure temperature at the site of importance to the organism—in the organism itself, the temperature here (in the case of land organisms) bears little relationship to the meteorological temperature. The latter can serve only as a rough estimate. Accordingly, it is considerably harder to establish a relationship between the distribution of an animal or plant and the temperature than, for example, to learn how distribution depends on the salinity of the available water.

As an illustration of the difficulty of such judgements, consider the following example. In Bavaria, the part of West Germany with the most continental climate, the nightingale is found in only a very few places where the climate is especially mild. It is of common occurrence in many other parts of Germany. About April 20 the nightingale returns from its wintering grounds; it cannot tolerate nighttime frost, and hard frosts are generally to be expected in Bavaria until the beginning of May. By contrast, the hoopoe is a regular inhabitant of many locations in Bavaria. It requires considerably higher temperatures than does the nightingale. It arrives later (not before the beginning of May), when the night frosts in general have stopped and the approaching continental summer brings very warm days and very small amounts of rain. In most other parts of Germany the hoopoe is considerably less common, because the summer temperatures are lower and the precipitation greater. It is not clear whether the influence of temperature here is direct or is exerted by way of another factor such as food supply. Temperature affects all chemical processes as formulated in Van t'Hoff's Law: a 10°C increase in temperature accelerates a chemical reaction by a factor of 2-4. We say that this chemical reaction has a Q_{10} of 2-4. The biochemical reactions of organisms are naturally subject to this law. Given that the temperature can fluctuate over a wide range during the course of a day, it is understandable that during evolution all organisms have developed mechanisms that liberate them, to a greater or lesser extent, from this temperature dependence. The warm-blooded animals have gone furthest in this respect, and will be discussed separately. The ectothermic organisms, with body temperatures that can change over a wider range (microorganisms, plants, and poikilothermic animals), seem at first glance to be obliged to follow even the most erratic changes in environmental temperature. But in fact they too have developed a lavish array of regulatory mechanisms, and in many respects even these organisms are evidently unaffected by temperature.

The most illustrative example is the physiological clock of plants and animals, which runs with a period of about 24 h regardless of the prevailing temperature. It is self-evident that this should be so; a physiological clock that ran faster at high than at low temperatures would offer no advantage in the struggle for existence and would therefore never have become established in evolution. But this example demonstrates that temperature compensation is possible even in ectothermic organisms. If we want to know more about the general aspects of such compensation, we must ask how widespread these phenomena are and explore the underlying physiological mechanisms.

It is astonishing that true compensation seems never to have become a property of the processes of development. All ectothermic organisms appear to develop in dependence on Van t'Hoff's Law (Fig.4.1). And this dependence is even more extensive than is evident in the figure.

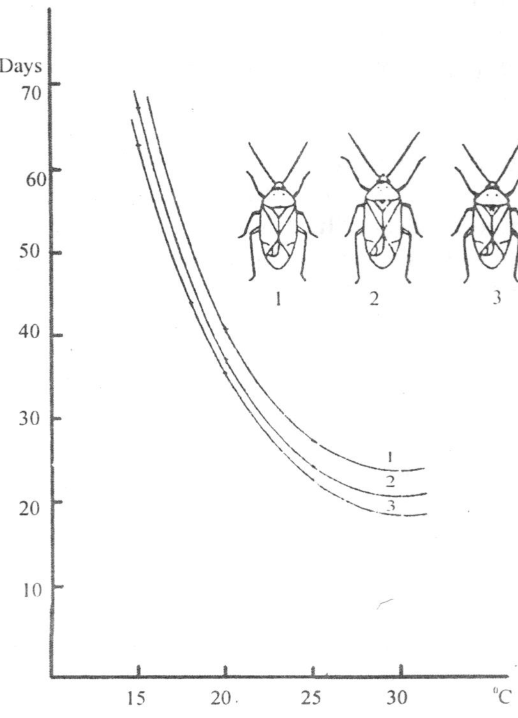

Fig. 4.1 Temperature-dependence of the duration of development of three species of the bug Lygus (*1* maritimus; *2* pratensis: *3* rugulipennes). Growth and temperature are related identically, in principle, in all heterotrophic ecothermic organisms. The product of duration of development (*T*) and the temperature (*t* in degrees above the null point t_o for development of a species) is constant (the thermal constant *K*); $Tu=t_o=K_2$. The null point for development can be computed from experiments at two temperatures, by the formula

$$t_0 = t_1 - \frac{T(t_1 - t_2)}{T_1 - T_2}$$

Homeotherms and Poikilotherms

When examining the relationships between organisms and environmental temperature, it is usual to subdivide organisms. One possible division is between the 'warm-blooded' and the 'cold-blooded'. However, these terms are subjective; a more satisfactory classification divides organisms into *homeotherms* and *poikilotherms*. As environmental temperature rises, homeotherms maintain an approximately constant body temperature, while the body temperature of poikilotherms varies with environmental temperature. One problem with this classification is that even classic homeotherms such as mammals and birds experience periods of reduced temperature (e.g. during hibernation) while some poikilotherms (e.g. Antarctic fish) experience only small body temperature variations because their environmental temperature remains constant. Also, many poikilotherms are capable of some body-temperature regulation. A better distinction between organisms is described below.

Ectotherms and Endotherms

A clearer classification of the relationships between organisms and environmental temperature is to subdivide organisms into *ectotherms* and *endotherms*. Endotherms regulate their body temperature by producing heat within their own bodies; ectotherms rely on an external heat source. This represents a distinction between birds and mammals (endotherms) and all other organisms. Although there are numerous exceptions to this (fox example, some reptiles and insects can elevate their body temperatures to facilitate activity) the distinction is nevertheless valuable. Ectotherms and endotherms differ in the extent to which they are able to maintain a constant body temperature. Over a certain temperature range (*the thermoneutral zone*) and endotherm consumes energy at a basal rate. However, at environmental temperatures further and further away from this zone, the endotherm consumes more and more energy in maintaining a constant body temperature. Moreover, even in the thermoneutral zone endotherms typically consume energy much more rapidly than ectotherms. Endotherms produce heat at a rate controlled by the brain. They usually maintain a constant body temperature between 35°C and 40°C, and therefore they tend to lose heat to the environment. However, this loss is moderated by insulator material (fur, fat or feathers) and by controlling blood flow near the skin surface. To increase heat loss, mechanisms such as panting are used. These mechanisms enable a high degree of control of body temperature to be maintained, enabling a consistency of peak performance. The price paid is high energy expenditure.

Heat Exchange

All organisms gain heat from and lose heat to their environment as well as producing heat. Almost all ectotherms modify heat exchange using the avenues of heat exchange shown in Fig.4.2. Among the mechanisms used, some are fixed properties of particular species and some are behavioural responses, while others are more sophisticated behavioural patterns; other mechanisms are aspects of their physiology. Despite these mechanisms, the body temperature of an ectotherm varies significantly with environmental temperature for three main reasons:

- the ability of many ectotherms to regulate temperature is very low;
- ectotherms are to some extent always dependent on the external source of heat;
- energy must be expended to modify the heat budget. The extent to which an organism regulates its temperature will therefore be a compromise between cost and benefit.

Temperature Thresholds

There are three main temperature ranges of interest: very low, very high and the temperatures in between. The most dangerous thing about high temperatures is that they lie only a few degrees above the animal's metabolic optimum, a result of the physio-chemical properties of their enzymes. High temperatures may therefore lead to enzyme inactivation or the unbalancing of components of metabolism (e.g. respiration proceeding faster than photosynthesis and leading to starvation). However, the most frequent effect of high temperature on organisms is *dehydration*. One of the problems with dehydration in animals is the loss of the ability of the animal to cool itself due to a reduction in the volume of blood reaching the body surfaces. All terrestrail ectotherms must conserve water, but at high temperatures rates of water loss can be lethal. Plants living in hot environments may suffer severe water shortage and as a consequence they cannot use the latent

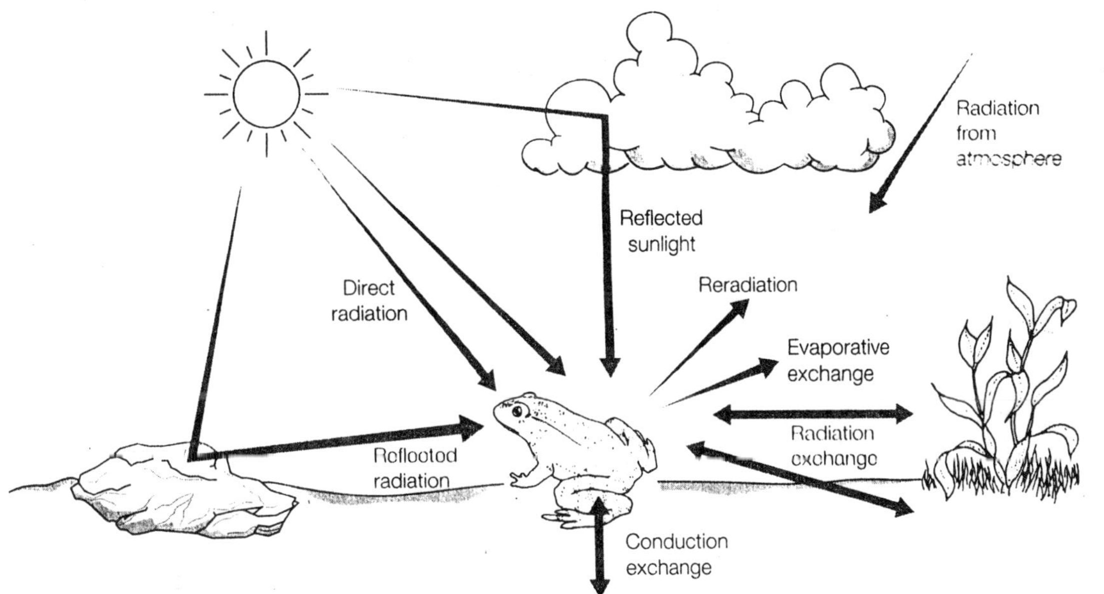

Fig. 4.2 The avenues of heat exchange between an ectotherm and the physical environment. Redrawn from Hainsworth, F.R., *Animal Physiology*, 1981, Addison Wesley Longman.

heat of evaporation of water to keep leaf temperature down. In such plants (e.g. desert succulents) the risk of overheating can be minimized by a low surface-to-volume ratio. Even if atmospheric humidity is high, the risk of overheating exists through reduced evaporative cooling. Nevertheless, most problems with overheating are found in hot, dry environments. There may be stages in an organism's life cycle that are particularly resistant to the effects of high temperatures. This is true of dormant structures such as bacterial endospores and seeds of plants, mainly as a result of their naturally dehydrated state.

At low temperatures, there are large differences between the tolerances of differing species associated with the processes of freezing, chilling and hardening. Many species are killed by temperatures below –1°C due to the damaging effects of ice-crystal formation within cells. Those species that survive lower temperatures do so because they have evolved mechanisms that prevent ice-crystal formation within cells. The crystals may eigther damage cell integrity or absorb water leaving a concentrated solute suspension which may be lethal. Some ectothermic animals which are exposed to low temperatures accumulate solutes which act as an antifreeze, preventing crystal formation. Low temperature tolerance in plants is almost always associated with a period of acclimatization or hardening. Also, resistance to freezing injury changes according to the plant's stage of development. Most seeds are resistant to low temperatures. Similar examples can be found among ectothermic animals; those that live through freezing winters do so at a resistant, dormant stage of their life cycle. Even at temperatures above freezing, metabolic reactions slow down and may almost cease. Ectothermic animals become moribund and cease to carry out basic maintenance functions. This may weaken an organism making it more susceptible to sources of mortality. Specifically, plants are liable to injury by chilling at temperatures around 10°C, brought about probably through disruption of membrane structure. There are also many animals which are susceptible to chilling damage, especially those that are not usually exposed to low temperatures.

RESPONSES TO TEMPERATURE

Temperature and Rates of Enzyme Reaction

In ectotherms the metabolic rate is relatively slow at low temperatures and more rapid as the environment becomes warmer. Endotherms buffer their internal organs from fluctuations in the environmental temperature, and hence do not exhibit such effects. The increase in metabolic rate with temperature can be described by a *temperature coefficient* (Q_{10}), which is given by:

$$Q_{10} = \frac{\text{metabolic rate at body temperature } T^{\circ}C}{\text{metabolic rate at body temperature } (T-10)^{\circ}C}$$

The value of the coefficient indicates the increase in reaction rate caused by a 10°C rise in temperature, and is commonly about 2.0, although Q_{10} is not constant across all temperatures, showing deviations towards the upper and lower thermal limits of an organism.

Rates of Development and Growth

Within the nonlethal temperature range the most important effect on ectotherms of temperature is likely to be its effect on the rate of development and growth. When rate of development is plotted against body temperature there exists an extended range of temperatures over which the relationship is linear. Figure 4.3 shows the development of the cabbage white butterfly, *Pieris rapae,* from egg hatch to pupa. As most organisms spend all of their time below their nonlinear high temperature zone, it is assumed that the rate of development rises linearly with temperature above a *developmental threshold temperature.* The consequence of this relationship is that, unlike ourselves and endotherms, octotherms cannot be said to require a certain length of time for development. What they require is a combination of time and temperature, *physiological time.* The importance of this concept lies in the ability to understand the timing of events, and thus population dynamics. In practice, however, there are difficulties in determining an organism's physiological time-scale, mainly due to the effects of fluctuating temperatures. The linear relationship itself is never more than an estimate, and there are problems with the monitoring of body temperature in the field.

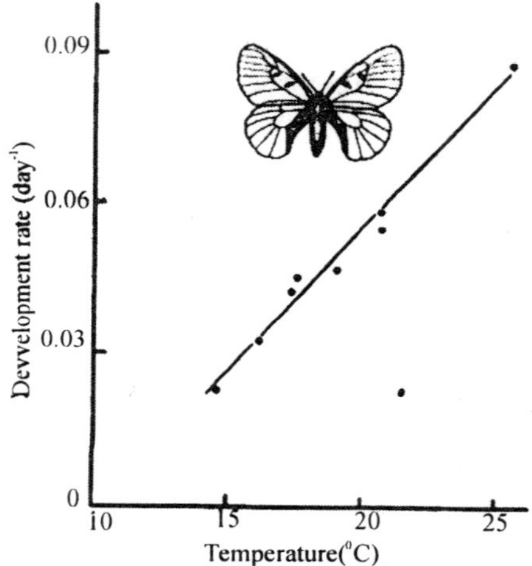

Fig. 4.3. Development of the cabbage white butterfly, Pieris rapae, from egg hatch to pupa requires 174 day-degrees above a threshold temperature of 10.5°C.

Acclimation and Acclimatization

Temperature may also act as a stimulus, determining whether the organisms will begin development. For many plants a period of chilling or freezing is necessary before germination. For example, seeds of winter wheat only develop and flower if they have been pre-chilled. This phenomenon—the induction of flowering by low temperatures—is called *vernalization*. Temperature may also interact with other stimuli (e.g. photoperiod) to break dormancy. The responses of an ectotherm to temperature are based on the temperatures the organism has experienced in the past. Exposure of an ectotherm to a relatively high temperature for several days can shift its entire temperature response upwards; conversely, several days' exposure to low temperatures can shift its response downwards. This process is known as *acclimation* if these changes are laboratory induced or *acclimatization* if they occur naturally. This process takes time. The variations that occur within species are relatively slight compared with interspecific differences. For example, some mosses and lichens can withstand temperatures of 70°C, while there are bacteria which can live and reproduce at temperatures above 100°C. However, within species there are differences in temperature responses between populations from different locations, usually as a result of genetic differences rather than just acclimatization. In practice, geographic differentiation is usually limited enabling species to be described as having one temperature response.

TEMPERATURE AND SPECIES DISTRIBUTION

Species Distribution and Temperature

It is appropriate to begin this section with a definition of the potential role of temperature in determining the distribution and abundance of organisms. Variations can be defined under seven main headings (Table 4.1).

The distribution of the major biomes over the Earth represents a reflection of the major temperature zones. Similarly, changes in species with increasing altitude are again a reflection of changing temperatures. However, it is more difficult to attribute a role to temperature when considering single species. In certain cases, the distribution limits of a single species can be attributed to a lethal temperature which precludes the species' existence. For example, frost damage is probably the single most important factor in determining plant distribution. However, a more widespread type of relationship is one which has a close, but not ideal, correlation between the distribution limits of a species and an isotherm. An *isotherm* is a line on a map that joins places having the same mean temperature at a particular time of the year. Whilst some correlation can usually be found, only limited significance can be attached to the isotherm in terms of a complete explanation. Even every mobile animals such as birds may have their distributions closely linked to temperature, as in the case of the eastern phoebe (*sayornis phoebe*), a migratory bird of eastern and central North America. The wintering population of the eastern phoebe is confined to that part of the US in which the mean minimum

Table 4.1 Variations in the Temperature profile of regions

Variation	*Description*
Latitudinal/seasonal	These two variations cannot be separated. The tilting of the Earth to the sun results in generalized temperature zones, with the hottest temperatures occurring in the middle latitudes (38°C) rather than the equator (35°C).
Altitudinal	Superimposed on these broad geographical trends is the influence of altitude. There is a drop of 1°C for every 100 m increase in altitude in dry air, and a drop of 0.6°C in moist air as a result of the adiabatic expansion of air.
Continentality	The effects of continentality are largely attributable to different rates of heating and cooling of land and sea. Land surfaces reflect less heat than water so quickly warm up, whilst also cooling quicker. The sea therefore has a moderating effect on coastal regions.
Microclimate	On the small scale, local variations can give rise to microclimate variations. For example the air temperature in a patch of vegetation can vary by 10°C over a vertical distance of 2.6 m from the soil surface to the top of the canopy
Depth	Depth, either in soils or water can have two effects on temperature fluctuations. Firstly, the fluctuations are dampened, and secondly, they lag behind surface fluctuations. The strength of these effects increases with depth and decreases with the thermal conductivity of the medium (low in soil, higher in water). A meter below the soil, daily temperature fluctuations are damped out completely.
Diurnal	The daily rhythm of solar radiation causes changes in temperature.

January temperature exceeds –4°C. The close correlation of the bird's winter range to this isotherm probably relates to its energy balance. An organism's limit to distribution is determined not by lethal temperatures, but by conditions that make it a poor competitor. Competition is not the only biological interaction that combines with temperature to limit distribution. Many animals have a distribution which correlates with temperature and the occurrence or quality of their food.

Variations in temperature may also be intimately associated with another environmental condition or resource such that the two are inseparable. The most widely used example of this is the relationship between relative humidity and temperature. For aquatic organisms the relationship between temperature and dissolved oxygen concentration is important. The solubility of oxygen decreases with increasing temperature. There exists a close correlation between these environmental factors and species distribution patterns, in which it is impossible to separate the effect of temperature and oxygen concentration.

Evolved Response to Temperature

As already discussed, the effects of temperature on individuals may be moderated by acclimatization or by evolved differences. These factors also influence distribution and abundance. For example, endothermic animals from cold climates tend to have shorter extremities (ears and limbs) compared with animals from warmer climates (*Allen's rule*). Also, there is a tendency for birds and mammals to be larger

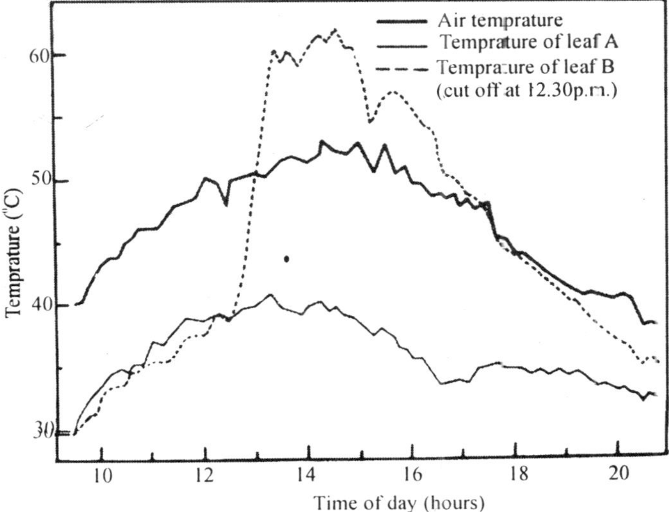

Fig. 4.4 The temperature of a leaf can be well below that of the air, as a result of transpiration and the associated cooling. A cut-off leaf is rapidly heated by sunlight, because of lack of water.

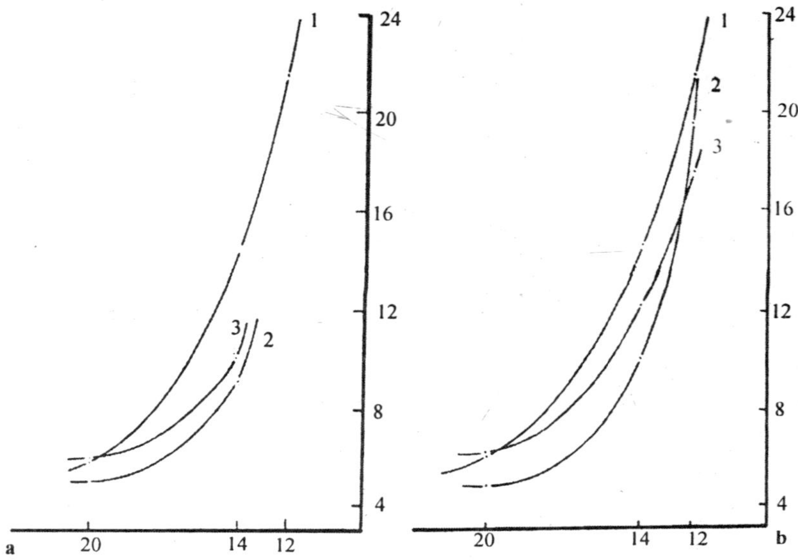

Fig.4.5 **a,b.** Time elapsed from emergence of the imago to the first egg-laying by Drosophila sp., as a function of temperature. Ordinate: time in days: abscissa: temperature. *1*constant temperature; *2* temperature fluctuating about the corresponding mean during the day; *3* computed curve for the fluctuating temperature. **a** Amplitude of the temperature cycle 8°C, **b** amplitude 5°C. At the lower temperatures development takes less time when the temperature fluctuates than at the corresponding constant temperature.

in colder areas (*Bergmann's rule*). The explanation in both cases is that endothermic organisms in colder climates should have a smaller surface area relative to volume across which they lose heat. While Allen's rule has widespread applicability, Bergmann's rule is much less universal, probably due to the number of other factors which affect body size, but it is a valuable predictor at the intraspecific level.

However, there are a few notable phenomena that have not yet been satisfactorily explained.

1. The productivity of many plants is increased when the temperature fluctuates (Fig. 4.6). Perhaps low nighttime temperatures act to prevent high respiratory losses. Such

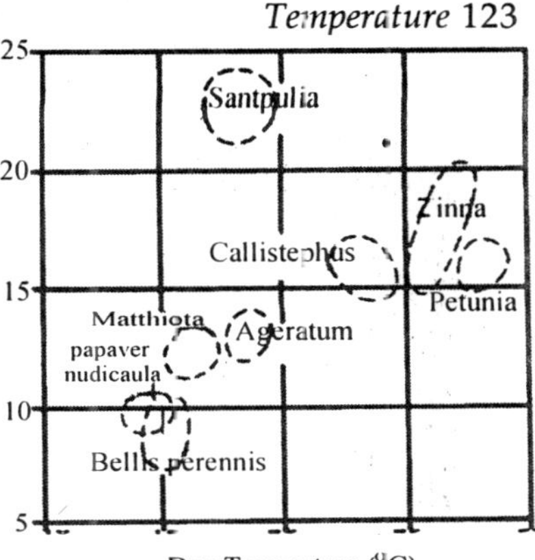

Fig. 4.6 Many ornamental plants develop best when the night-time temperature is below that during the day. The most favourable temperature difference is about 5°-10°C. The African violet is an unusual exception.

an explanation could also apply to the diurnal vertical migrations of planktonic animals, which when they are not feeding sink to the colder levels in the water.

2. In many of the ectothermic animals that have been studied the rate of reproduction is greatly increased in fluctuating temperatures (these include the rotifer Brachionus and various terrestrial insects (Fig. 4.7). Parasitic wasps of the genus Trichogramma raised under fluctuating-temperature conditions have appreciably greater effects, when they are set free for the control of the host insects, than those raised in constant temperature. So far it has not been possible to decide whether such examples represent a general rule, nor do we know enough about their physiological bases. In view of the ecological significance of such an increase in the production of organic matter or of eggs, this phenomenon definitely deserves further study.

Finally, temperature almost always acts in concert with wind and moisture (rain). Their effects are practically inseparable. Indeed, it seems almost miraculous that we are able to say anything at all about the influence of temperature on animal distribution. In the area of forest entomology it has proved useful to construct climate curves based on both precipitation and temperature. Merkel (1977), for example, used such climate curves to study a bark beetle. Temperature was automatically recorded on an hourly basis; the time when it exceeded 7°C, the null point for movement and feeding by the

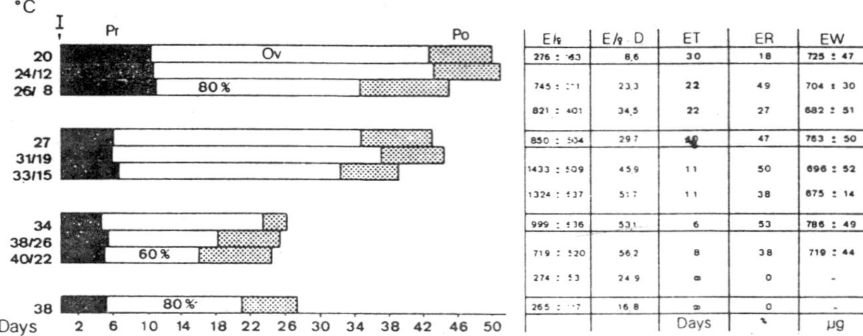

E/♀	E/♀ D	ET	ER	EW
276 ± [illegible]	8.6	30	18	725 ± 47
745 ± [illegible]	23.3	22	49	704 ± 30
821 ± 401	34.5	22	27	682 ± 51
850 ± 504	29.7	[illegible]	47	763 ± 50
1433 ± 509	45.9	11	50	696 ± 52
1324 ± 437	[illegible]	11	38	675 ± 14
999 ± 436	53.1	6	53	786 ± 49
719 ± 520	56.2	8	38	719 ± 44
274 ± 53	24.9	∞	0	-
265 ± [illegible]	16.8	∞	0	-
		Days	%	µg

Fig. 4.7 Fertility of Gryllus bimaculatus under conditions of constant and diurnally fluctuating temperatures. *I* imaginal molt; *Pr* pre-oviposition period; *Ov* oviposition period; *Po* Post-oviposition period; *E* number of eggs per female; *E/D* number of eggs per animal per day; *ET* time until emergence of the larvae; *ER* emergence rate; *EW* weight at emergence.

beetle, were added and the effective temperature sum per day and month was calculated. Plotting these sums on the *x* axis of a coordinate system and the sums for precipitation on the axis gives the climate curve (Fig. 4.8). When there is heavy rainfall the curve rises steeply, and when the weather is dry it is nearly parallel to the *x* axis. By entering observations of stages in the life of the beetle in this graph, one obtains data specific to the measurement site which can be compared with results similarly obtained at other locations or in other years. Using this method, researchers have developed considerable insight into the dependence of insects on the climate in their habitats. But even this procedure is not entirely satisfactory, because depending on the general state of the weather the beetles can require different integrated temperatures. For example, the large bark beetle Ips typographus does not appear until the temperature at the site where it has spent the winter rises above 7°C. By this time the surface of the soil has reached 10°-20°C. The beetles then begin feeding under the bark; to do so, they require temperatures between 12° and 19°C. The temperature sums necessary for maturation of the gonads vary depending on the weather. The breeding flight of the mature beetle can occur only if the air

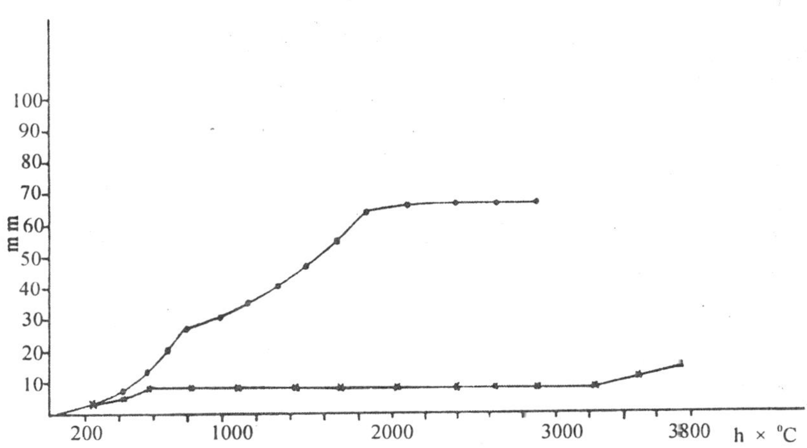

Fig. 4.8 Example of a way in which the climate at two different sites can be described. Ordinate: precipitation in mm; abscissa: integrated temperature (hours × °C). The curves are cumulative, with the months plotted in succession. In the first month the two curves coincide (the weather was the same at the two sites); during the following months one curve rises sharply (rain and low temperatures) whereas the other is flat (dryness and warmth). To make the differences more distinct, one can take as a reference level the null point for development of a species.

temperature is at least 20°C. The body of the animal must have reached at least 23°C, which can happen at an air temperature of 20°C if the weather is sunny but requires 23°C air temperature under overcast skies.

Another approach to the difficult problem of temperature in terrestrial habitats is to determine the effective mean temperature (eT) by the method of Pallmann *et al.* (1940). This procedure is based on the fact that sucrose in aqueous solution is broken down by hydrogen ions to form glucose and fructose. In a buffered solution, with constant hydrogen-ion concentration, the reaction is temperature-dependent. The degree of sucrose inversion can readily be monitored polarimetrically, providing a convenient measure of the temperature situation in a particular period of time. The advantage of the method lies in the fact that the reaction rate, like the growth processes of ectotherms, rises exponentially with temperature. Sucrose inversion ought therefore to give a very precise measure of temperature-dependent biological processes. But in spite of this obvious advantage the method has not been widely adopted, no doubt, for fear of overgeneralization. That is, there are no really well documented relationships between organic life and the temperatures recorded in this way. Exemplary studies in which this method is compared with other would be most informative.

In water things are simpler. A general heating by the sun's radiation is impossible because infrared light does not penetrate the water far enough. It is almost completely absorbed in the upper millimeters, which may become distinctly warmer than the rest of the body of water. If black bodies are present in this uppermost layer they become quite warm. By this means Aedes larvae can develop very rapidly in the spring, even in ponds still partly covered by ice.

Only one ecologically fundamental process is clearly not covered by the above considerations—photosynthesis in green plants. Photosynthesis is not a simple biochemical process; to a considerable extent it involves photochemical events and is thus not so dependent on temperature. The Q_{10} of photosynthesis is less than 2, distinctly lower than that of biochemical reactions in general (with a Q_{10} of 2-4). Hence photosynthesis can operate at low temperatures, and at higher temperatures does not increase to the same degree as respiration or any of the functions of animals and microorganisms. Figure 4.9 illustrates a basic ecological phenomenon: in the moist, warm tropics dead organic matter is rapidly decomposed, whereas in cool regions decomposition occurs slowly. However, the production of organic matter by the green plants is not very different in the two locations. This fact was remarked upon by Darwin in his book about the voyage of the Beagle. In comparing the jungles of the Amazon and of Tierra del Fuego, he writes that the tropical rainforests appeared to him as a symbol of life, full of power and growth; the forests at the southern tip of the continent seemed like a symbol of death, full of dead tree trunks, branches and twigs. Forests with dead trees are characteristic of temperate and cool zones, because at low temperatures the dead wood remains intact for a long time. In a tropical rainforest fallen trees decay so rapidly that they are hardly noticeable. From the different slopes of the two lines in Fig. 4.9, we would also predict that plant productivity at consistently high temperatures cannot be much higher—and in certain circumstances may be even less—than at temperatures in the intermediate range. At high temperatures the steady losses due to respiration are so large that the balance between photosynthesis and respiration may show a deficit. Again, of course, the temperatures of interest are not those in the meteorological reports. If the plants have access to enough water, massive transpiration can hold the leaf temperature well below that of the surrounding air. Warming by heat radiation, such as occurs in animals, plays no role in the leaves of plants; essentially all the infrared radiation passes through the leaf or is reflected by the chlorophyll.

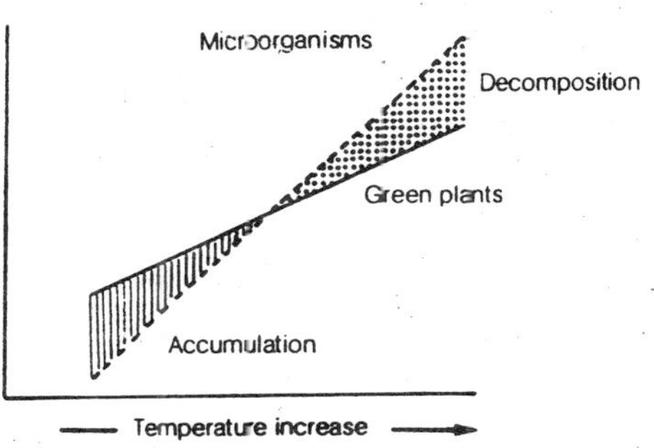

Fig. 4.9 The influence of temperature on production by green plants and on the processes of decomposition by respiration in animals and microoganisms. In a tropical warm-moist climate plant matter is decomposed more rapidly than it is synthesized; the layer of humus on the ground is but a thin film covering the mineral soil, and in a few months it can be completely eroded. In cool regions the humus layer is potentially much thicker.

So far we have been considering only the period in which plants and animals are active. In climates with a cold winter the organisms face the problem of surviving freezing temperatures—and sometimes temperatures well below the freezing point. Many animals can avoid exposure by moving to parts of the water or soil that do not freeze, but many others cannot do this; in the permafrost regions of the arctic such a strategy is entirely impossible.

Very few organisms can tolerate freezing of their body fluid or of their cells. At the lowest temperatures which they can survive, nearly all organisms manage to keep their internal milieu in the liquid state. This fact, together with the photochemcial nature of photosynthesis, explains the ability of green plants to assimilate carbon dioxide with a positive balance even in winter. Cold-weather assimilation has been demonstrated in the coniferous forests of the taiga and even more strikingly in antarctic lichens, but it also occurs elsewhere —for example, in the high-mountain plant Ranunculus glacialis. How is this prevention of freezing achieved? The organisms produce substances that lower the freezing point, and accumulate them in their body fluids and cells. A particularly well known antifreeze agent in animals is glycerol, those of plants include the sugar hamamelose. There are a number of chemically similar substances that tend to produce the same effect. Because of these, some beetles can tolerate temperatures of –80°C or lower. When an organism is so well adapted it would seem irrelevant whether the winter temperatures stay above zero or fall to -20° or -30°C. In fact there is an additional complication; at temperatures between the freezing point and about +10°C ectothermic animals and plants lacking green parts use up a great deal of energy in respiration; in the temperature range 6° -10°C many animals can even move actively about. But only predatory organisms can, under certain conditions, cover losses, since the nutrients in animal tissue are readily utilizable. Herbivores and detritus feeders cannot eat enough to replace the lost energy at those temperatures. It is considerably better for them if the winter temperatures remain well below the freezing point, so that none of their stored energy is used up. The paucity of species in maritime regions (for instance, Schleswig-Holstein or the British Isles) is in part explainable on this basis; the winter is not warm enough for feeding, but not so cold that animals can retain all of their stored energy. In this regard continental climates are more favourable. The great variety of insects in the northern United States and central Russia is partially due to this effect. On the other hand, the very warm summers of continental regions are naturally more favourable to ectothermic animals than the rather cool summers of maritime regions with the same average annual temperature (Fig. 4.10).

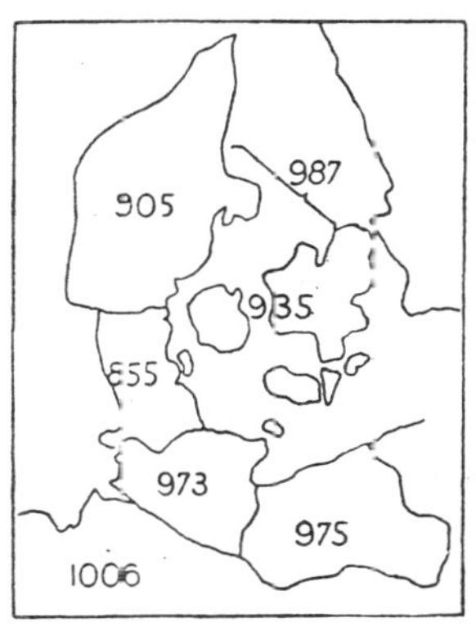

Fig. 4.10 The number of species of wild flowering plants in northern Germany and Denmark; it is lowest in the region with the most pronounced maritime climate. The regions with many species have areas of continental climate, so that they also include continental species.

Bony fishes in the ocean face a special problem in surviving the winter. As mentioned previously, their osmotic pressure is lower than that of their surroundings. The water in the open ocean freezes at a temperature of –1.7 °C, whereas a fish would freeze at only –1°C. Here we have yet another price paid for adaptation to a new habitat. Arctic and antarctic fishes solve the problem by moving to deeper levels during the winter or by raising the osmotic pressure of the blood to approximately that of the medium. In so doing, however, they evidently enter a sort of dormant state in which they take hardly any food. Species less well adapted in some situations—for example, when ice crystals are stirred into deep water by violent storms —die in massive numbers.

Warm-blooded animals are in a quite different situation. These are organisms with greatly elevated metabolic rate and concomitant high energy consumption and heat production. The heat is a side product of metabolism which may be given off from the body or, with suitable insulating mechanisms, can be retained. By changing the insulation properties of the body surface as environmental conditions change, such an animal can achieve a constant body temperature between 35° and 42°C, depending on the species. When the external temperature is very high cooling mechanisms involving the evaporation of water come into play. As a result the animals are largely independent of the surrounding temperature; they can remain fully active even at low temperatures and colonize regions from which most ectothermic animals are excluded. Such regions include the antarctic continent and to a considerable extent, the tundra of the far north (around latitude 80°N), where the temperature is so low and solar radiation so slight that ectothermic herbivores are practically negligible. The price the warm-blooded animals have paid is that they require much more food than ectothermic animals of equal size. This requirement is in fact entirely a consequence of their temperature. Extrapolating the energy consumption per unit time of a crayfish at 20° - 39°C, one obtains precisely the energy consumption of a warm-blooded organism that weighs the same. But the advantages of warm-bloodedness are easy to see. Food that is hard to utilize fully (all plant matter!) is processed a good deal more successfully in the warm digestive tract of such animals (in some cases with the aid of specially adapted microorganisms) than by any ectothermic animal. A warm-blooded herbivore thus exploits its food more thoroughly than does an ectotherm. On the other hand much of the energy obtained is lost in maintaining the necessary temperature. In effect, then, an ectothermic organism produces more animal matter than a warm-blooded animal does, for a given amount of plant matter consumed.

It is not surprising that attempts have been made to reduce this high energy requirement. The technique of insulation has been brought to perfection by some organisms; arctic animals consume hardly any more energy when the environmental temperature is low than when it is high (Fig. 4.11). But warm-blooded animals, like ectotherms, still face the problem of minimizing the loss of energy during the least favourable seasons. A warm-blooded animal that simply took shelter when the weather was bad would lose so much energy because of its ongoing high metabolism that it could survive only briefly. The conspicuous phenomenon of bird migration, and the comparable migrations of bats, show one way out of this dilemma. Another solution is periodically to reduce body temperature, as hibernators do. In the dormant state these animals can last out the unfavourable season without much loss

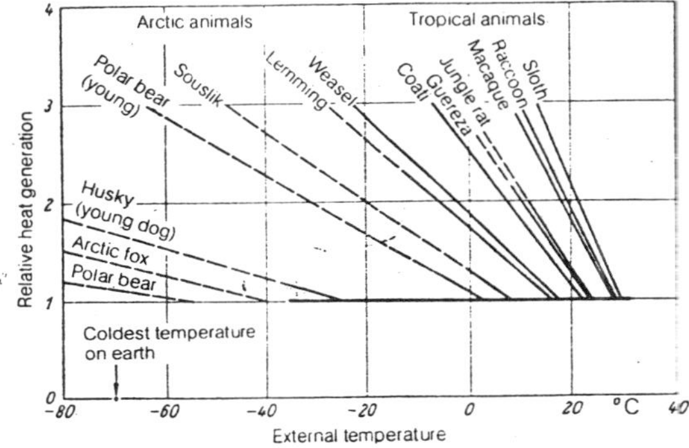

Fig. 4.11 The heat generated (energy consumption) by arctic and tropical animals at rest, as a function of the external temperature. The minimal turnover is set at 1 in each case. - measured mean values; —extrapolated values.

of energy—but they lose the advantage of warm-bloodedness, that of being constantly prepared for activity. Hibernating mammals need special hiding places, since they cannot defend themselves against predators. Small animals with very high energy requirements cannot even afford a long rest during the night; shrews are active at regular intervals without regard to the day/night cycle. Hummingbirds can greatly alter their body temperature during a 24-h period. Those in the high Andes enter a sort of 'hibernation' (torpor) every night. Many insect-eating birds can go into torpor for a few days when food is scarce. This is especially true of the young birds (swifts and swallows). Other birds adopt a less extreme form of this strategy; tits can lower their temperature slightly during the night in the arctic winter, and thus save energy. The basal metabolic rate of birds in general falls by about 20% during the resting period. Aquatic mammals are particularly vulnerable during the arctic winter. The extremities of the body (the feet of birds, and the snout, eyes, and fins of seals) are quickly cooled because of the high heat conductivity of water. In fact, these parts of the body are better described as ectothermic than as warm-blooded. Their enzyme patterns are adapted to these conditions. Tuna are comparable in this respect; the core of the body is homoiothermic. They have no swim bladder, so that their musculature is constantly in motion. The active swimming of these fish by day and night provides sufficient energy that with a certain amount of insulation the interior of the body can be kept at a constant temperature. But the gills present a problem. These thin-skinned structures must be exposed to the open water so that oxygen exchange can occur, and the blood flowing through them is cooled. At low temperatures hemoglobin takes up more oxygen than it can bind at high temperatures. Mammalian hemoglobin would take up so much oxygen that, once inside the body, the oxygen would come out of solution and cause embolism. The hemoglobin of tuna is specially adapted in a way that makes its oxygen affinity independent of temperature. Only by such means could a gill-breathing organism be effectively warm-blooded. Just as the specialized hemoglobin made possible this temperature compensation, so enzymes in the feet, fins, snout, and eyes of seals and sea birds allow these organs unimpaired function even though their temperature may be between 2° and 6°C.

In view of all the different temperature effects that have been listed here, it is evident that apart from a certain large-scale temperature dependence hardly any specific predictions can be made on the basis of the available climatological data. To predict temperature relationship accurately, one would have to take account of several aspects in addition to the mean annual temperature—the difference between winter and summer temperature, the daily amplitude of temperature fluctuation, and the elevation of the sun, which limits the degree to which organisms can be warmed by direct heat radiation. The topography of a region is also a factor. A south-facing slope in the northern hemisphere receives considerably more solar radiation than a slope with northern exposure. The south-facing slope is therefore dryer, and during the summer cooling by evaporation is less, which accentuates the difference in temperature. Finally, where diurnal temperature fluctuations are concerned, it is necessary to consider the ability of very many animals to seek out favourable positions. Because not all animals can do this, and no plants can, inferences about the species composition of a habitat based on climatic data are extraordinarily problematic. For example, the flora of the tundra in the far north (the region around latitude 80°N is barely distinguishable from that of subarctic montane regions (southern Norway, 60°N). Because of difference in sun elevation and the associated difference in the intensity of incident radiation, however, there are marked differences in the fauna of the two regions. These are especially apparent among the herbivores and species with special forms of behaviour by which they make direct use of the sunlight. The preferred temperatures established in the laboratory give only an approximate indication of the distribution of a species in nature.

Such experiments play a central role in ecology, however, so that they deserve a brief discussion here, in the context of temperature preferences. The first question to consider is whether the preferred temperature corresponds to the optimum. This is by no means necessarily the case. Behavioural research has shown that brooding geese presented with eggs of different sizes will choose one much larger than their own. Here the preferendum is far from the optimum And we know that similar discrepancies can occur in the realm of ecology. The Mediterranean cricket Gryllus bimaculatus has a temperature preferendum in the region of 34°C. But the optimum for this species is much lower, between 25° and 31°C. This range is so wide because the measured optimal differ, depending on the function being analyzed—egg production, the quantity of reserve substances in the eggs, the rate of development to the imago stage, the final size of the imago, or the level of mortality. But in the life of the animal there are no distinct optima. The best indication one can give of the effective optimum temperature is a range of temperatures within which all the vital functions can operate more or less efficiently. An added complication is that in the field the temperature fluctuates. Such fluctuation has a positive effect on egg production. However, the crickets do not (as tiger beetles do) seek out locations at different temperatures at different times of day. Finally, the food supply can modify the effect of temperature. We may well ask whether animals live within their optimum range at all. To what extent may other factors encountered under field conditions force them into a range of temperatures we would not consider optimal? We cannot answer this question here, but will return to it in the discussion of competition.

Light

Light is the factor that permits life to exist on earth at all. But it is difficult to distinguish among various habitats on this basis, for there is enough light to support photosynthesis over the earth's entire surface. Very few habitats—the ocean depths, the deep zones of some inland waters, and caves—do not receive enough light for plants to be productive. Everywhere else in the biosphere there is light in adequate amounts and with the correct spectral composition. It is only natural, then, that the quantitatively significant production of organic matter everywhere is in the final analysis based on the same mechanism—photosynthesis by means of chlorophyll. The different kinds of chlorophyll, as far as we know, hardly differ at all in productivity.

As an initial consideration, we can take it that light energy is a factor in excess. Plants make use of but a small fraction of the incident radiation; the efficiency of light utilization is always less than 5% and usually 1% of the available energy (Table 4.2). It follows that productivity within a habitat is usually not dependent on the amount of light—there is always enough. The level of production is dictated by temperature and water supply, as well as by the presence of minerals. Therefore it must be possible to calculate the level of production possible Table 4.3. Under the same conditions of soil, nutrient availability, and climate all plant communities must in theory exhibit roughly the same production, and observations have shown that in fact they do. Studies carried out in the Solling Project showed that the production of organic matter in a meadow, an area of coniferous forest, and an area of beech forest was quantitatively comparable.

This fact is hard to accept. We know, after all, that there are shade plants and sun plants. We know of plants that carry on photosynthesis during only part of the year-spring flowers, for example, which later die back into the ground. How can we reconcile such things with the above claims regarding uniform productivity and a single basis biochemical process?

Table 4.2 Production by a field of maize, as an example of the calculation of primary production

Total dry weight of the maize from 0.4 ha (=1 acre)(10,000 maize plants)	6,000 kg
Ash (inorganic components) subtracted	– 300 kg
Total weight of organic components	5,700 kg
Their equivalent in glucose	6,700 kg
Plus organic matter lost by transpiration appropriate to the season (expressed as the glucose equivalent)	2,000 kg
Total weight of the glucose formed by 0.4 ha maize	8,700 kg
Energy required for synthesis of 1 kg glucose	3,300 kcal
Energy required for synthesis of 8,700 kg glucose	ca.33 x 10^6 kcal
Total solar energy available to 0.4 ha	2,040 x 10^6 kcal

% utilization of the available energy = $\frac{33,000,000 \times 100}{2040,000,000}$ =1.6%

Table 4.3 Theoretical maximum yields under the geographical conditions in various habitats, and the theoretically calculable utilization by a warm-blooded animal.

Theoretically possible annual maximum yield (kg dry matter ha, from de Witt as cited by Baeumer, 1971)	*Location*	*Number of sheep theoretically supportable by this amount of energy (per ha per year)*
25,000	Stockholm	68
30,000	Berlin	80
51,000	Puerto Rico	140
57,000	Tropical Australia	156

The only difference between shade leaves and sun leaves is that the former contain more chlorophyll; the same is true of shade plants as compared with sun plants. The greater amount of chlorophyll compensates for the lower relative light intensity. Greater amounts of chlorophyll are obtained at a price; there is room for them in the leaf only if the thick epidermis and cuticle are eliminated. Therefore shade plants are very vulnerable to drought. In their normal habitat this is irrelevant, because shade is practically always associated with ample moisture. The essential point here is that the mechanism of photosynthesis is identical to that of the sun plants. The assertion that productivity of different plant communities is similar under similar conditions does not refer to single species within the communities; no one could maintain that productivity is the same among different species. Differences at this level are inevitable in view of the variations in length of growing season, which is controlled by other factors. The comparison here is between one plant community composed of many species and another of equally complex composition. The overall production in such a community results from the activity of different plants occurring at different times; but because the underlying biochemical system is the same in all cases, the total production of the community per unit area and per unit time is the same. To recapitulate: the uniformity in the productivity of different plant communities under the same conditions is ultimately based on the fact that the amount of chlorophyll exposed to the light is independent of the species composition of the community.

The modern high-yield varieties developed for agricultural purposes produce no more organic matter than the wild varieties, but what produced is differently distributed. Instead of an extensive root system and strong stems more is produced of the part usable by humans—the kernels of grain, for example. There is one exception in principle to the uniform-productivity rule—one which leads to a slightly but distinctly increased productivity rule—one which leads to a slightly but distinctly increased productivity. Whereas the first product of normal photosynthesis is a triose, a sugar with three carbon atoms (hence the term "C_3 plants" for those with this form), in a number of plants the CO_2 taken up is first coupled to phosphoenol-pyruvate—a C_3 molecule—to form oxalacetate, a C_4 molecule. Since the first product of photosynthesis here is a C_4 molecule, these are called C_4 plants. Later the oxalacetate is split up again; the carbon dioxide becomes available to the normal photosynthetic apparatus and is further processed in the normal pathway of photosynthesis, described previously. This procedure uses up more energy than normal photosynthesis, and it is probably for this reason that C_4 plants are restricted to regions where solar radiation is very intense. But it conveys advantages. The limiting factor for the plants is carbon dioxide. Phosphoenolpyruvate (PEP) carboxylase has a higher affinity for carbon dioxide than the ribulose-diphosphate (RuDP) carboxylase that acts as the acceptor for the CO_2 molecule in normal photosynthesis. As a result, more carbon dioxide can be fixed per unit time and productivity can be greater than that possible with normal photosynthesis (Fig. 4.12 and 4.13a,b). Examples of plants with this kind of photosynthesis can be found among the grasses and dicotyledons (e.g. Atriplex) in very sunny regions which (usually) have an irregular water supply. The familiar crops maize and sugarcane are in this group, as well as aggressive, wide-spreading tropical weeds. Because the difference between the two photosynthesis types is not clear-cut but gradual (all plants can bind slight amounts of carbon dioxide to PEP), a number of attempts are currently being made to find and cultivate varieties of other crop plants with C_4 photosynthesis. The superiority of C_4 plants over C_3 plants is based on yet another principle: C_3 plants have so-called photorespiration. That is, they lose considerable quantities of organic matter by respiration even during the day—in fact, the daytime respiration level is about 5 times as high as that at

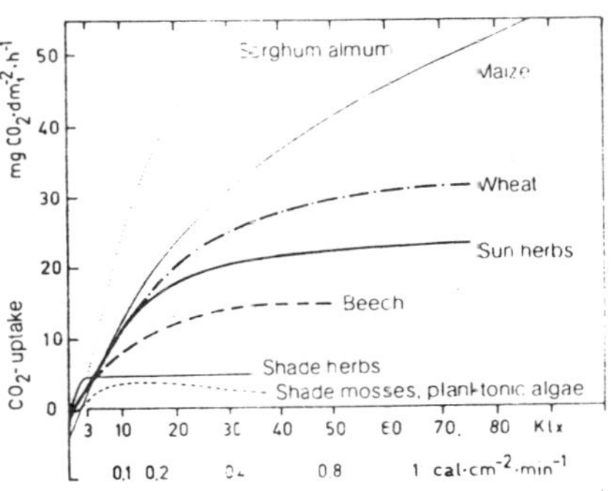

Fig. 4.12 Light-dependence of net photosynthesis in different plants with the natural amount of CO_2 available. The C_4 plants sorghum and maize are more productive than the C_3 plants.

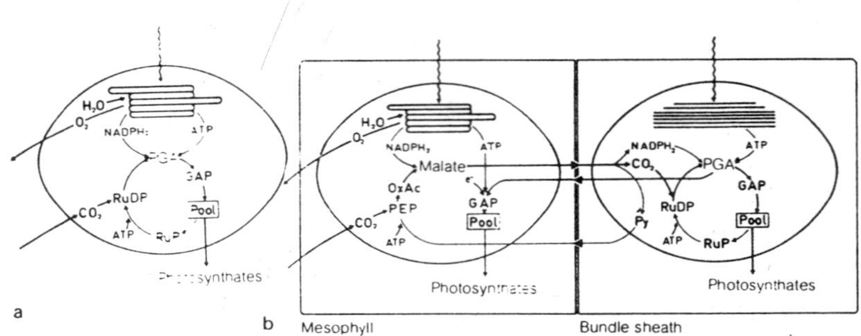

Fig. 4.13 **(a)** Simplified diagram of CO_2 fixation and the production of photosynthates by C_3 plants. **(b)** C_4 plants. *PEP* phosphoenolpyruvate: *RuDP* ribulose diphosphate; *PGA* 3-phosphoglyceric acid; *GAP* 3-phosphoglycer-aldehyde; *Pool* intermediate C_3 to C_7 compounds; *RuP* ribulose-5 phosphate; *OxAc* oxaloacetate; *Py* pyruvate. The regeneration of *PEP* from *PGA*, in which water is given off, is not shown.

night. The enzymes involved are not located in the mitochondria, but in the very small peroxysomes. For a long time there has been speculation about the significance of this photorespiration, which cannot be found in C_4 plants. Now it has turned out that RuDP carboxylase, which couples carbon dioxide to the RuDP molecule, often 'confuses' oxygen with carbon dioxide, especially at high oxygen concentrations. When this happens the ribulose diphosphate is oxidized to form phosphoglycerate and phosphoglycolate. The latter is oxidized to glyoxylic acid: the hydrogen peroxide produced in this reaction is converted to water by peroxidase. During photorespiration—the oxidation of glycolate—the plant obtains no energy; the process represents a pure waste. The entire pathway is probably explicable only as a "historical relict." It originated at a time when the oxygen concentration was still significantly lower and that of carbon dioxide higher than today. The C_4 plants, with their higher specific affinity for carbon dioxide, have solved the problem of CO_2 deficiency, but the C_3 plants have not. On the other hand, the C_4 plants require greater amounts of energy for their actual photosynthesis—a higher level of solar radiation per unit time. For this reason they cannot spread into regions too far from the equator.

A very similar modification, but complicated by introduction of a timing factor, is found in many succulent desert plants (Fig. 4.14). Their stomata are opened only at night; at this time they bind carbon dioxide to PEP as the C_4 plants do. Most of the oxalacetate is converted to malate and stored in the vacuoles. As it accumulates during the night, the pH of the vacuole contents falls dramatically with the increasing concentration of organic acids. In the morning the stomata close, and carbon dioxide is split off from the malate and enters the normal photosynthetic pathway. The pH of the cell sap again

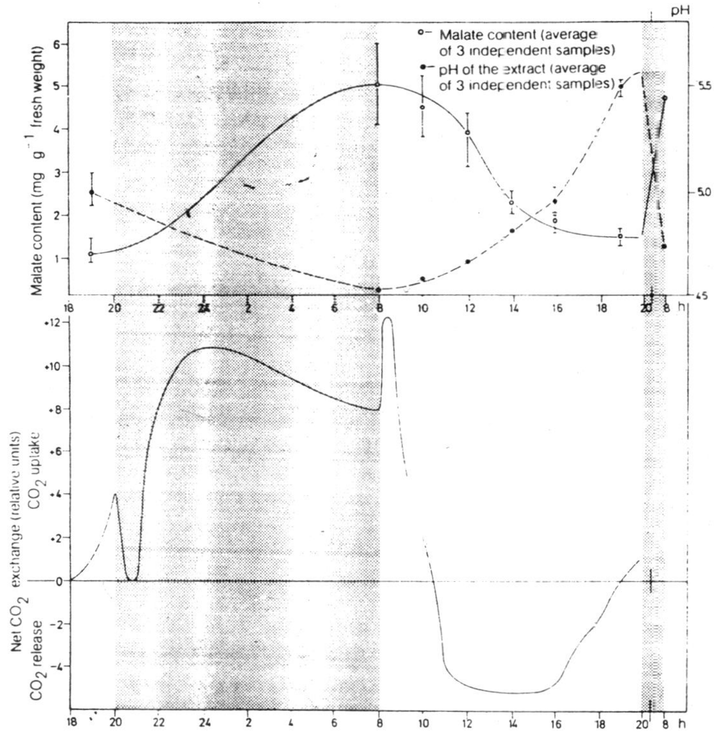

Fig. 4.14 Changes in the malate content and pH within the vacuole of Tillandsia usneoides in the course of a day.

approaches the alkaline range (diurnal acid rhythm). By keeping the stomata closed in the daytime the plant loses no water vapour during the period of greatest heat and low relative humidity—and for desert plants, water conservation is important.

It may be that the C_4 pathway of photosynthesis gives still another selective advantage. In the high-performance plants conversion of the C_4 acids to carbon dioxide and pyruvate occurs in specific bundle-sheath cells that are indigestible by ectothermic herbivores. Grasshoppers that eat such grasses excrete undigested bundle-sheath cells. These plants are therefore a relatively unprofitable source of food for such animals. The pronounced daily rhythm of acidity in the succulents may also provide protection from plant-eating animals, but this possibility has not yet been documented.

As far as ecology is concerned, a crucial aspect of productivity is the site of incorporation of the organic matter produced by a vascular plant—does the new substance take the form of nectar, seeds, leaves, roots of wood? At present there is no satisfactory answer to this fundamental question. Evidently the different plants vary in this regard; annuals differ from perennials. In addition, there are endogenous control mechanisms, most conspicuous and self explanatory in biennial plants. And within a given species "resource allocation" depends very much on external factors. A well-fertilized and well-watered plant develops only a small root system, whereas under water stress or conditions of mineral scarcity a larger, and usually deeper, root system is elaborated. Flowers and seeds account for a smaller proportion of the total growth of a plant when water is abundant than when it is in short supply. This may be an absolute rule for some species; these form flowers and seeds only under conditions of water stress. Finally, the growth of a plant is critically affected by the amount of light received (photomorphogenesis, the regulatory hormone phytochrome).

These considerations lead us to the effects of light animals. For animals, light is almost never a source of energy; it is most significant with respect to orientation of an animal in space (via the eyes) and in time. Light thus appears not to be a factor necessary for animal metabolism, so that one would expect there to be no problem in raising animals in permanent darkness if other conditions were favourable. Such experiments have been done with only a few species, and most of them have failed. Drosophila, terrestrial chironomids and cicadas, for example, have very high mortality rates in maintained darkness. No one knows why.

As a model of the ecological effect of light in the context of the physiology of metabolism, consider Vitamin D. This is taken up as 7-D-hydrocholesterol; in the skin, by the action of short-wavelength sunlight, it is converted into cholecalciferol. This conversion does not occur if insufficient light reaches the skin. When the supply of Vitamin D is inadequate the skeleton is incompletely ossified, leading to the typical symptoms of rickets in children. Rickets is found predominantly at northern latitudes, where the winters are long and dark and UV radiation is weak even in summer. In tropical regions with intense sunlight the illness is essentially unknown. Dark-skinned humans living in the north are particularly subject to rickets, because most of the limited radiation available is absorbed by their skin pigment and cannot be used for conversion of the precursors into the vitamin. Conversely, people with light skins exposed to the strong tropical sunlight are in danger of excessive ossification. The problem of insufficient irradiation can be simply solved by taking vitamin pills, but so far no means have been found by which overproduction of Vitamin D can easily be prevented.

This example suggests that the distribution of animals may be closely related to light, but there is as yet practically no confirming evidence. It also appears plausible that animals of valleys and the high mountains might differ in their radiation requirements and tolerance. Such a possibility has often been proposed, but again evidence is almost entirely lacking. Gluck (1979), however, has shown that there is a clear correlation between the locations where birds breed successfully and the amount of light falling on their nests. He found that the nests of chaffinches, goldfinches, and hawfinches were located where the total irradiation was relatively high, whereas greenfinches, serins, and linnets choose nest sites with a distinctly smaller amount of light.

Some animals—unlike plants—are active at night. The cause of nocturnal activity is frequently the higher relative humidity that prevails at night. In this case, then, light acts solely as a timing signal.

5

Inter-Relationships

Populations of organisms interact by depressing one another's food supply (competing), by eating one another (predation), by benefiting one another (mutualism), or by helping without being helped (commensalism). Predators and prey influence one another's abundances in the short run and also influence one another's evolution, as illustrated by mimicry. Competing organisms exclude one another from some habitats, but many species with similar resource requirements do live together. Mutualisms commonly involve plants because of the importance of animals as transporters of gametes and seeds. The role of an organism in its community is its niche. More species of most taxa live together in tropical environment than in temperate ones.

This chapter deals with predator-prey interactions, competition, commensalims, mutualism, ecological niches, and patterns of species richness.

Populations of organisms interact with one another in complex and varied ways that can be grouped into a few major types (Table 5.1). If two species use the same resources and those resources are in insufficient supply for their combined needs, the species are called *competitors* and their interactions constitute *competition*. If one organism uses another as its source of energy, the eater is called the *predator* and the eaten its *prey*. Their interactions are known as *predator-prey interactions*. In some interactions both participants benefit, in which case we say that they as *mutualists* and that their interaction constitutes a *mutualism*. If one participant benefits but the other is unaffected, the interaction is a *commensalism*. If one participant is harmed but the other is unaffected, the interaction is an *amensalism*.

Table 5.1 Types of Ecological Interactions between Species

	Effects on Organism 2		
Effects on Organism 1	*Benefit*	*Harm*	*No effect*
Benefit	Mutualism	Predation or parasitism	Commensalism
Harm	Predation or parasitism	Competition	Amensalism
No Effect	Commensalism	Amensalism	Neutralism

These types of interactions blend into one another because the strengths of interactions vary from strong to weak and because many cases in nature do not fit these categories neatly. For example, two species of flour beetles living together in a container of flour eat the same food but also act as predators on one another's eggs and pupae. Birds and large predatory insects, such as praying mantids, eat the same insects, but the birds may prey upon the mantids as well. Often it is clear that one participant benefits, but it is difficult to tell whether the weak effect on the other is positive, neutral, or negative.

Predator-prey Interactions

Predators and prey come in a variety of sizes and shapes that make different types of interactions distinctive. The relative sizes of predators and their prey determine how the prey are acquired. If the predator is much larger than its prey, the prey are handled in bulk. The world's largest predators—the baleen whales—feed on very small prey, which they filter from the water. Predators that are only moderately larger than their prey usually pursue and capture them as individuals. Many predators are the same size or smaller than their prey and feed upon them internally or by attaching to them externally. They are usually called *parasites*. If they are about the same size as their prey, such that one or a few predators totally consumes a prey item, they are known as *parasitoids*. Most species of insects are attacked by parasitoid wasps or flies.

Prey types also differ in the extent to which they are defended against their predators. Sessile prey usually defend themselves chemically (by possessing toxic chemicals) or physically (by having hard shells, spines, thorns, and so on). Motile prey depend more heavily on their ability to escape from their predators. Prey are also defended by camouflage, which makes them difficult for predators to find.

Prey also differ in the speed at which their populations recover after they have been reduced by predators. For animals such as barnacles on a rocky shoreline, each wave brings a new and undiminished supply of food. Many prey, however, reproduce only once during the year, and their populations do not recover until the next breeding season. How rapidly populations of prey renew determines the kinds of behaviour predators can employ. Sessile predators, such as web-spinning spiders, mussels, and barnacles, occur only where prey populations can renew rapidly.

Types of Predators and Prey

Filter feeders are found in several animal phyla (brachiopods, mollusks, various worms, chordates). They feed upon prey much smaller than themselves that are suspended in the water or air. Every filter feeder has an apparatus with which it filters prey from the medium, and its structure determines the upper and lower size limits of prey that can be captured. Smaller prey pass through the filter and larger ones bounce off it. For filtering to work, the surrounding medium must move through the filter. This can be accomplished either by movement of the medium itself (e.g., water currents or wind) or movement of the animal through the medium. An important feature of filter feeding is that most of the work of feeding is expended *before* the prey contact the filter. The predator cannot save very much energy by rejecting prey. In fact, it must usually expend energy to remove a prey from the filter. For this reason most filter feeders are very nonselective in their foraging. They accept virtually all prey that are captured by their filter. Differences in the rate of feeding are due primarily to where the filter is located.

Detritivores (decomposers) feed upon the dead bodies and waste products of organisms at some point in their passage from freshly dead to complete conversion to inorganic molecules. Detritus is often patchily distributed in space and time. The deaths of bears in a forest are rare events but create large and rich patches of detritus. Falling leaves create a much more widespread food source for organisms such as worms, slugs, millipedes, and insects, but it may be carried by wind and water currents and concentrated in places where movement of the medium is reduced. Detritus begins its existence with the physical and chemical defenses possessed by the living organisms of which it was a part. These defenses are reduced by the attacks of the first detritivores, so that later consumers of detritus do not have to deal with those defenses. Plant detritus contains large quantities of cellulose and is harder to break down than animal detritus. Except for the bones and shells, the bodies of animals decompose rapidly.

Parasitoids and parasites, being no larger than their prey, are often able to meet all of their needs by feeding upon a single prey item. Therefore, choice of food is made only once or few times during a life cycle, usually by the egg-laying female. An individual parasite is usually deposited as an egg in food-rich environment, but prey have defensive reactions to the presence of a parasite and may be able to destroy it. The ability of the host to defend itself depends upon the number of parasites attacking it and upon its condition. If the host is already weakened by stresses imposed by the physical environment or shortage of food, the parasites are more likely to gain a foothold. Bark beetles attack pine trees by tunneling into the nutritive layers just below the outer bark. The tree defends itself by exuding the sticky pitch for which pines are famous. The more beetles that attack, the greater the probability that they will overcome the defenses of the tree. Each colonizing beetle releases a powerful pheromone that attracts other individuals to the same tree. If there are enough beetles in the general area, a mass attack results that may kill the tree.

Herbivores eat the tissues of plants. They have large quantities of food available to them, but much of it is of very poor quality. Wood consists of cellulose and is impregnated with lignins (which are even more difficult to digest). Very little wood is consumed before the plant that produced it dies or is about to die. If this were not true there would be no forests in the world. Plants also produce a variety of other tissues, such as roots, leaves, flowers, fruits, and seeds, that differ in their chemistry, size, structure, and pattern of production. The organisms that directly eat terrestrial plants are a rich and diverse group and are divided into many different types of animals specialized for utilization of these different plant tissues.

Typical predators, often called *carnivores* are the ones we usually think of when predation comes to mind. They are generally moderately larger than their prey (but not necessarily so—wolves are smaller than moose and caribou), pursue and capture their prey as individuals, and eat many individual prey items during their lives. The number may be very large indeed. A small bird in a temperate forest during the winter must find an average-sized prey (insect or seed) every few seconds to maintain itself.

Foraging Theory

Foraging theory has been developed to help us determine the rules by which predators select where to forage and which prey to take from among the array of potential prey they encounter. All such theories involve guessing the 'goal' of the predator, that is, what is being favoured by natural selection. Possible 'goals' are maximization of energy intake per unit time spent foraging, minimizing the actual time spent pursuing and handling prey, minimizing risk of starvation during an ensuing period (such as night) when foraging may not be possible, or maximizing intake of some critical vitamin or mineral. Each may be appropriate in different situations.

Several mathematical formulations of theories for typical predators that pursue and handle prey items as individuals but search simultaneously for many different prey types have been developed. For predators that select prey items so as to maximize their rate of energy intake, an interesting result emerges. Whether or not a prey item is included in the diet of the predator does not depend upon the prey item's own abundance, but depends on the abundance of prey *types* that yield more energy per unit time spent capturing and eating them. If the top-ranked prey type is sufficiently abundant, the predator captures more energy per unit time spent foraging by taking only that prey type and ignoring all others. As the abundance of the top-ranked prey type decreases, the predator adds lower-ranked prey items to its diet in the order of the energy per unit time that they yield.

Many tests of this theory have been performed. A two-prey case was examined using great tits in experiments in which birds were presented with food on a conveyor belt. The two prey types were whole mealworms and pieces of mealworms. It took the tits approximately the same amount of time to handle both types. By knowing the energy contents, handling times, and rate of arrival of whole mealworms on the conveyor belt, the abundance of whole mealworms at which tits should shift from taking both prey types and concentrate only on whole mealworms can be calculated (Fig. 5.1). At lower arrival rates of whole mealworms, the bird did take both prey types, and it also nearly stopped eating pieces of mealworms when whole mealworms were more common. However, it did not shift at exactly the predicted density and it never totally stopped taking pieces of mealworms. Several factors could explain the deviations from the predicted results. First, the bird might have had difficulty remembering precisely how many captures it had made. Second, because mealworms arrived randomly, several rather long time intervals between arrivals occurred, even when the mean arrival rate was high enough that the bird should have ignored mealworm pieces. Finally, the bird might have made mistakes in identification. Nonetheless, the behaviour of the experimental bird, and others like it, suggests that prey are being selected by rules similar to those postulated in the theory. These birds appear to be foraging so as to maximize their rate of energy intake.

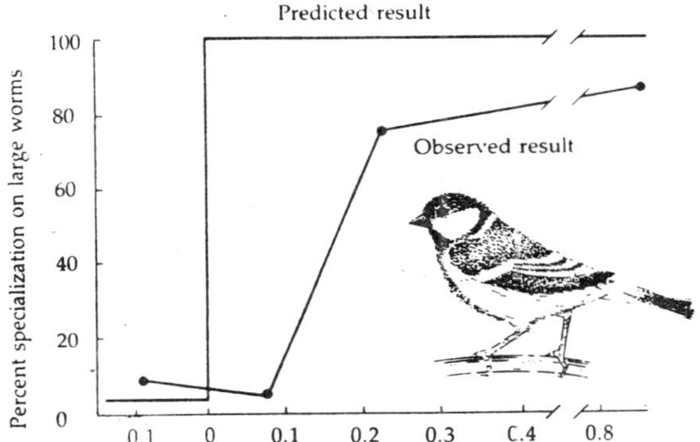

Fig. 5.1 Rate at which large worms arrive (numbers indicate extra payoff for eating large worms)

Short-term Consequences of Predator-prey Interactions

Predators, by eating some of their prey, immediately reduce the size of the prey population, but this is offset by reproduction of the prey. How numbers of predator and prey fluctuate is seen most readily by considering a single predator species eating a single prey species in a homogeneous environment. We will assume that the prey species grows according to an S-shaped curve that, when plotted as a population rate, yields a curve of the form of an inverted U. The height of the curve represents the number of prey that could be cropped by a predator without reducing the size of the prey population. An individual predator searching for only individuals of that prey species will find enough to eat if its encounter rate with the prey is above a threshold value. Below that threshold, it will lose weight and eventually starve. Nevertheless, the predator continues to eat the prey and can drive its populations to low levels before starvation or emigration of the predator allows to prey population to recover. Therefore, predator-prey interactions tend to lead to oscillations in population densities of both species.

Many laboratory and field studies have confirmed the reality of oscillations in populations of predators and prey. In nature, the best-known regular population cycle occur in small mammals and their predators at high latitudes. Populations of Canadian lynxes fluctuate in a 9-11 year cycle following a similar cycle in the population of their principal prey, snowshoe hares. The causes of the hare cycle are not fully known but are believed to involve both predation by lynxes and changes in the quality of the plants upon which the hares feed. A 3-4 year cycle is observed in populations of lemmings and their chief predators, snowy owls, jaegers, and Arctic foxes. However, not all predator-prey interactions yield conspicuous oscillations.

In heterogeneous environments, prey are more vulnerable to predators in some patches than in others. In the absence of predation, prey would be distributed over those patches in accordance with the availability of food and other resources. As predation increases, prey are removed selectively from those patches where the predators are most efficient at harvesting them. A striking example of this is provided by the interaction between St. John's wort (*Hypericum perforatum*) and a beetle (*Chrysolina quadrigemina*) that preys upon it. St. John's wort, an annual plant introduced into California from Europe, increased rapidly and soon dominated vast expanses of range land. Because it is much less desirable as a forage species for cattle than the grasses it replaced, attempts were initiated to control the weed. Spectacular success was achieved by the introduction of *Chrysolina,* which has all but eliminated St. John's wort in sunny areas. However, the plant remains common in shadier areas where the beetle does not do well.

Patchiness can also occur in time. The cactus *Opuntia* sp., introduced into Australia from North America, also spread rapidly and took over vast expanses of valuable sheep grazing land in Australia. It was controlled by the introduction of a moth (*Cactoblastis cactorum*) that completely destroys patches of *Opuntia* when the egg-laying females find them. However, new patches of cactus are formed from seeds dispersed from existing clumps by birds. These flourish until they are found and destroyed by the moths. Today, there is fairly constant low population of both *Opuntia* and *Cactoblastis,* if an average is taken over large areas; but on a local scale there are vigorous oscillations that result in the extermination of first the prey and then the predator, which has eaten itself out of food.

Plant Defenses

The most obvious ways in which plants limit the activities of predators, in this case herbivores, are *morphological defenses*. Thorns, spines, and prickles play an important role in discouraging browsers, although some are able to overcome these obstacles. The defensive role of plant hairs, especially those that have a glandular, sticky tip, is less obvious. However, the enormous abundance of insects and the scale at which they operate as herbivores provide some indication of why hairs of various kinds are important to the plants that possess them. The strengthening of plant parts by depositing silica in the leaves, is an element in the protective system of some plants, such as grasses. If enough silica is present in their cells, the plants may simply be too tough to eat.

Significant as these morphological adaptations are, the chemical defenses that occur so widely in plants are even more crucial. As research in this area proceeds, we continue to be astonished by the number and diversity of these features. Most animals can manufacture only 8 of the 20 amino acids. The exact number varies from species to species and even during the life span of an individual in some species. All of the other amino acids are required and must be obtained from plants. By restricting the amount of one or more of the amino acids that animals require, plants can limit their nutritional suitability for animals. As a result, they may be protected from herbivores.

Best known and perhaps most important in the defenses of plants against herbivores are *secondary chemical compounds*. These are distinguished from *primary compounds*, which are regularly formed as components of the major metabolic pathways, such as respiration. Virtually all plants, and apparently many algae as well, contain very structurally diverse secondary compounds that play a defensive role. Estimates suggest that perhaps 50,000 to 100,000 different secondary plant compounds may exist; some 15,000 of these have been characterized chemically.

We now know a great deal about the roles of secondary compounds in nature. As a rule, different kinds of compounds are characteristic of particular groups of plants. For example, the mustard family (Brassicaceae) is characterized by a group of chemicals known as mustard oils. These are the substances that give the pungent aromas and tastes to such plants as mustard, cabbage, watercress, radish, and horseradish. The same tastes that we enjoy signal the presence of chemicals that are toxic to many groups of insects.

Another group of plants, the potato and tomato family (Solanaceae), is rich in alkaloids and steroids, complex ring-forming molecules that exhibit an enormous range of structural diversity. Similar compounds occur widely among flowering plants. Sometimes the compounds are involved in more complex defensive system: for example, plants of the milkweed family (Asclepiadaceae) and the closely related dogbane family (Apocynaceae) tend to produce a milky sap that deters herbivores from eating them. In addition, these plants usually contain cardiac glycosides molecules named for their drastic effect on heart function in vertebrates.

One of the best known plant groups with toxic effects includes poison ivy, poison oak, and poison sumac (*Toxicodendron* species). All contain a gummy oil called urushiol, which causes a severe rash in susceptible people. Urushiol can persist for years on clothes or other objects. There are at least 140,000 cases of this rash annually that cause absence from work in the United States, and many more are not reported. Approximately one out of two people is at least moderately sensitive to these plants. Immediate washing with soap will remove the excess urushiol and prevent its further spread, but the reaction of exposed body parts seems to be immediate.

Many angiosperms are protected by a rich and varied chemical arsenal. Mustard oils, alkaloids, steroids, and other classes of secondary compounds either are toxic to most herbivores or disturb their metabolism so greatly that they are unable to complete their development normally. As a consequence, most herbivores tend to avoid the plants that possess these compounds. The pattern of occurrence of such chemicals has had major consequences on the evolution of both plants and herbivores, especially insects, that feed on them.

The Evolution of Herbivores

Associated with each family or group of plants protected by a particular kind of secondary compound are certain groups of herbivores that are able to feed on these plants, often as their exclusive food source. How do these animals manage to avoid the chemical defenses of the plants, and what are the evolutionary precursors and ecological consequences of such patterns of specialization?

As a starting point, we can offer the observation that some herbivore groups feed on plants of many different families, others on plants of just a few families. In general, those herbivores that feed on a restricted array of plant families, perhaps only one, feed on a plant groups in which certain kinds of secondary compounds are well represented. For example, larvae of cabbage butterflies (subfamily Pierinae) feed almost exclusively on plants of the mustard and caper families, as well as on a few other small families of plants that are characterized by the presence of mustard oils. Similarly, caterpillars of monarch butterflies and their relatives (subfamily Danainae) feed on plants of the milkweed and dogbane families. No other groups of butterflies have larvae that feed on plants of the families that are well protected by secondary plant compounds. This indicates the efficiency of the chemicals in retarding feeding by most kinds of butterflies. Other kinds of insects that feed on these plant families are also usually specialized within their groups.

We can explain the evolution of three particular patterns as follows. Once the ancestors of the caper and mustard families acquired the ability to manufacture mustard oils, they were protected for a time against most or all herbivores that were feeding on other plants in their area. The plants that contained the mustard oils apparently succeeded very well as a result of this particular defense. These plants and their descendants evolved and migrated, eventually giving rise to the thousands of species of mustards and capers that now grow worldwide.

At some point, certain groups of insects—for example, the cabbage butterflies—developed the ability to break down mustard oils and thus feed on these plants without harming themselves. Having acquired this ability, the butterflies themselves had registered an important evolutionary "breakthrough." They were able to use a new resource without having to face competition for it from other herbivores. Often, in groups of insects such as cabbage butterflies, sense organs have evolved that are particularly sensitive to the secondary compounds that their food plants produce. Clearly, the relationship that has formed between cabbage butterflies and the plants of the mustard and caper families is an example of coevolution.

Chemical Defenses in Animals

Some groups of animals that feed on plants rich in secondary compounds receive an extra benefit, one of great ecological importance. When the caterpillars of monarch butterflies feed on plants of the milkweed family, they do not break down the cardiac glycosides that protect these plants from herbivores. The cardiac glycosides in the leaves that the caterpillars eat are concentrated, stored in fat bodies, and passed through the chrysalis stage to the adult and even to the eggs. All stages of the monarch life cycle are protected from predators by incorporation of the same compounds synthesized by the plant to deter herbivores. A bird that eats a monarch butterfly quickly regurgitates it and in the future avoids the kind of conspicuous orange-and-black pattern that characterizes the adult monarch. Locally, however, some birds appear to have acquired the ability to tolerate the protective chemicals and eat the monarchs.

Insects that feed regularly on plants of the milkweed family are, in general, brightly coloured. Among them are brightly coloured cerambycid beetles, whose larvae feed on the roots of the milkweed plants; bright blue or green chrysomelid beetles; and bright red true bugs of the order Hemiptera. In some parts of the world, there are also bright red grasshoppers and other very obvious insects. These herbivores clearly are 'advertising' their poisonous nature by their bright colours, using an ecological strategy known as *warning coloration*. Similar relationships occur in marine communities. A recent investigation of coral reef invertebrates at Lizard Island, on the Great Barrier Reef off the northeast coast of Australia, revealed that of the common and easily visible reef invertebrates, three-fourths were toxic to fish; of the protectively camouflaged invertebrates, only one-fourth were toxic.

Insects that eat plants lack specific chemical defenses are seldom brightly coloured. In fact, many of these insects are *cryptically* coloured—coloured to blend in with their surroundings and thus be hidden from predators.

Insects such as the larvae of cabbage butterflies are able to feed on plants with chemical defenses because of their ability to break down the toxic plant molecules, but are cryptically coloured. Because the toxic chemicals are broken down, they do not provide protection for the herbivore.

Some marine animals—such as certain nudibranchs—acquire defensive chemicals or defensive cells from their prey. Hydroids often provide stinging cells to animals that graze on them. Off the coast of Panama, the large nudibranch *Aplysia* grazes selectively on red algae of the genus *Laurencia*, which is protected by elatol, a powerful inhibitor of cell division. Perhaps this is the reason that few fish feed on *Aplysia*. There is an intensive investigation of marine animals, algae, and flowering plants for new drugs against cancer and other diseases, or as sources of antibiotics. It holds great promise because of the enormous diversity of chemical compounds that occur in these organisms.

Animals also manufacture and use a startling array of substances to perform a wide variety of defensive functions. Venomous snakes, lizards, and fishes are well known; in addition, bees, wasps, predatory bugs, scorpions, spiders, and many other arthropods have chemicals that they use to defend themselves and to kill their prey.

Various chemical defenses are found among the vertebrates, as well. The dart-poison frogs of the family Dendrobatidae produce toxic alkaloids in the mucus that covers their skin.

Some of these toxins are so powerful that a few micrograms will kill a person if injected into the bloodstream. Most of the over 100 species of these frogs are brightly coloured; they are favourites in zoos and aquaria for that reason.

Aposematic Colouration

Warning colouration, as discussed previously, is also known as *aposematic colouration*. Such colouration is characteristic of animals that have effective defense systems. This includes not only poisons, but also stings, bites, and other means of repelling predators. An organism possessing such a system will benefit by advertising the fact clearly, perhaps by showy colours that are not normally found in that particular habitat. The distasteful or poisonous individual wants to avoid an attack; otherwise, the individual runs the risk of being killed while protecting itself. An individual protected in this way has a selective advantage over others that are not protected.

The animals that display aposematic colouration must occur at relatively high densities if the system is to be effective. If genetically related individuals are similarly coloured and live in the same vicinity, the selective advantage is obvious. Predators will tend to avoid such individuals.

Such animals tend to live together in family groups, unlike those that tare cryptically coloured. If camouflaged animals lived together in groups, one might be discovered by a potential predator, offering a valuable clue to the presence of others.

Mimicry

During the course of their evolution, many unprotected (nonpoisonous) species have come to resemble distasteful ones that exhibit aposematic colouration. The unprotected mimic gains an advantage by looking like the distasteful model. Two types of mimicry have been identified: Batesian mimicry and Muellerian mimicry.

Batesian Mimicry: *Batesian mimicry* is named for W.H. Bates, a British naturalist who first brought it to general attention in 1857. Bates discovered that if the unprotected animals are present in numbers that are low relative to those of the species that they resemble, they will be avoided by predators. If the unprotected animals are too common, or do not live among their models, many of them will be eaten by predators that have not yet learned to avoid the individuals with those particular characteristics.

Many of the best-known examples of Batesian mimicry occur among butterflies and moths. Obviously, predators in systems of this kind must use visual cues to hunt for their prey; otherwise similar patterns of colouration would not matter to potential predators. There is also increasing evidence indicating Batesian mimicry can also involve nonvisual cues, such as olfaction, although such examples are less obvious to humans.

The groups of butterflies that provide the models in Batesian mimicry are, not surprisingly, members of groups whose larvae feed on only one or a few closely related plant families. The plant families on which they feed are strongly protected chemically. The model butterflies take poisonous molecules from these plants and retain them in their bodies. The mimic butterflies, in contrast, belong to groups in which the feeding habits of the larvae are not so restricted. As caterpillars, these butterflies feed on a number of different plant families, but not those protected by toxic chemicals.

One often-studied mimic among North American butterflies is the viceroy, *Limentitis archippus*. This butterfly, which resembles the poisonous monarch, ranges from central Canada south through much of the United States east of the Sierra Nevada and Cascade Range into Mexico. The larvae feed on willows and cottonwoods, and neither the larvae nor the adults were thought to be distasteful to birds, although recent findings may dispute this. Interestingly, the larvae of the viceroy are camouflaged on leaves because they resemble bird droppings, while the distasteful larvae of the monarch are very conspicuous.

Muellerian Mimicry: Another kind of mimicry, *Muellerian mimicry*, was named for the German biologist Fritz Mueller, who first described it in 1878. In Muellerian mimicry, several unrelated but protected animal species come to resemble one another. Thus, different kinds of stinging wasps have yellow-and-black striped abdomens, but they may not all be descended from a common yellow-and-black striped ancestor. In general yellow-and-black and bright red tend to be common colour patterns that warn predators relying on vision that such animals are to be avoided. If animals that resemble one another are all poisonous or dangerous, they gain an advantage because a predator will learn more quickly to avoid them.

In both Batesian and Muellerian mimicry, mimic and model must not only look alike but also act alike if predators are to be deceived. For example, the members of several families of insects that resemble wasps (Figure 5.2) behave surprisingly like the wasps they mimic, flying often and actively from place to place.

Theory of Competition

Two species may *overlap* in their utilization of resources without competing if the quantity of resources is large enough that the presence of one species does not reduce the other's energy intake.

Competition that occurs because resources are depleted is known as *exploitation competition*. It results in lower encounter rates with prey items and, hence, lower growth and reproductive rates. Competition may also result from direct behavioural interactions: *interference competitio*. Behaviourally more complex animals are more likely to engage in direct behavioural interference, whereas simpler animals usually affect one another through exploitation. Plants usually compete by exploitation.

Competing organisms may coexist, although at lower population densities than either would achieve in the absence of the competitor; or one species may exclude the other. Which of these two outcomes occurs depends on the efficiency with which each species exploits the environment (the carrying capacity of the environment for them) and the effects individuals have on members of the other species compared to the effects they have on individuals of their own species (conspecifics). Coexistence is favoured when the addition of an individual lowers the success of conspecific individuals more than it lowers the success of individuals of the other species. The reverse-*inter*specific competition being stronger than *intra*specific competition—favours elimination of one species by the other. Intraspecific competition is greater than interspecific competition when there are differences in the resources exploited by individuals of the two species. If two species are forced to compete in a simple laboratory situation where only a single resource is provided, one out-competes the other, a process known as *competitive exclusion*. The first such

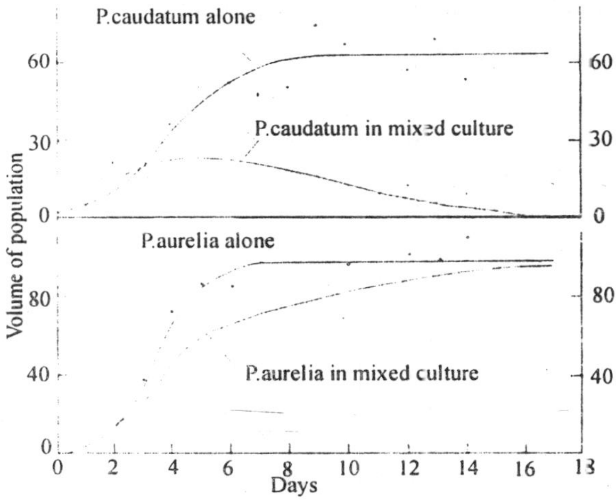

Fig. 5.2 Competition between two similar protists
Paramecium caudatum and P. aurelia have very similar food and temperature requirements. When grown alone in containers with a constant food supply, the population of each species grows in an S.shaped curve until it reaches the carrying capacity of the environment (black and solid colour curves). However if they are grown together in the same container, one species soon eliminates the other (gray and light colour curves). In the experiments shown here, P. aurelia eliminated P.caudatum (under somewhat different conditions the reverse happens). Because of the very small size of these organisms, population sizes are expressed here as volume rather than numbers of individuals. Notice that in these experiments the species that maintained the largest population when grown alone (P. aurelia) was also the competitive winner. This is not always the case in competition experiments, however.

experiments were performed by the Russian ecologist G.F. Gause and reported in an influential book, *The Struggle for Existence,* published in 1934.

Competitive exclusion can be avoided even in very simple laboratory environments by providing a bit of environmental heterogeneity such that there are some places where one species does better. For example, *Paramecium bursaria* and *P. aurelia* form stable mixtures in the laboratory under certain conditions because *P bursaria,* which has symbiotic algae within its cell, is able to feed in the deoxygenated water at the bottom of the containers whereas *P. aurelia* is unable to do so.

Competition in Nature

Nature is vastly more complex than simple laboratory environments, and species often coexist even when they overlap greatly in their use of resources and are known to be competing. The most direct way to demonstrate competitions to remove one species and observe an increase in the population of another species. This is done by everyone who weeds a garden. There is a drop in productivity of desired plants when gardens are subjected to the periods of neglect that most of us impose on them. A well-known generalization among agronomists (students of crop plants) about plant competition is the *self-thinning rule,* which states that the number of individuals decreases under competition but not as fast as the mean weight of the individuals increases, so that the weight of the population as a whole increases. Thinning occurs more rapidly on fertile plots because each individual grows more rapidly and competition is more quickly manifested (Fig. 5.3).

Rapid response to the removal of competitors has been demonstrated among the common meadow vole (*Microtus pennsylvanicus*)—a grassland species—and the red-backed vole (*Clethrionomys gapperi*) and the deer mouse (*Peromyscus maniculatus*)—both normally woodland species. Animals were confined in enclosures 0.2 hectares in size, each of which was half woodland and half meadow. When any one of the three species was alone in an enclosure, it occupied both habitat types. *Microtus* caused restriction of both the other two species to the woodland and immediately adjacent grassland; but if

Microtus was removed, the others expanded back into the grassland. These changes are not due to exploitation but to direct aggressive interference by the larger *Microtus*.

Indirect evidence of competition is provided by habitat restrictions or expansions by species in the presence or absence of a presumed competitor. The chaffinch (*Fringilla coelebs*) is one of the most abundant birds in Europe, being found in both coniferous and broad-leafed woodland. At some undetermined time in the past, chaffinches invaded the Canary Islands off the northwestern coast of Africa. They evolved there in isolation for

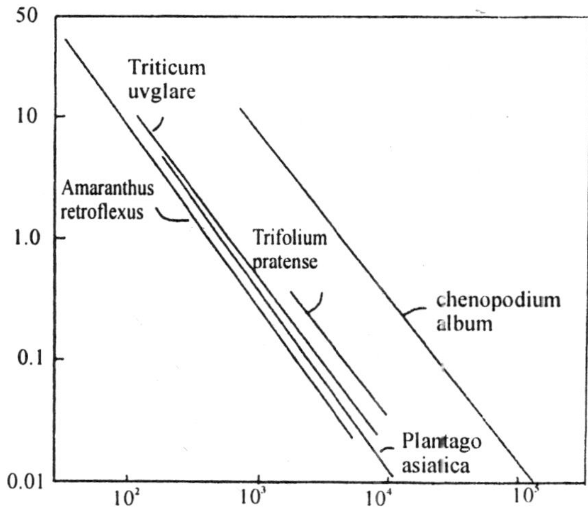

Fig. 5.3 Self-thinning among annual plants

As populations of annual plants grow from tiny seedlings to larger mature plants, individuals compete with one another for resources and some die (self-thinning). Many experiments with such growing plants have shown that there is a predictable relationship between the density of the plants and the size of the survivors. Studying the curves from right to left, this figure shows that the number of individuals decreases more slowly than the mean weight of survivors increases so that the total weight of the population increases as the density decreases due to competition. A number of different species of annual plants are plotted here, but they all show exactly the same relationship—that is, the lines all have the same slope.

long enough that when a second invasion of chaffinches occurred they were isolated reproductively from them. Today on the Canary Islands, there are two species of chaffinch: the blue chaffinch (descended from the early colonization) and the common chaffinch. On the Canaries, however, the blue chaffinch occupies the coniferous forests and the common chaffinch is restricted to broad-leafed forests. This is the only place that the common chaffinch does not occur in coniferus forests and the only place where it coexists with another species of chaffinch. It is reasonable to attribute this, and other similar cases of habitat restriction, to competition.

Ecologists have long been interested in how similar two species can be and still coexist. There is no single value to the amount of overlap in resource utilization compatible with coexistence because more overlap can be tolerated if resources are stable in abundance than if they fluctuate widely. In addition, interference competition may be very effective even when there is relatively low overlap in resource utilization between two species. Finally, the long-term consequences of competition are not necessarily the same as the short-term ones.

In a patchy environment, differences among competitors make them at least slightly different in their abilities to exploit prey in different patches. This allows one competitor to depress prey populations in some patches to such low levels that individuals of the other species cannot forage there profitably. The competitors then segregate by habitat types, each concentrating its foraging where it is most efficient. This tendency can be reinforced by behavioural interactions of the kind seen between *Microtus* and *Clethrionomys* and *Peromyscus,* either before or during depression of the prey. Habitat segregation is reinforced every generation because wandering individuals, particularly juveniles, explore different patches but find some less profitable because of the presence of the competitors.

Evolutionary changes gradually occur as each species adapts better to the restricted set of environmental patch types it actually occupies. These changes make each species less able to exploit the preferred patches of the other and reduce the chances that they will immediately invade the preferred patches of the other even if the competitor is removed.

We have discussed predation and competition as if they were very different processes, but they are intimately interrelated. Competitors are really rival predators that share some important resources and thereby influence one another. Their mutual influences may also affect their relationships with still other predators. For example, a competitor may force an animal to forage in more dangerous places, thereby exposing it to its own predators. Similarly, a predator, by preying preferentially upon a dominant competitor, may create openings for species that would otherwise be competitively excluded. The seastar *Pisaster ochraceous* is an abundant predator on rocky intertidal communities on the Pacific coast of North America. Its preferred prey are mussels (*Mytilus californianus*) which, in the absence of seastar predation, dominate a broad belt of the intertidal zone, pushing out other competitors for space. *Pisaster,* by heavily cropping mussels, creates bare spaces that are occupied by a variety of other species, some of which are shown. The actual structure of rocky intertidal communities is the result of interactions between competition and predation.

MUTUALISM

Mutualism is an association between two or more species in which all derive benefit in feeding or in some other way. The term symbiosis has often been applied to this relationship, but *symbiosis* properly refers to the intimate association of two or more dissimilar organisms, regardless of benefits or the lack of them, and hence includes mutualism, commensalism, and parasitism.

Mutualism, as is true also with commensalism and parasitism, may be *facultative,* when the species involved are capable of existence independent of one another, or *obligate,* when the relationship is imperative to the existence of one or both species. Considerable study and experimentation is sometimes required to decide whether a particular relationship is facultative or obligate, or even whether it is truly mutualistic.

Mutualism in plants is demonstrated in the association of fungi and algae to form lichens, of nitrogen-fixing bacteria with the roots of legumes, and of fungal mycorrhizae with the roots of many flowering plants.

There are many intimate relations between plants and animals. Mutualism is suspected in the presence of photosynthetic algae cells in the protective ectoderm of green hydra, and those associated with turbellarians, mollusks, annelids, bryozoans, rotifers, protozoans, and the egg capsules of salamanders. The algae give off oxygen, benefiting the animals, which in turn supply carbon dioxide and nitrogen to the plants. The thick growth of algae often found on the carapace of the aquatic turtles is important mostly as camouflage for the turtles. Certain beetles, ants and termites cultivate fungi for food. Bacteria in the caeca and intestine of herbivorous birds and mammals aid in the digestion of cellulose. The cross pollination of flowers by the agency of insects and birds seeking nectar and pollen is of such great importance that many structural adaptations in both plants and animals fit the one to the other to ensure the success of the function.

Animals, especially birds and mammals, are of great importance as agents of plant distribution. Seeds, fruits, even entire plants become attached to feathers or fur, or ingested seeds are eaten and eliminated unharmed with the feces. When bare seeds are eaten they are usually macerated, digested, and entirely destroyed unless they have very hard coats. But fruits are fed upon primarily for pulp, and most of the seeds pass through the alimentary tract unharmed. Animal transportation of ingested seeds is perhaps the most important means by which fruit species are dispersed. Furthermore, germination of the seeds is frequently improved by mechanical abrasion in the stomach and thinning of the seed coat by digestive juices, making them more permeable to water and oxygen. Germination of acorns and nuts is improved if they are buried in the ground rather than left lying on the ground surface. Squirrels, chipmunks, wood rats, and some birds, particularly jays and woodpeckers, cache acorns and nuts as a winter food supply, hiding them in cavities and nooks or burying them in the soil. Perhaps most are recovered and eaten; Cahalane (1942) found that 99 per cent of the acorns buried by the fox squirrel in a locality where the animals were numerous were recovered by the animals, largely through the sense of smell. One per cent of the thousands of nuts produced by it during the lifetime of a tree that are buried but not recovered would be adequate to ensure the continuance of the forest. Invasion of oak and hickory trees into sandy areas is greatly accelerated by, and is sometimes dependent on, this coaction of squirrels, and the dispersal of forests up the slopes of mountains against gravity may also depend in large part on transportation of the heavy seed by animals. The interesting concept involved here is that plants have evolved fruits and nuts that are highly attractive to animals as food substances. However, the production of prodigious numbers of fruits and nuts during their lifetime ensures that at least some will escape consumption and will be more widely and effectively dispersed.

Large populations of such herbivores as rabbits and deer sometimes do considerable damage to new propagation of herbs and trees, but the effects of over-browsing cannot be dismissed as all bad. Removal of the lower branches of established trees by deer may not seriously affect the vigour of the trees; indeed, such pruning may actually increase their value as lumber. Deer pawing the leaf-litter may thereby plant, so to speak, some seeds that would not otherwise become established. Thinning dense stands of young trees may allow residuals to grow more rapidly; much of new growth is doomed anyway because of root competition, and shading cast by established trees. Some species of shrubs and trees actually produce more annual growth under heavy than light browsing. Other species, however, may be killed when small by heavy browsing, although they tolerate considerable browsing when mature. Detrimental effects of both browsing and grazing become evident in an area in the form of excessive invasion of new species which are little used as food, and disappearance or stunting of the food species that are desirable.

Some tropical acacias have evolved foliar nectaries or other food bodies as well as enlarged hollow stipules, spines, or other structures to attract stinging ants. In return, the plants obtain protection from herbivorous mammalian and insect enemies.

Interspecific mutualism is nicely demonstrated by the flagellate *Trichonympha,* an obligate anaerobe in the gut of several species of wood-eating termites (but not in the family Termitidae) where it digests cellulose. *Trichonympha* and related species also occur in the alimentary tract of the wood-eating roach *Cryptocercus.* The termite and roach reduce the wood to small fragments, passing them through the alimentary canal to the hindgut, where the protozoans digest the cellulose, changing it into sugar. The host benefits the protozoa by removing harmful metabolic waste products and maintaining anaerobic conditions in the intestine.

The ruminant stomach and the horse caecum contain enormous numbers of ciliates and bacteria, some of which digest cellulose. The microorganisms reproduce the equivalent of their biomass each day. Digestion of these organisms provides the host with about 20 per cent of its nitrogen requirement.

COMMENSALISM

Commensalism defines the coaction in which two or more species are mutually associated in activities centering on food and one species, at least, derives benefit form the association while the other associates are neither benefited nor harmed. It is often difficult to establish definitely the nature of the relations between species; and phenomena considered at one time to be commensalism have been later found to be parasitism or mutualism. The concept of commensalism has been broadened, in recent years, to apply to coactions other than those centering on food; cover, support, protection, and locomotion are now frequently included.

The remora fish are remarkable for having the spinous dorsal fin modified to form a sucking disk on top of the head by means of which they become attached to the body of the shark, swordfish, tunny, barracuda, or sea turtle. They are of small size and are not burdensome to the host. The host benefits the remora, however, for when the host feeds, the scraps of food floating back are swept up by the remora. Many small animals become attached to the outside of larger ones, such as the protozoans *Trichodina* and *kerona* on *Hydra,* vorticellids on various other aquatic organisms, branchiobdellid annelids on crayfish, and so on. Commensals may also be internal; consider, for instance, the harmless protozoans that occur in the intestinal tract of mammals, including man.

The fellow-bellied sapsucker drills holes into the phloem of trees from which oozes a flow of sap. The woodpecker uses the sap for soaking or mixing with the insects that it captures for food or may drink the sap separately. The sap and the insects that it attracts are also fed on extensively by other birds, especially hummingbirds, and by other insects and a few mammals.

The pitcher of the pitcher plant found in bogs furnishes a breeding site or home for certain species of midge flies, mosquitoes, and tree toads. Many kinds of microorganisms, both plant and animal, live in the canal system of sponges.

The nest of one species often furnishes shelter and protection for other species as well. Ant nests may contain guest species of various other insects. Large hawk nests sometimes have nests of smaller species tucked in their sides, some birds place their nests close to wasps, bees, or ants for the protection offered by these insects. Woodchuck burrows are used also by rabbits, skunks, and raccoons, especially in the winter. During dry periods the water in crayfish burrows, a meter below the ground surface, often teem with entomostraca.

COMMUNITY ORGANIZATION

The final stage in the evolution of cooperation is the biotic community. Analogous to a multicellular individual, the community is composed of organic units, in this case organisms and species rather than cells and tissues. It has a definite anatomy in its stratification, niches, and food chains. The community, too, is a thing born, and it exhibits the same characteristics of growth and maturity as do individuals. There is succession of stages to the climax community like the series of instars in the life cycle of an insect. If the community is injured, it heals the wounds in its structure through secondary succession. The community is self-sustaining in that it absorbs energy from the sun and metabolizes it at various trophic levels in order to do work. There is division of labour, analogous to the functions of the various organs in the body of a single individual; plant species manufacture the food that animals need, and dominant species create environment conditions within the community suitable for the existence of other species. There is transmission of stimuli, intercommunication, between individuals and species by voice, odour, sight and contact. There is control over the numbers of individuals of each species in the balance of nature. The result is that the biotic community is a highly integrated recognizable unit in which species exhibit various degrees of interdependency. The existence of each component depends to a certain extent on cooperation between them all, so that the community functions as if it were an organic entity. These complicated inter-relationships have come about through evolution because of their survival value for the component species involved.

PARASITISM

Parasitism is the relation between two individuals wherein the *parasite* receives benefit at the expense of the *host*; parasitism is therefore a form of disoperation. Parasitism is mainly a food coaction, but the parasite derives shelter and protection from the host as well. A parasite does not ordinarily kill its host, at least not until the parasite has completed its reproductive cycle. Were the parasite to kill its host immediately on infecting it, the parasite would be unable to reproduce and would quickly become extinct. The balance between parasite and host is upset if the host produces antibodies or other substances which hamper normal development of the parasite. In general the parasite derives benefit from the relation while the host suffers harm, but tolerable harm.

Classification

Parasites are commonly classified as *ectoparasites*, those which live on the outside of the host, and *endoparasites*, those which live in the alimentary tract, body cavities, various organs, or blood or other tissues of the host. Ectoparasites may be parasitic only in the immature stages—the hairworm larvae, parasitic in aquatic insects; only the adults parasitic—fleas, on birds and mammals; or both larvae and adults may be parasitic—the blood-sucking lice and flies, biting lice, mites, and ticks that occur on birds, mammals, and sometimes reptiles, and the monogenetic trematodes on fish. Similar relations obtain among endoparasites, although it is more common to have all stages parasitic: entozoic amoebae, trichomonad flagellates, opalinid ciliates, sporozoans, pentastomids, nematodes, digenetic trematodes, acanthocephalons, cestodes, and some copepods.

Animals may also be parasitic on plants. Nematodes infest the roots of plants. Galls are formed by wasps or gnats especially on oaks, hickories, willows, roses, goldenrods, and asters. Mites stimulate formation of witches' brooms in hackberry. A variety of insects the larvae of which are leaf miners, wood borers, cambium feeders, and fruit eaters, should be included here. Plants themselves may be parasites either on other plants or on animals. Bacteria and fungi are among the most important disease-producing organisms in animals.

Social parasitism describes the exploitation of one species by another, for various advantages. Old World cuckoos and the brown-headed cowbird of North America do not build nests of their own; rather, they deposit their eggs in the nests of other species, abandoning eggs and young to the care of foster parents. The bald eagle sometimes robs the osprey of fish that it has just caught. One species of ant waylays foraging workers of another species and snatches away the food they are transporting, or the robber species may deliberately rob another nest of food. Some species of ants make slaves of the workers of other species. Various other types of dependency of one species on another have evolved, not only between ants, but also in other social insects, such as termites, wasps, and bees. Social insects are apparently the only animals other than man to have succeeded in domesticating other species, and of cultivating plants, particularly fungi, for food.

Organisms that produce disease generally fall into one or two categories. They are either present in the body at all times but not normally virulent or they are normally absent but are virulent from the moment the host is infected by them. Even the healthiest animals chronically entrain many parasites and noxious organisms in the body, but these organisms wreak overt harm only when they become unusually abundant, when virulent mutant strains develop, or if, for one or another reason, the host's vitality and resistance decline to the point where the host is no longer able to withstand the effects of their presence. Any animal suffering an unusually heavy infestation of parasites will show the tax thus put upon its vitality as a loss of vigor and weight, decreased growth rate, and low resistance to vicissitudes of its natural environment. Normally, a more or less mutual tolerance exists between host and parasite such that the demands of the parasite are in equilibrium with the host's capacity to meet them. Host-tolerant parasites have been naturally selected for; mutant strains that are exceptionally virulent quickly die out because they kill the host, without which they cannot survive.

A single attack, even a mild one, of some diseases often confers a partial or complete *immunity* from further attacks of the same disease, even though the agent of the disease may still be carried in the body of the recovered victim. Immunity is an acquired physiological adaptation by which the immune is able to withstand the presence of an otherwise noxious organism, suffering little or no deleterious consequence of the presence. The fact of immunity is demonstrated when parasites not conspicuously harmful to their normal hosts are introduced into a species to which they are normally exotic. The novel host has had no prior occasion or opportunity to adapt immunitively to the alien parasite, and may sicken, even die, of the effect of the parasite's presence, the same effects in kind and intensity which the normal host easily takes in stride. For instance, a trypanosome that is a natural parasite in many of the larger wild mammals of Africa evokes no spectacular effects in its usual hosts. But when the parasite is vectored by the tsetse fly to man, it causes sleeping sickness; to cattle, nagana. *Zoonosis* is the study of diseases common to animals and man. Animals often serve as reservoirs for diseases organisms that may under proper circumstances be conveyed to man. The threat of plague is ever present is some parts of the world, where the organism prevails among the resident rodents.

PARASITOIDISM AND PREDATION

Some Diptera and Hymenoptera deposit their eggs in the immature stages of other insects; the larvae on hatching feed on the host until they are fully grown (Fig. 5.4). The relation of the larva to its host is frequently described as one of parasitism. But it is fundamentally different from parasitism in that the host generally dies of the larval depredations before the larva emerges, but the larva generally lives in spite of the host's death. The relationship resembles that of predator to prey, except that, unlike the true predator, the larva lives within the

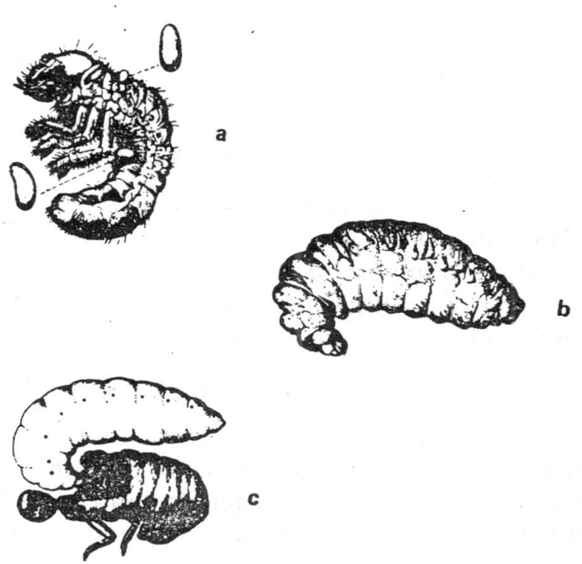

Fig. 5.4 Development of a parasitoid black digger wasp: (a) eggs in position on the host larva; (b) the developing larva; (c) the fully grown larva devouring the remainder of the host.

body of its prey and kills it slowly as it feeds, not suddenly before it feeds. Such larvae are, for these reasons, best thought of as *parasitoids*.

Parasitoids may in turn be infested with *hyperparasitoids*. In the Chicago, Illinois, region *Samia cecropia,* a saturniid moth, suffers the destruction of nearly 23 per cent of its cocoons by an ichneumonid parasitoid, *Spilocryptus extrematis,* which deposits an average of 33 eggs on the inside of each cocoon or on the surface of the larva. The host larva dies in a few hours after the parasitoid hatches, and the ichneumonid larva moves about freely, feeding on the cuticle or burrowing into the tissues to drink the body fluids. Another ichneumonid, *Aenoplex smithii,* was found as a secondary parasitoid, feeding on the larvae of *S.extrematis* in about 13 per cent of the cecropia cocoons infested by the latter species. A chalcidid, *Dibrachys boucheanus,* fed both upon *S. extrenatis* and, as a tertiary parasitoid, upon *A. smithii.* Another chalcidid, *Pleurotropis tarsalis,* infected cocoons containing *D. boucheanus* and eventually killed the larva as a quaternary parasitoid. To have five links in an inverted parasitoid food chain is perhaps unusual, but hyperparasitoidism is common and of importance in controlling the size and inter-relations of animal populations.

Predation is a form of disoperation, at least in point of immediate effects, since one animal kills another for food. Predation is important in community dynamics in so many ways that we will divide discussion of it to food coactions, productivity, and regulation of population size.

COMPETITION

Competition is the more or less active demand in excess of the immediate supply of material or condition exerted by two or more organisms. The materials and conditions sought by animals include food, space, cover, and mates. When these materials are in more than adequate supply for the demands of those organisms seeking them, competition does not occur; when they are inadequate to satisfy the needs of all the organisms seeking them, the weakest, least adapted, or least aggressive individuals are forced to do without, or go elsewhere. Competition may result in death for some competitors, but this is from fighting or being deprived of food or space rather than being killed for food as in predation, or by disease as in extreme parasitism.

Competition may be either direct or indirect. It is *direct* where there is active antagonism, struggle, or combat between individuals; *indirect,* when one individual or species monopolizes a resource or renders a habitat unfavourable to the establishment of other organisms having similar requirements. Direct competition, or *interference,* is evident in the fighting of bull seals for larger harems and of grouse for a better position in the social hierarchy; in chasing and colour displays (a sort of saber-rattling) by fish and birds for defense of territories; in the singing and calling of birds, some mammals, and frogs as bids for mates; and in the excretion of chemicals that affect the behaviour or health of other organisms.

Indirect competition, or *exploitation* is common among plants when certain species monopolize the water and nutrient resources of the soil or available light so that competing species cannot maintain themselves. Once an area is well saturated with established individuals to seek homes elsewhere, even in less favourable situations, than to intrude. To be successful by indirect competition, a species needs to get established in an area first, or if the invasion of various species is nearly simultaneous, then to have a more rapid rate of reproduction and growth, or a greater longevity, so as to utilize the resources of the habitat to the fullest possible extent. It is desirable, but not always possible, to distinguish between and quantify the relative roles of interference and exploitation when species compete.

Competition is usually keenest between individuals of the same species, *intraspecific competition,* because they have identical requirements for food, mates, and so on, and because they are more nearly equal in their structural, functional, and behavioural adaptations. *Interspecific competition* occurs where different species require in common at least some materials or conditions. The severity of competition depends on the extent of similarity or overlap in the requirements of different individuals and the shortage of the supply in the habitat. It is generally the case that the more unlike the kinds of competing organisms, the less intense the competition. Yet birds compete with squirrels for acorns, nuts, and seeds; insects and ungulates compete for food in grassland; the bladderwort plant competes with small fish for entomostraca and other plankton.

The immediate test of success in competition is survival; the ultimate test is leaving the largest number of established offspring. Aside from allelochemistry, which is only beginning to be understood, competition has five important effects in the animal community:

1. Establishment of social hierarchies;
2. Establishment of territories;
3. Regulation of population size;
4. Segregation of species into different niches;
5. Speciation.

Allelochemistry

Included here are the coactions whereby chemicals secreted by one organism affects the growth, health, or behaviour of other organisms. *Allelopathy* is produced in plants when toxins are liberated that inhibit seedling growth in the vicinity. This may affect succession of plant species, especially important in the early stages. Some pioneer species in the abandoned field sere produce substances inhibitory to nitrogen-fixing and nitrifying bacteria. This retards invasion of other species that require higher nitrogen concentration in the soil. Volatile inhibitors are generally more prevalent than water-soluble ones, and relatively more prevalent in arid than humid climates. Antibiotics produced by bacteria, fungi, actinomycetes, and lichens are widespread in nature and may be one of the reasons why bacteria pathogenic to man cannot multiply well in within any species is so wide that strong individuals of one species may be despotic over weak individuals of another even though the majority of individuals in the first species are submissive.

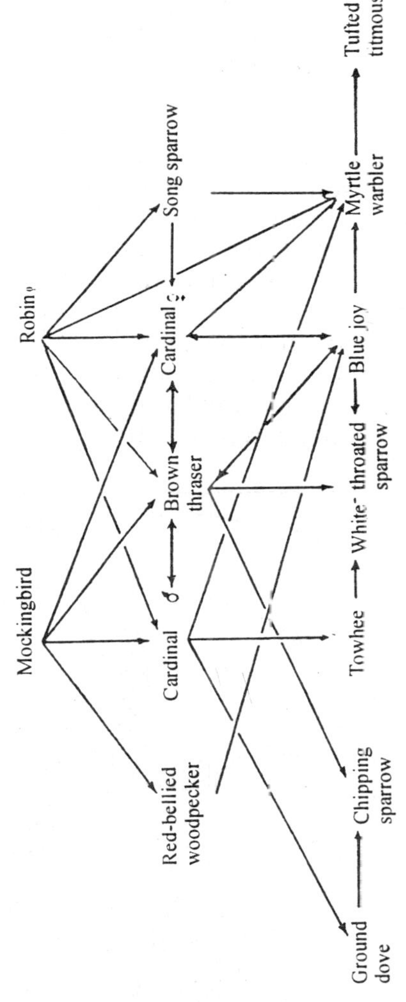

Fig.5.5 Social hierarchy among species of birds visiting a feeding station during the winter.

However, the sharptailed grouse is usually dominated over the ring-necked pheasant, and the latter is usually dominated over the prairie chicken. The manner in which different species fit into a social hierarchy may be a key to structure and organization of communities.

Territory and Home Range

The establishment of territories, especially during the breeding season, is another expression of despotism, but a special one in that it determines the spatial relations between motile animals. A *territory* is any area defended against intruders. It may be the entire home range over which the animal is active, or only a small portion around the nest (Fig. 5.6). Although many animals tend to be gregarious during the non-breeding seasons, they frequently take up isolated positions and become intolerant of the close presence of others when undertaking reproduction. A *home range* is that area

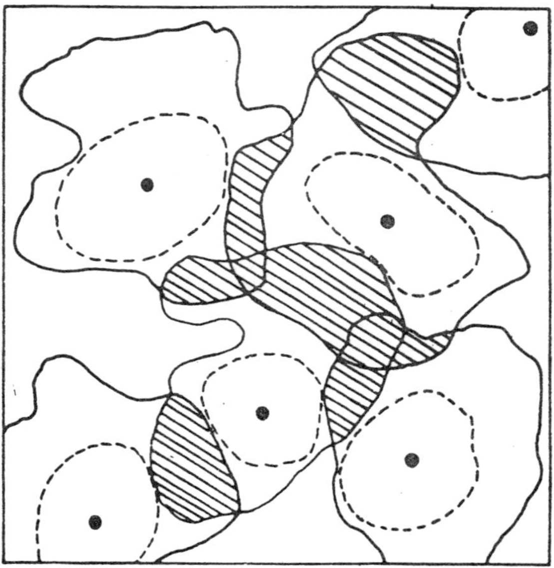

Fig. 5.6 Theoretical relation between home ranges (area enclosed within solid lines) and territories (area enclosed within broken lines). The black dots represent nesting sites.

regularly traversed by an individual in search of food and mates, and caring for young but is not defended (Fig.5.7).

The establishment of territories is best developed in birds, but also occurs in some other vertebrates, including man and certain invertebrates (marine limpets, wood ant, dragonflies, killer wasps, field crickets, and others). There is increasing evidence that most adult animals, except for small aquatic species, establish home ranges, if not territories, at least during the breeding season (snails, crayfishes, fish, toads, salamanders, uta lizards, turtles, mammals. Immature animals, species in migration, or shifting populations during the non-breeding seasons commonly do not have definite areas to which they confine their activities. An area should not be called a territory unless one can ascertain that it is defended against intruders of the same species or at least exhibits exclusive possession in the presence of others. A territory is usually defended by

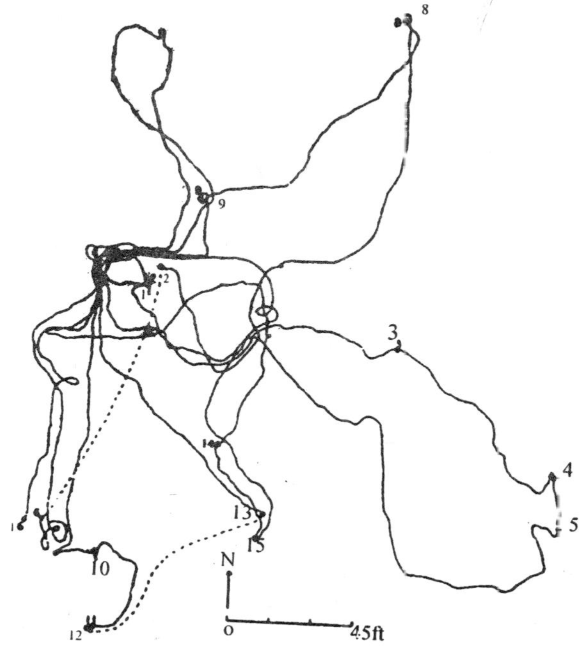

Fig. 5.7 Travels of a box turtle over its home range during a week, July 7-14.

individuals, but it may be defended by a closely integrated group of individuals, even over a period of years, as in lemurs. The size of home ranges and territories varies proportionally to weight (W) in lizards ($W^{0.95}$) birds ($W^{1.16}$), and mammals ($W^{0.69}$). Larger animals have higher metabolic requirements and therefore need larger areas over which to hunt for food.

Territoriality has become so ingrained in the behaviour of some types of animal that simple advertisement of possession constitutes adequate defense. Such advertisement takes the form of song or other vocal expression in birds, some mammals, and some frogs, or the deposition of scent or chemical cues (pheromones), as in many mammals and insects. If an intruder persists in invading a territory, however, the owner will variously display bright threatening colouration, scold or growl, give chase, or actually engage in physical combat.

Maintenance of a definite territory has several benefits: a definite breeding location in which the nest can be confidently established and protected is afforded; it aids the acquisition of mates; it ensures an area of sufficient size to provide food both for the adults and, later, for the young; and it frees the possessor of the onus of despotic interference by other individuals. The extent to which these advantages are attained varies with the species. Although competition for territory is most keen between individuals of the same species, it also occurs between different species with similar requirements for food and reproduction. A home range, on the other hand, only provides a breeding location. Possession of territory lessens the pressure of competition during the reproductive period, particularly for the female, when the entire energy and attention of animals needs to be devoted to the production of offspring.

6

Eco Systems

An eco system is composed of all the living organisms in an area plus the surrounding physical environment with which they interact. Because no sharp boundaries exist between different ecosystems, studies must necessarily be limited both in area and content. After the purpose of a study is defined, arbitrary limits must be set. While it would be impossible for you to study the entire ocean ecosystem, you could investigate specific aspects affecting the decline of one fish population. This might include food supply, predators, and environmental pollutants.

The purpose of thinking in terms of an ecosystem is to link obligatory, interdependent, and causal relationships that form the whole—much as individual ingredients go together to make a cake. Shifts in the constituents of either change the end product.

Biosphere is the term used to denote the collective ecosystems of the world. Biosphere studies are difficult to undertake because of the magnitude of interactions involved, yet examination of carbon dioxide levels, radiation changes, and energy flow must be applied to the biosphere as a whole.

Ecologists refer to all living organisms in a given ecosystem as a *community*. Life within a forest is called a forest community. Sometimes we study subdivisions of a community, such as the plant, vertebrate, mammal, or rodent communities. Life above ground and in the soil may be called terrestrial communities, while plants and animals in bodies of water may be called aquatic communities.

Within the community, organisms are grouped into populations. A *population* is a group of one species of organisms occupying a particular space at a particular time. Each population has characteristics, such as birth rate, death rate, age distribution, and genetic composition, which no individual in the population has by itself. Population characteristics are frequently expressed statistically. The ecologist uses populations and communities to tell us more about the ecosystem.

TROPHIC LEVELS

There are two major ways in which organisms derive energy; *autotrophs* pick up energy from the sun and nonliving sources, whereas *heterotrophs* eat living matter or, in the case of saprophytes and decomposers, dead material originally derived from living autotrophs. The transfer of food energy from the source, in plants, through herbivores to carnivores occurs through the *food chain*. Two examples of food chains are:

Example 1	*Example 2*
Sun	Sun
Producer	Producer
Corn	Pine tree
Primary consumer	Herbivore
Corn earworm/caterpillar	Aphid
Secondary consumer	Carnivore
Ichneumonid wasp/parasite	Spider
Tertiary consumer	Secondary carnivore
Spider	Chickadee

Every step in the food chain is termed a *trophic level*, and *herbivores* and carnivores feed at different trophic levels. *Omnivores* feed on at least two trophic levels. Each pair of trophic level is connected by a chain known as a *link*. Thus, the trophic level of a species is one more than the chain length. Charles Elton (1927) was one of the first to explain the

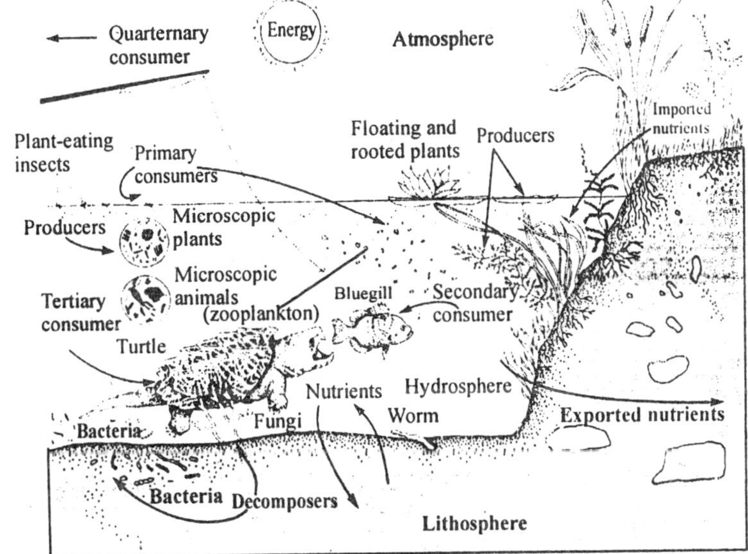

Fig. 6.1 : Producer, consumer, and decomposer in a pond ecosystem. Each of these roles is filled by a number of different organisms. For example, additional secondary or tertiary consumers might be water snakes, snapping turtles, and various birds of prey. Although ecosystems are often thought of as closed systems, none really is. Typically, both living and nonliving things are imported and exported.

importance of food chains to ecology. He pointed out that most chains are actually fairly short and contain no more than five or six links. Some of the most important links in the food chain are those involving detritivores. It is generally underappreciated that much primary production is not consumed by herbivores but dies and rots on the ground to be consumed by detritivores. Bacteria and fungi are thus common constituents of many food chains. Often the rate at which mineral resources such as phosphorus and nitrogen, locked up in dead material, are released and again made available for growing plants is controlled by the rate of action of the decomposers.

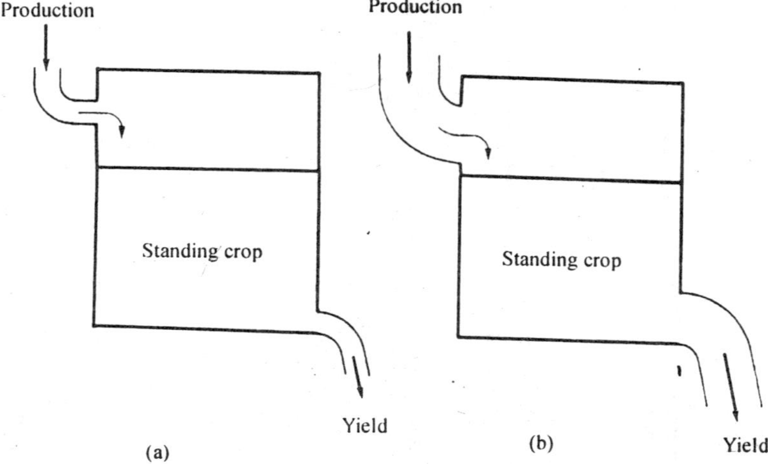

Fig. 6.2 Hypothetical illustration of two equilibrium communities (where input equals output). **(a)** Low input, low output, slow turnover. **(b)** High input, high output, rapid turnover. Standing crop is not related to production or yield because turnover time for all systems is not a constant.

Food Webs

Few ecosystems are so simple that they are characterized by a single, unbranched food chain. Several types of primary consumers usually feed on the same plant species; for example, one may find several insect types feeding on one tree and some vertebrate grazers. Also, any species of primary consumer usually eats several different plants. Such branching of food

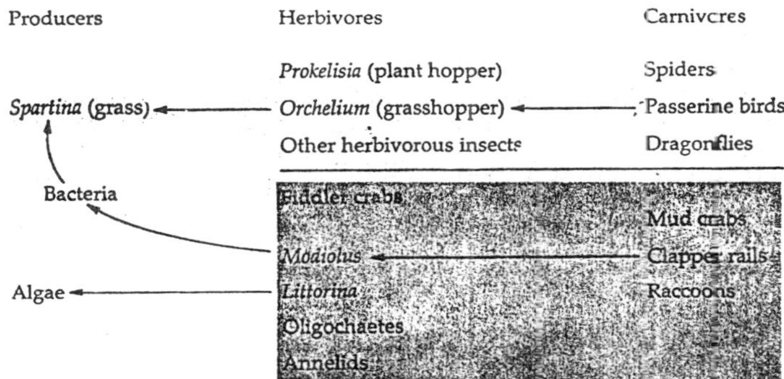

Fig. 6.3 Food web of a Georgia salt marsh with groups listed in their approximate order of importance. Often Spartina grass is the only plant in this food web. Detritivores and the species that feed on them (under the solid line) constitute a large fraction of the total species. Note that although newer studies have revealed the presence of additional insect herbivores, the general pattern remains unaltered.

chains occurs at other trophic levels as well. For instance, frogs eat several different types of insect species, which also may be eaten by different types of birds. Owls may eat primary consumers such as field mice and also prey on higher level organisms like snakes. It is more correct, then, to draw relationships between these plants and animals, not as a simple chain but as a more elaborate interwoven food web. The classification of organisms by trophic levels is one of function rather than species. For example, male horseflies are herbivores, feeding on nectar and plant juices, whereas females are blood-sucking ectoparasites.

Most food webs are imperfectly known because they are so vast. Ecologists have recognized three kinds of food web:

a Source webs—one or more kinds of organisms and the organisms that eat them, their predators, and so on.

b Sink webs—one or more kinds of organism, the organisms they eat, their other prey, and so on.

c Community webs—a group of species within a defined area or habitat (like the pitcher plant). Most food-web theory employs these webs. More than 200 such webs have been reported in the literature.

Food webs can form a useful starting point for the analysis of ecosystem organization. The relative complexity of a food web can be denoted by a measure known as *connectance*, where

$$\text{connectance} = \frac{\text{actual number of interspecific interactutions}}{\text{potential number of interspecific interactions}}$$

so that for n species

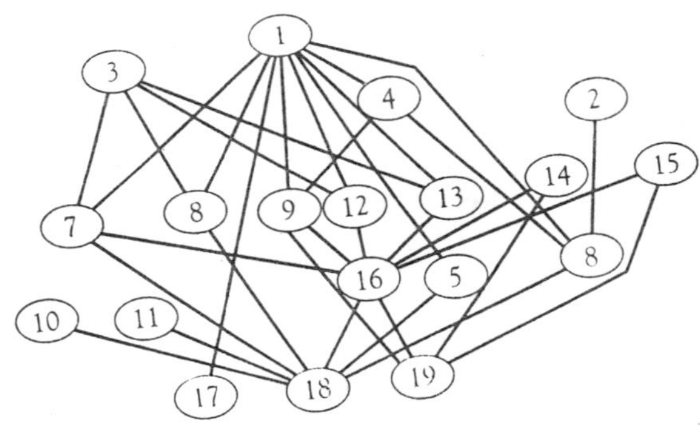

Fig. 6.4 A food web of the insects in the pitcher plant Nepenthes albomarginata in West Malaysia. Each line represents a trophic linkage; predators are higher in the figure than their prey. Key: (1) Misumenops nepenthicola; (2) Encyrtid (near Trachinaephagus); (3) Toxorhynchites klossi; (4) Lestodiplosis syringopais; (5) Megaselia sp. (nepenthina); (6) Endonepenthia schuitemakeri; (7) Triperoides tenax; (8) T. bambusa; (9) Dasyhelea nepenthicola; (10) Nepenthosyrphus sp.; (11) Pierretia urceola; (12) Culex curtipalpis; (13) C. lucaris; (14) Anotidae sp. 1; (15) Anotidae sp.2; (16) bacteria and protozoa; (17) live insects; (18) recently drowned insects; (19) older organic debris.
Number of top predators is 7: species 1, 2, 3, 10, 11, 14, 15.
Number of basal species is 3: species 17, 18, 19.
Number of intermediate species is 9: the rest.
Number of linkages: intermediate to top 14; basal to top 4; intermediate to intermediate 8; basal to intermediate 7.

the number of potential interspecific interactions

$$= \frac{n(n = 1)}{2}$$

The number of links per species is called linkage density, *d*. In Figure 6.4, there are 19 "species," 33 actual interactions, (18 x 18)/2 = 171 potential interactions, and a connectance of 0.19. Linkage density = 33/19 = 1.74.

A number of generalizations have been made about food webs and some of these are summarized by Pimm, Lawton, and Cohen (1991):

1. Cycles are rare. That is species A eats species B, species B eats species C, and species C eats species A. Cannibalism is a cycle in which one species feeds on itself.
2. The average proportion of top predators (nothing preys on them), intermediate species (with both predators above and prey below), and basal species (autotrophs) remains constant in webs regardless of the number of species.
3. The proportion of trophic links between top predators and intermediates, intermediates and intermediates, basal species and intermediates, and basal to top remains constant.
4. Linkage density of often constant, except for webs with large numbers of species. Links per species is scale invariant and is roughly equal to two.
5. Connectance remains constant as the number of species in the food web increases. This is the opposite of Cohen's (1990) scale invariant law (#4). Imagine an insectivorous bird that feeds in two communities, A and B. A has twice the number of insects than B does. Martinez (1992) argues that a bird would eat twice as many species of insect in community A as in community B. Because this would likely apply to all species in the community, so connectance changes as species richness increases.

6. Omnivory is rare.
7. Contrary to intuition, food-chain lengths do not differ greatly among ecosystems with different primary productivities. This was elegantly shown by Pimm and Kitching (1987), who fertilized certain habitats, increasing primary productivity, but noted no increase in numbers of trophic links. However, Schoener (1989) presented evidence that food-chain lengths are linked to the amount of productive space, a quantity that combined productivity with area (or volume) occupied by the food web.
8. There are only slight differences in chain lengths where consumers are vertebrates, compared with where they are all invertebrates.
9. Chain length is smaller on smaller islands.
10. Chains are shorter in areas with frequent natural or experimental disturbances.
11. Chains are shorter in two-dimensional habitats such as grasslands than three-dimensional habitats such as forests or reefs.
12. The ratio of the number of prey to the number of predators is about one to one.

Energy Flow in Ecosystems

When a primary consumer eats a producer organism and is itself then eaten by a secondary consumer, we say that energy is *flowing through* the ecosystem. As the energy available in the chemical bonds of organic molecules moves through an ecosystem, it is gradually degraded to nonuseable form. Thus, *energy flow is a one-way process,* and as we will see shortly, this means that ecosystems are unable to function unless there is constant input of energy from an external source.

Food Chains and Food Webs

One way to follow energy flow through an ecosystem is to identify the sequences of organisms through which the energy moves. Such a sequence is known as a *food chain*. A simple food chain involves a producer and a primary consumer (herbivore):

Sunlight → Plants → Medow mouse

A longer chain would involve a secondary consumer (carnivore):

Sunlight → Plants → Meadow mouse → Weasel

In the same ecosystem, a still-longer food chain involving a *teritary consumer (top carnivore)* might be as follows:

Sunlight → Plants → Grasshopper → Meadowlark → Cooper's hawk

Each step involves a transfer of energy from one feeding group, or link, to another. This type of food chain is called a *grazing food chain* because it involves the consumption of green plants by herbivores, which in turn support additional feeding links of carnivores and top carnivores.

Not all of the biomass produced in an ecosystem is channeled into grazing food chains. In fact, most plants and animals die without being eaten, thus contributing to *detrital (detritus-based) food chain*.

Furthermore the grazing food chain may be tied into the detrital food chain. This happens when a consumer in the grazing food chain feeds on some organism in the detrital food chain, such as when a meadowlark eats a spider.

Not all producer organism are eaten by primary consumers and not all primary consumers are eaten by secondary consumers. In fact, most of the organisms at each feeding level die without being eaten and their bodies become part of the system's detritus. Thus, a major portion of the energy incorporated at a given feeding level is not utilized by organisms of the next feeding level. An important characteristic of food chains is that there is less energy available at successively higher feeding levels.

Another significant portion of the energy consumed by organisms at each feeding level is not digested and absorbed and thus simply passes from the body as feces. Of the portion that is absorbed, some is lost to the system as the heat of respiration. Heat is a by-product of metabolic processes such as

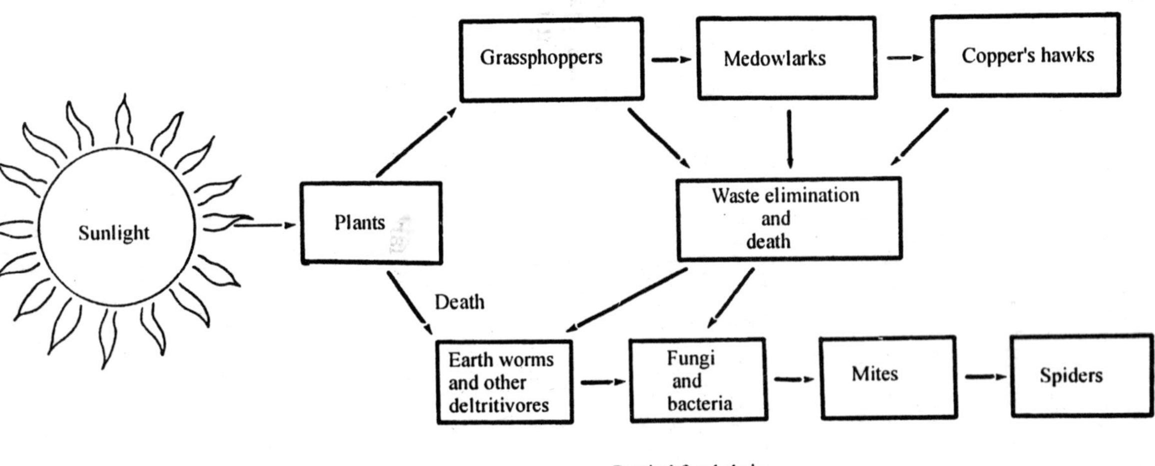

Fig. 6.5 Examples of grazing and detrital (detritus-based) food chains in an ecosystem. The situation is even more complex than shown here because dead detrital food chain organisms also contribute to detritus. Furthermore, organisms in grazing food chains can consume organisms in detrital food chains, as when a meadowlark eats a spider.

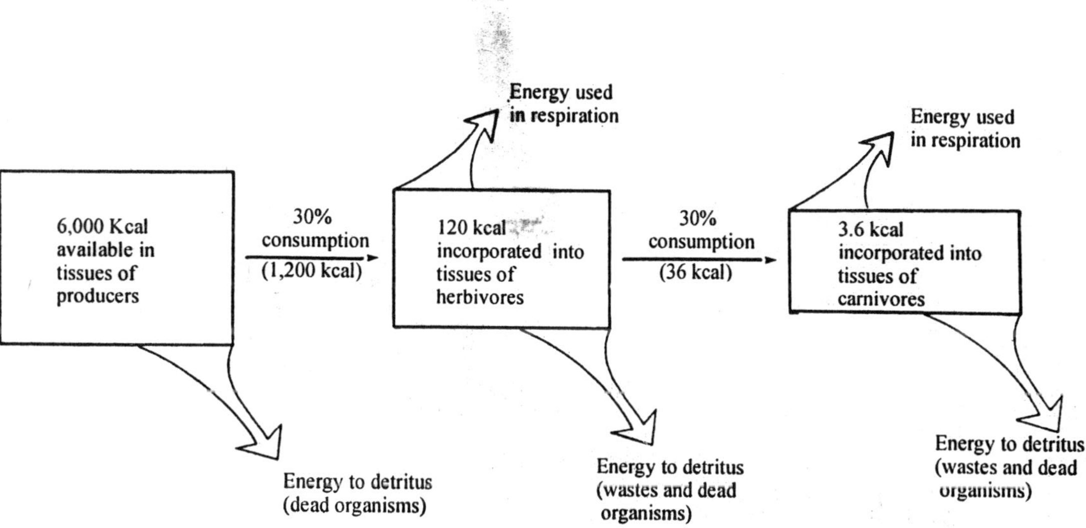

Fig. 6.6 A diagram showing an example of energy relationships in a grazing food chain.

the Krebs cycle reactions and electron transport systems which are involved in cellular energy conversions and various forms of cell work. Potential energy available in the chemical bonds of food molecules that is assimilated but not lost during respiration is incorporated into the tissues of organisms.

On the average, in grazing food chains only about 10 per cent of the energy ingested by the organisms at a particular feeding level is incorporated into the tissues of those organisms. This unavoidable inefficiency in energy transfer between feeding levels explains why food chains are relatively short—at the most, four or five links—and why mice are more common than weasels, foxes, or hawks. This also explains why more people could be fed on grain than on the meat of grain-fed animals.

A food chain represents just one path of energy flow through an ecosystem. However, ecosystems ordinarily contain numerous food chains, each linked to others to form complex *food webs*.

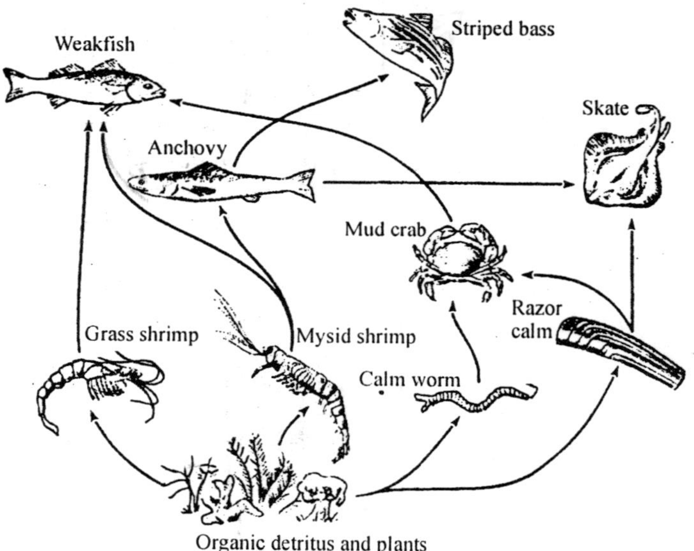

Fig. 6.7 A food web (highly simplified) of a shallow coastal saltwater ecosystem.

Trophic Levels and Ecological Pyramids

If we group all the organisms in each part of a food web according to their general source of nutrition, we can divide the food web into *trophic* (feeding) *levels*.

Biologists have agreed to assign producer organisms to the first trophic level, primary consumers (herbivores) to the second trophic level, secondary consumers (carnivores) to the third trophic level and tertiary consumers (top carnivores) to the fourth trophic level.

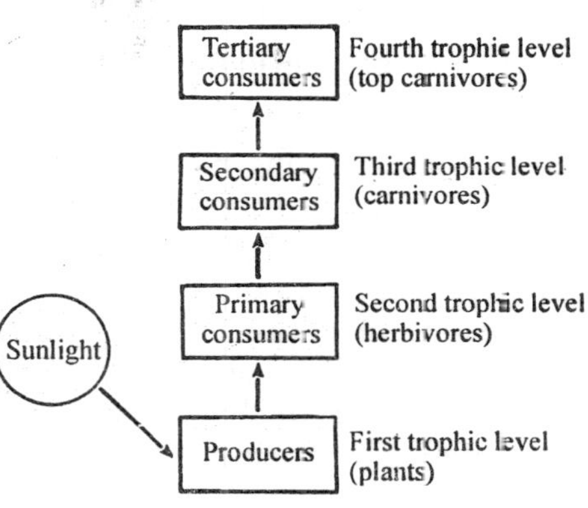

Fig. 6.8 Diagrammatic representation of trophic levels in an ecosystem.

The trophic structure of an ecosystem can be summarized in the form of *ecological pyramids*. The base of each pyramid represents the producers of the first trophic level, while the apex represents tertiary or higher-level consumers; other consumer trophic levels are in between.

There are three kinds of pyramids. One is based on the numbers of organisms at each trophic level and consequently is known as a *pyramid of numbers*. For most ecosystems,

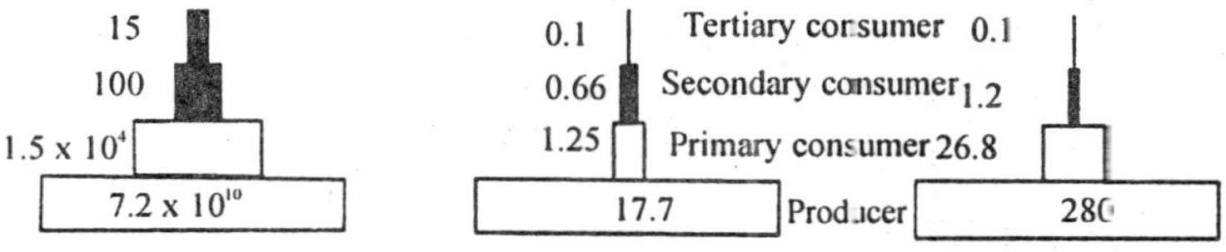

Fig. 6.9 Ecological pyramids based on data for an experimental pond (a) Pyramid of numbers (individuals/m^2), (b) Pyramid of biomass (dry g/m^2), (c) Pyramid of energy (dry mg/m^2/day).

pyramids of numbers are right side up because numbers of organisms decrease at successively higher trophic levels. However, there are some ecological systems for which the pyramid of numbers is inverted. For example, in a system involving a single tree and all of the insects that feed on it, the base of the pyramid, which represents the producer trophic level (the tree), would be smaller than the part representing the consumer trophic level (the many insects).

A second type of pyramid is a *pyramid of biomass*. Biomass is the weight of living material. By weighing plants in sample plots in a field or forest at the end of the growing season, biologists can estimate the biomass of plants in that area. By multiplying the number of animals by the average weight per individual, they can estimate the biomass of each of the consumer trophic levels. In general, the biomass of producers is much greater than the biomass of herbivores, and the biomass of herbivores is much greater than the biomass of carnivores. However, this is not always the case. In some aquatic ecosystems, the pyramid of biomass may be inverted because of the rapid turnover of the small, short-lived, single-celled algae that dominate the producer trophic level. At a particular time in such systems, consumer biomass may exceed producer biomass. However, if biomass is measured over a period of time, producer biomass will be greater than consumer biomass.

A third kind of pyramid is a *pyramid of energy*. The pyramid of energy is always upright because, as we mentioned earlier, a given trophic level has a smaller energy content than does the trophic level immediately below it. When studying trophic relations in ecosystems, a pyramid of energy is more meaningful than a pyramid of numbers or biomass, but the information necessary to construct a pyramid of energy is more difficult to obtain.

Primary Production

Through the process of photosynthesis green plants utilize the sun's energy to convert carbon dioxide and water into carbohydrates. The total amount of energy incorporated by producer organisms during photosynthesis (in many ecosystems chemesynthesis also makes atleast a small contribution to primary production) is called *gross primary production*. We use the word 'primary' because we are referring to the producer trophic level. We should note as well that ecologists also use the term *productivity* interchangeably with production.

Some of this incorporated energy must be utilized by the producers themselves in respiration to support their own metabolism. What is left after the requirements of respiration have been met is called *net primary production,* the energy that is stored by producer organisms. Arithmetically, net primary production is equal to gross primary production minus respiration. When the energy intake (gross primary production) of plants just equals the amount they use in respiration, there is no net primary production. Thus, the less energy plants require for their own metabolic needs, the more energy is left to accumulate as net production.

Since production is a rate function, it is always expressed on a per unit time basis. Gross and net primary productions are most often expressed as dry organic matter (g/m^2/yr) or as calories (kilocalories/m^2/yr).

At any particular time, the amount of organic material present, the *biomass (standing crop),* can be measured directly, but production, being a rate function, is more difficult to determine. Nevertheless, ecologists have succeeded in estimating the primary production in many ecosystems. We can compare primary production in several different ecosystems. The least productive are open oceans and deserts. The productivity of open oceans is limited by a lack of nutrients because light energy is absorbed by water before it can reach the nutrient-rich bottom waters. In contrast, deserts have low production because of a lack of water, even though they may be nutrient rich. Among the most productive natural ecosystems are estuaries, marshes, coral reefs, and moist tropical forests. Their net primary production ranges between 10,000 and 20,000 kcal/m^2/yr. Some agricultural systems also fall in this range. These ecosystems have available an abundance of nutrients and moisture, as well as favourable growing seasons.

Secondary Production

Net primary production represents the amount of energy potentially available to consumers. Although some of this organic matter is ingested by primary consumers relatively soon after it is produced, the greater portion of it is not utilized immediately. Some of it is unavailable or inaccessible—for example, most net primary production stored as wood in trees becomes available only when trees die and decompose. Eventually, though, almost all net production is processed by consumers and/or decomposers.

A large portion of the material that is available as a result of net primary production and that is eaten by consumers, as we said earlier, passes through their digestive systems without being absorbed. Grasshoppers, for instance, assimilate only about 30 per cent of the food they consume; the rest is lost as feces and becomes part of the detritus. Some of the organic matter that is assimilated by consumers is used to provide energy to meet the everyday metabolic demands that ensure normal body maintenance. The assimilated material left over after maintenance costs have been met goes into body growth and reproduction. The organic matter that is a accumulated during growth and reproduction is known as *secondary production*. Thus, we can generalize to say that secondary production refers to the rate at which consumers in an ecosystem store energy.

The amount of secondary production varies widely among consumers. Homeotherms use about 98 per cent of the energy they assimilate for their metabolic processes, and only 2 per cent goes into secondary production. Among poikilotherms, especially invertebrates, about 75 per cent of the assimilated energy is used for metabolism, leaving about 25 per cent available for secondary production.

The Pond Ecosystem

Let us consider the pond as a whole as an ecosystem, leaving the study of populations within the pond for the second section of this book. The inseparability of living organisms and the nonliving environment is at once apparent with the first sample collected. Not only is the pond a place where plants and animals live, but plants and animals make the pond what it is. Thus, a bottle full of the pond water or a scoop full of bottom mud is a mixture of living organisms, both plant and animal, and inorganic and organic compounds. Some of the larger animals and plants can be separated from the sample for study or counting, but it would be difficult to completely separate the myriad of small living things from the nonliving matrix without changing the character of the fluid. True, one could autoclave the sample of water or bottom mud so that only nonliving material remained, but this residue would then no longer be pond water or pond soil but would have entirely different appearances and characteristics.

Despite the complexities, the pond ecosystem may be reduced to the several basic units:

1. *Abiotic substances* are basic inorganic and organic compounds, such as water, carbon dioxide, oxygen, calcium, nitrogen and phosphorus salts, amino and humic acids, etc. A small portion of the vital nutrients is in solution and immediately available to organisms, but a much larger portion is held in reserve in particulate matter (especially in the bottom sediments), as well as in the organisms themselves. As Hayes (1951) has expressed it, a pond or lake "is not, as one might think, a body of water containing nutrients, but an equilibrated system of water and solids, and under ordinary conditions nearly all of the nutrients are in a solid stage." The rate of release of nutrients from the solids, the solar input, and the cycle of temperature, day length and other climatic regimes are the most important processes which regulate the rate of function of the entire ecosystem on a day-to-day basis.

2. *Producer organisms.* In a pond the producers may be of two main types: (1) rooted or large floating plants generally growing in shallow water only and (2) minute floating plants, usually algae, called *phytoplankton (phyto*

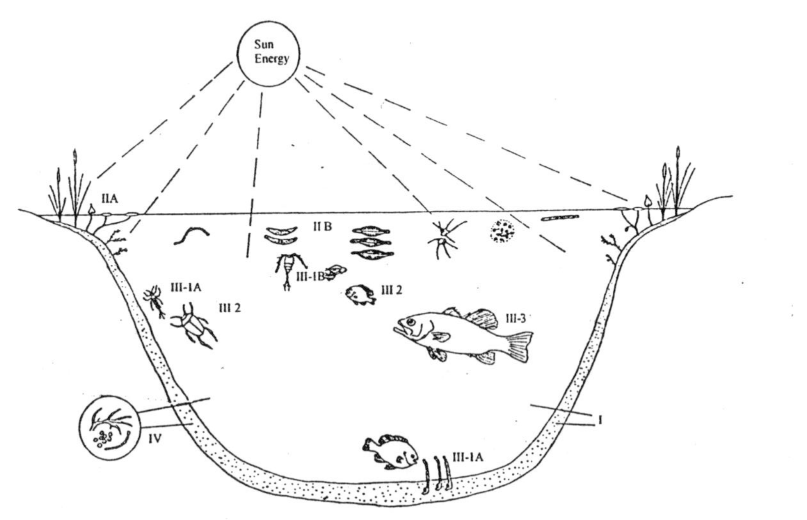

Fig. 6.10 Diagram of the pond ecosystem. Basic units are as follows: I, abiotic substances—basic inorganic and organic compounds; IIA, producers—rooted vegetation; IIB, producers—phytoplankton; III-1A, primary consumers (herbivores)—bottom forms; III-1B, primary consumers (herbivores)—zooplankton; III-2, secondary consumers (carnivores); III-3, tertiary consumers (secondary carnivores); IV, saprotrophs—bacteria and fungi of decay. The metabolism of the system runs on sun energy, while the rate of metabolism and relative stability of the pond depend on the rate of inflow of materials from rain and from the drainage basin in which the pond is located.

= plant; *plankton* = floating), distributed throughout the pond as deep as light penetrates. In abundance, the phytoplankton gives the water a greenish colour; otherwise, these producers are not visible to the casual observer, and their presence is not suspected by the layman. Yet, in large, deep ponds and lakes (as well as in the oceans) phytoplankton is much more important than is rooted vegetation in the production of basic food of the ecosystem.

3. *Macroconsumer organisms,* animals such as insect larvae, crustacea, and fish. The primary macroconsumers (herbivores) feed directly on living plants or plant remains, and are of two types, namely *zooplankton* (animal plankton) and *benthos* (=bottom forms) paralleling the two types of producers. The secondary consumers (carnivores) such as predaceous insects and game fish feed on the primary consumers or on other secondary consumers (thus making them tertiary consumers). Another important type of consumer is the *detritivore,* which subsists on the 'rain' of organic detritus from autotrophic layers above.

4. *Saprotrophic organisms.* The aquatic bacteria, flagellates, and funge are distributed throughout the pond, but they are especially abundant in the mud-water interface along the bottom where bodies of plants and animals accumulate. While a few of the bacteria and fungi are pathogenetic in that they will attack living organisms and cause disease, the great majority begin attack only after the organism dies. When temperature conditions are favourable, decomposition occurs rapidly in a body of water; dead organisms do not retain their identification for very long but are soon broken up into pieces, consumed by the combined action of detritus-feeding animals and microorganisms, and their nutrients released for reuse.

The partial stratification of the pond into an upper 'production' zone and a lower 'decomposition - nutrient

Table 6.1 Daily community metabolism in the water column of a pond as indicated by mean oxygen changes at successive depths

	O² Change (GMS/M³)			
Depth	*Light bottle*	*Dark bottle*	*Gross production (GMS O_2/M³)*	*Community respiration (GMO²/M²)*
Top m³	+3	−1	4	1
2nd m³	+2	−1	3	1
3rd m³	0	−1	1	1
Bottom m³	−3	3	0	3
Total metabolism of water column, (gms O_2/m²/day)	–	–	8	6

regeneration' zone can be illustrated by simple measurements of total diurnal metabolism of water samples. A 'light-and-dark bottle' technique may be employed for this purpose, and also to provide a starting point for charting energy flow (one of the six processes listed in the ecosystem definition). Samples of water from different depths are placed in paired bottles, one of which (the dark bottle) is covered with black tape or aluminum foil to exclude all light. Other water samples are 'fixed' with reagents so that the original oxygen concentration at each depth can be determined. Then the string of paired dark and light bottles is suspended in the pond so that the samples are at the same depth from which they were drawn. At the end of a 24-hour period the string of bottles is removed and the oxygen concentration in each sample is determined and compared with the concentration at the beginning. The decline of oxygen in the dark bottle indicates the amount of respiration by producers and consumers (i.e., the total community) in the water, whereas oxygen changes in the light bottle reflects the net result of oxygen consumed by respiration and oxygen produced by photosynthesis, if any. Adding respiration and net production together, or subtracting final oxygen concentration in the dark bottle from that in the light bottle (provided that both bottles had the same oxygen concentration to begin with) give an estimate of the total or gross photosynthesis (food production) for the 24-hour period since the oxygen released is proportional to dry matter produced.

The hypothetical data in Table 6.1 illustrates the kind of results one might expect to get with a light-and-dark bottle experiment in a shallow, fertile pond on a warm sunny day. In this hypothetical case photosynthesis exceeds respiration in the top two meters and just balances it in the third meter (zero change in light bottle); below three meters the light intensity is too low for photosynthesis, so only respiration occurs. The point in a light gradient at which plants are just able to balance food production and utilization is called the *compensation level* and marks a convenient functional boundary between the autotrophic stratum (*euphotic zone*) and the heterotrophic stratum.

A daily production of 8 grams O_2 per m and excess production over respiration would indicate a healthy condition for the ecosystem, since excess food is being produced in the water column that become available to bottom organisms as well as to all the organisms during periods when light and temperature are not so favourable. If the hypothetical pond were being polluted with organic matter, O_2 consumption (respiration) would greatly exceed O_2 production, resulting in oxygen depletion and (should the imbalance continue) eventual anaerobic (= without oxygen) conditions which would eliminate fish and most other animals. In assaying the 'health' of a body of water we need not only to measure the oxygen concentration as a condition for existence, but also to determine rates of change and the balance between production and use in the diurnal and annual cycle. Monitoring oxygen concentrations, then, is one convenient way of "feeling the pulse" of the aquatic ecosystem. Measurement of "biological oxygen demand" (B.O.D.) is also a standard method of pollution assay.

Enclosing pond water in bottles or other containers such as plastic spheres or cylinders has obvious limitations, and the bottle method used here as an illustration is not adequate for assaying the metabolism of the whole pond since oxygen exchanges of bottom sediments and the larger plants and animals are not measured.

Freshwater Ecosystems

Freshwater ecosystems are divided into two general groups—lotic and lentic systems.

Lotic Systems: Lotic systems are characterized by following water and include brooks, streams, and rivers. Lotic systems differ in such factors as size, current velocity, water depth, and the type of bottom (mud, salt, stone, or gravel). Lotic systems are inhabited by organisms that are well-adapted to maintaining their position in flowing water. For example, fish that live in mountain streams, such as trout, have a streamlined shape. Invertebrates, such as caddis flies and mayflies, live beneath stones or attach themselves to submerged objects by means of sticky surfaces, hooks, and suckers. Algae grow tightly attached to the surface of rocks and may be encased in a slippery, gelatinous sheath.

The most important primary producers of lotic systems are algae, but in most streams the major energy source is organic matter that is carried in from surrounding terrestrial ecosystems. Many of the stream inhabitants feed on *drift*, which consists of organisms and detrital material carried downstream by the current.

However, in places where surface gradients are not very steep, streams and rivers flow very slowly, and they may have characteristics similar to those of lakes. Phytoplankton (a collection of small, photosynthesizing organisms), along with rooted vascular plants, are likely to be the dominant producers. Free-swimming invertebrates are usually abundant, and fish such as bass are more common than trout or other fish that tend to inhabit swiftly flowing streams.

Lentic systems: In contrast to lotic aquatic ecosystems, *lentic* systems are characterized by relatively still water. Lakes and ponds are examples of lentic ecosystems.

Vertical stratification of temperature, light, and oxygen often occurs in lentic systems and these factors influence the distribution of organisms. For example, temperature strongly affects the structure of lakes and ponds in northern temperate regions. During the summer, intense solar radiation absorbed by surface water so that a layer of warmer, lighter (less dense) water 'floats' on top of heavier (more dense), cooler water. Since the warm upper layer does not mix with the colder lower layer, a one of steep temperature decline, called the *thermocline*, develops between the two layers. This thermocline marks the boundary between the upper warm layer (*epilimnion*) and the deeper cold layer (*hypolimnion*). The well-lighted epilimnion, where most of the phytolankton grows, is well aerated due to photosynthetic oxygen production by plants and continual mixing by the wind. The hypolimnion, however, may be deficient in oxygen due to bacterial decomposition. Furthermore, because sediments and organic matter accumulate on the bottom, the deeper waters are relatively high in nutrients, while the nutrients of the upper waters become depleted due to their uptake by phytoplankton during the growing season.

During the fall, the upper water cools to a temperature below that of the deeper waters, becomes more dense, and thus sinks to the bottom, while the water below, which is now relatively warmer and therefore lighter, rises to the top. These temperature effects, together with strong autumn winds, bring about the *fall overturn*, a mixing of the previously separate layers.

As winter approaches, the water in lakes and ponds cools. When water reaches its freezing point, ice formation begins at the top of the water because ice is less dense than cold water (water reaches its maximum density at 4°C). As ice forms it floats on the more dense cold water. This freezing pattern leaves unfrozen water beneath the ice, which has an insulating effect that retards further cooling of the system. This permits aquatic organisms to live through the winter in the water beneath the surface ice.

In the spring, after the ice has melted, strong winds and temperature changes produce a *spring overturn*, which completely mixes the waters of the lake or pond. Then, as the summer sun warms the water, it becomes vertically stratified once again.

Vertical stratification and seasonal changes in the temperature of a pond or lake influence the seasonal distribution of fish and other aquatic organisms. For example, cold-water fish live in the deeper water in the summer, but move to the upper waters in the winter.

Horizontal zonation is another characteristic of lentic systems. One important horizontal zone in lakes and ponds is the open water that is dominated by phytoplankton and zooplankton (a collection of small floating animals that feed on phytoplankton). Thus, open water is characterized by grazing food chains.

Near the shore is a *littoral zone* that is dominated by floating and emergent vegetation rooted in the bottom. Some of the animals adapted of life in the littoral zone include *Hydra*, dragonflies, frogs, blackbirds, and muskrats. If the littoral zone is well developed, it is a source of considerable detrital material, and detrital food chains are very important in the functioning of the lake. Nutrients move from littoral plants to bottom sediments and then, after decomposition, become available to the phytoplankton of the open-water areas.

Characteristic *benthic* (bottom-dwelling) organisms such as worms, snails, clams, and some insect larvae live in the bottom 'ooze' of a lake.

Although lakes are sometimes regarded as self-contained ecosystems, they are strongly influenced by the input of nutrients from surrounding watersheds through precipitation, drainage of surface water, and leaching of nutrients from the soil, groundwater, and detrital material.

We can classify lakes and ponds by their nutrient status. *Oligotrophic* (nutrient-poor) lakes are characterized by low organic matter content, low nutrient-release from bottom sediments, and low productivity of phytoplankton. Such lakes are usually situated in nutrient-poor areas. On the other hand, *eutrophic* (nutrient-rich) lakes are characterized by high organic matter content, high release of nutrients from bottom sediments, high productivity of phytoplankton, and often a well-developed littoral zone. Such lakes are usually situated either where naturally nutrient-rich water drains into them or where nutrient enter from agriculture areas or human settlements. Oligotrophic lakes can become eutrophic through large inputs of nutrients. This is usually the fate of originally oligotrophic lakes after they become polluted by sewage and other wastes.

Wetlands are another kind of lentic ecosystem and occur in areas where the water tables at or above the level of the ground most of the year. In some cases, wetlands occur where lake and pond basins have filled in with sediments. Wetlands include marshes, swamps, and bogs. *Marshes* are wet 'grasslands' dominated by emergent vegetation such as cattails, bulrushes, and sedges. *Swamps* are wetlands dominated by woody vegetation such as buttonbush, willow, swamp oak, and cypress. *Bogs* have wet mats of vegetation, often dominated by *sphagnum* moss. Acidic, decaying material beneath the surface of bog forms peat.

Marine Ecosystems

Oceans, like lakes, have horizontal and vertical zonation. The horizontal zones of oceans include a littoral zone made up of coastal waters along shorelines, a *neritic zone* consisting of the relatively shallow areas over continental shelves, and the *oceanic zone* made up of the deeper, main ocean basins.

Each of these horizontal zones is also stratified vertically. The *benthic zone,* the ocean bottom, contains an assemblage of benthic organisms adapted to bottom-dwelling life. The water above the benthic zone constitutes the *pelagic zone*. It is inhabited by planktonic organisms, which are carried about by water currents, and by *nektonic* (free-swimming) organisms, that move actively from place to place.

Oceanic Zones

The open waters of the oceanic zone are generally regarded as being relatively unproductive. This is because photosynthesis occurs only in the *photic zone,* the upper 200 meters or so where light penetrates. There is usually little circulation between the

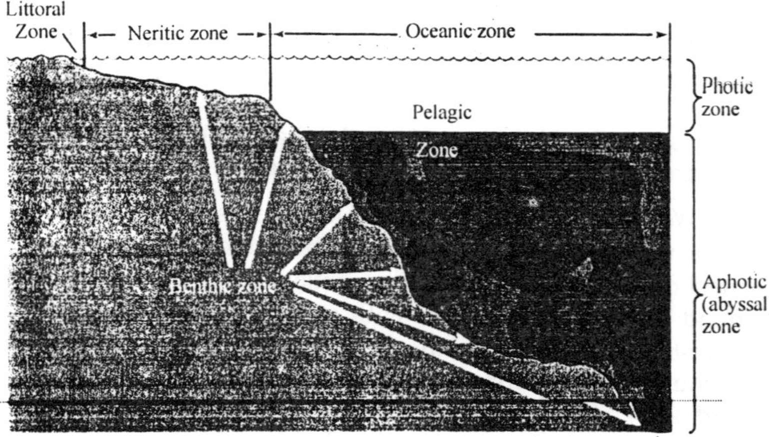

Fig. 6.11 Diagrammatic section of an ocean showing the major horizontal and vertical zones.

photic zone and the nutrient-containing bottom waters, so the growth of photosynthetic organisms in the photic zone tends to be nutrient-limited. In the deep, dark *aphotic (abyssal) zone* where nutrients are available, photosynthesis is virtually impossible.

Neritic Zones

The situation is different, however, in certain areas of relatively shallow water in the neritic zones over continental shelves. Such waters tend to be stirred by wave action and as a result are highly productive. High productivity also results where cold and warm ocean currents meet and produce upwellings that bring nutrient-rich bottom waters to the surface. This stimulates high levels of phytoplankton production, which in turn supports food chains composed of zooplankton, fish, and other consumers.

Littoral Communities

Littoral communities are very diverse. They include intertidal zones, mudflats, estuaries, and salt marshes.

Many organisms are adapted to live in *intertidal zones*, the [illegible]eas [illegible] covered and uncovered by the ebb and flow of tides. For almost twelve hours of each twenty-four-hour period, organisms in intertidal zones are exposed to the heat and drying effects of the sun; during the remaining time, they are under water. Some intertidal communities are constructed around large brown algae (kelp), which are well adapted to withstand the rigours of the intertidal environment. These large algae provide habitats for numerous other organisms. Other shoreline communities, such as *mudflats*, literally teem with life.

On the other hand, sandy beaches and rocky shores constitute rather severe environments for organisms. Many of the organisms on a sandy beach spend much of their time below the sand, becoming active and rising to the surface only during the period of high tide. On rocky shores, organisms such as rockweed, barnacles, and periwinkles are attached tightly to the rocks and thus can withstand the action of waves. When they are exposed during low tide, many of these organisms close their shells, which prevents their drying out. At high tide they reopen their shells and feed on detrital material and small organisms carried by the water.

Estuaries are the areas where rivers flow into the sea. As you might imagine, the salinity (saltiness) of estuaries is intermediate between those of fresh water and seawater. In estuaries certain interactions between outflowing river water and inflowing tides tend to create vertical water movements

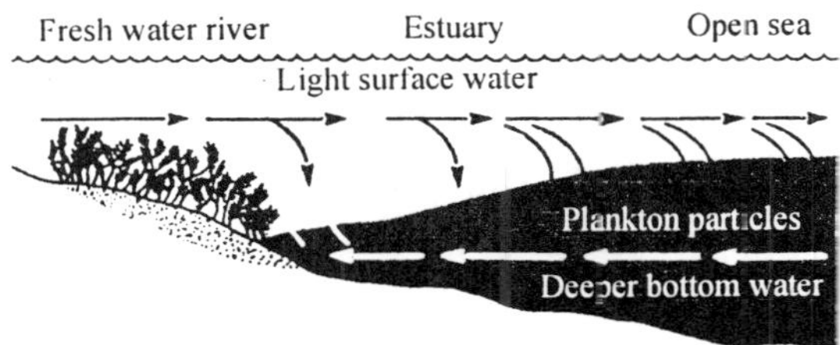

Fig. 6.12 Mechanisms by which nutrients are conserved and retained in an estuary. Lighter (less dense), fresh water from rivers flows seaward on top of denser seawater flowing in from the ocean. Nutrients that have settled to the bottom are carried toward land. Many of these nutrients are taken up by plants in the estuary.

that keep nutrients in circulation. As a result, estuaries are highly productive, and serve as nursery grounds for a variety of organisms, including many kinds of invertebrates, fish, and fish-eating birds.

Salt marshes are often associated with estuaries and are dominated by salt marsh cordgrass, a plant that is highly adapted to a saline environment. In addition to being an important wildlife habitat, salt marshes contribute nutrients and detrital material to estuaries.

Coral Reefs

Coral reefs are large formations that arise in warm ocean areas as a result of the continual deposition of calcium carbonate by coral animals and coralline algae over thousands of years. Living corals and the coralline algae that live among them occupy only the upper layer of the reef, but they grow on the accumulated remains of their predecessors.

There are a number of kinds of coral reefs, including some that form rings around tropical islands. Others form long reefs along shorelines. For example, the Great Barrier Reef along the northern portion of Australia's eastern coast stretches for 2000 kilometers, making it large enough to be seen from the moon.

Coral reefs form only in warm, shallow waters (usually no deeper than 50-60 meters) that are very clear. In fact, reef-building corals cannot grow in turbid water (water that contains large amounts of suspended material). The reason for this is at least in part due to the symbiotic algae (called *zooxanthellae*) that live within the cells that line the corals digestive tracts. Although corals are carnivores that feed on organisms they obtain from the surrounding water, they depend upon the photosynthetic activity of their algal symbionts for a large part of their energy requirements. Because zooxanthellae require light for photosynthesis, the vertical distribution of reef-building corals is restricted to relatively shallow, sunlit waters.

The photosynthetic activity of reef algae, especially the symbiotic and coralline algae, is great enough to make the coral reef one of the most productive types of ecosystems. The reef community that lives in and around the coral includes a diverse array of organisms, all of which find abundant food and living places within the reef ecosystem.

Terrestrial Ecosystems

We can categorize *terrestrial* (land-based) ecosystems on the basis of their most common types of mature plant communities and their climates. The major types of terrestrial ecosystems are called *biomes*.

Tundra

There are two types of *tundra*—Arctic Tundra, which occurs in a band across the northern latitudes of the world, and Alpine Tundra, which occurs at high altitudes on mountains at various latitudes around the world.

Arctic Tundra

Lying largely north of latitude 60°N and encircling the top of the earth is a vast, treeless region called the *arctic tundra*. The arctic tundra biome is dominated by mosses, lichens, sedges (grasslike plants with solid stems), grasses, and lowgrowing shrubs. The growing season (from the last killing frost of spring to the first killing frost of all) is short—approximately sixty days or less. Winters are extremely cold, and the precipitation, falling mostly as rain in summer and fall, is usually less than 25 cm a year. However, because of poor drainage, low temperatures that reduce evaporation and the presence of a permanently frozen soil (permafrost) a short distance below the surface, the land is wet and covered with numerous ponds, lakes, and bogs.

Net annual primary production in the arctic tundra is low because the growing season is short and nutrient availability is poor. Nutrient levels are also low because decomposition is extremely slow in soil that is very cold most of the year. In much of the arctic tundra the most important herbivores are lemmings, small rodents related to meadow mice. In other parts of the tundra, must ox and caribou are the major herbivores. The tundra's herbivores support a number of carnivores, including snowy owls, foxes and weasels.

Alpine Tundra

Above the timberline in the higher mountains of the world is the *alpine tundra*. Unlike arctic tundra, alpine tundra lacks a permafrost or has only some parts that are permanently frozen. Alpine tundra is characterized by moderate precipitation, widely fluctuating temperatures, and high ultraviolet radiation. The vegetation of alpine tundra, in contrast to that of arctic tundra, includes more grasses and sedges and fewer shrubs and lichens; animal communities are also different.

Coniferous Forest (Taiga)

South of the arctic tundra lies a worldwide belt of *coniferous forest (taiga)*. Here, the vegetation is dominated by spruce, fir, and pine trees. The climate is characterized by cool summers, cold winters, and a growing season of about 130 days. Precipitation ranges between 40 and 100 cm per year, with much of it coming as heavy snow. Soils in the coniferous forest biome are acid and infertile and have a thick litter of slowly decomposition needles. Much of the precipitation moves through the soil, carrying with it important nutrients. Although the growing season may be short and nutrients in poor supply, annual net production is rather high, between 2,000 and 3,000 kcal/m^2. This is true in part, because the evergreen nature of the needles permit photosynthesis on any favourable day during the year.

Many of the coniferous tree species that comprise the taiga are accumulator species; that is, they remove large quantities of elements from the soil, especially calcium, nitrogen, phosphorus, and potassium. Nutrient accumulation occurs, in part, because the needles are held on the tree for a long period of time and not dropped seasonally like the leaves of deciduous trees. Large portions of these accumulated nutrients are not recycled until trees die or are consumed by forest fires.

Mycorrhizae, which are symbiotic associations between certain types of fungi and the roots of coniferous trees, are very important in nutrient recycling in coniferous forests. Mycorrhizae extend into the slowly decaying litter of fallen needles, branches, and cones where decomposition is accomplished primarily by fungi. Mycorrhizae speed the transfer of nutrients from the litter to the plants.

Compared to temperate and tropical biomes, the coniferous forest has relatively few consumer species. However, the number of individuals within each consumer species tends to be large. Insects such as the spruce budworm and larch sawfly are herbivores that can defoliate trees over large areas when these insect populations occasionally grow very large. Moderately sized herbivores include the snowshoe hare, which is preyed upon by the lynx. Larger herbivores include the moose and woodland caribou, which are the prey wolves.

Temperate Deciduous Forest

South of the coniferous forest lies the *temperate deciduous forest*. In this biome, the dominant trees are *deciduous*, that is, they lose their leaves in the fall. Deciduous forest is found in areas of moderate climate with well-defined winter and summer seasons and relatively high precipitation (75 to 150 cm/yr). The growing season ranges between 140 and 300 days. The soil is relatively fertile and midly acid, and litter decomposes quite rapidly. Net primary production ranges from 4,000 to 8,000 kcal/m^2 a year, of which about two-thirds is stored in wood.

Production stored as wood forms a nutrient pool unavailable for short-term cycling, but some short-term nutrient cycling does take place through the decomposition of roots and litter. Nutrient cycling in deciduous forests is strongly influenced by the geology of the site on which the forest grows. For example, forests growing in soils over nutrient-poor bedrock store and cycle smaller quantities of nutrients than do forests growing in soils over rocks rich in calcium and magnesium.

Vertical stratification is well developed throughout the deciduous forest biome. The forest supports a great diversity of consumer organisms, the most numerous of which are insects. Insect populations seldom grow excessively large, but there is a notable exception. The gypsy moth, which was accidentally introduced into the northeastern United States, feeds on a wide variety of hardwood trees, especially oaks.

A major large herbivore of deciduous forests is the white-tailed deer. In areas of high population, white-tailed deer can influence the structure and development of a forest by overutilizing certain species of forest tree seedlings and other sprouts. Because of human activity, deer predators such as wolves and mountain lions have virtually disappeared from some areas.

Chaparral

In regions with hot, dry summers and mild, damp winters, the dominant vegetation is known as *chaparral*. Chaparral is characterized by low, shrubby vegetation with tough, waxy-coated leaves that are resistant to drought. Net production averages about 3,000 kcal/m^2 a year, with much of the production tied up in woody growth and litter. Periodic fires that sweep through the chaparral are important for nutrient cycling and stimulating new vegetative growth. Chaparral occurs in the southwestern United States, in Maxico, around the Mediterranean Sea, and in parts of southern Australia, South Africa, and South America.

Tropical Rain Forest

Tropical rain forest is located in the equatorial regions of Central America, central and northern South America, western Africa, and Indonesia. Tropical rain forests are found in areas where average temperatures are high (between 20° and 25°C) and precipitation is heavy (in excess of 200 cm per year). Temperatures fluctuate little, and the rainfall is evently distributed through the year. The forest is dominated by broadleaf, nondeciduous trees, which are part of a very rich diversity of plant species. The canopy of the forest is highly stratified and it provides habitats for tremendous variety of animal species. Because of the dense shade created by the towering trees, however, there is little undergrowth.

In such a warm, moist environment, organic matter decays rapidly, and only a small amount of litter accumulates on the forest floor. Nutrients are cycled directly from the litter to plant roots by way of mycorrhizae. Thus, most nutrients are stored in trees and nutrient reserves in the soil are small. Yet although the soil itself is relatively infertile, net production in a rain forest is very high because of high temperature, a twelve-month growing season, and the rapid recycling of nutrients from the litter.

Even though such soils are poorly suited for agriculture, tropical rain forests are being cleared and converted to agricultural use at an alarming rate. However, soil nutrients are so scarce in these areas that the soil seldom supports productive agriculture for long, and the cleared areas are usually abandoned after only a few years. If the cleared areas are small (less than 2-3 hectares), the rain forest will reoccupy them in time. Larger clearing operations, however, whether done for lumbering or agricultural purposes, are threatening to permanently destroy large areas of tropical rain forest in many parts of the world.

Bordering the tropical rain forest is the *tropical seasonal forest.* Not as lush as rain forests, tropical seasonal forests grow in regions where rainfall is seasonal with pronounced wet and dry seasons. Many trees in these forests are deciduous and lose their leaves during the dry season.

Tropical Savanna

Also associated with tropical regions is the *tropical savanna,* which is dominated by tall grasses and also supports scattered trees. In tropical Africa these trees are often flat-topped, thorny *Acacia* trees, while is South America they are mostly palms. Tropical savanna occurs in regions where temperatures are high and where rainy and dry seasons are pronounced. During the rainy season, precipitation ranges between 90 and 150 cm. The heavy rains of the wet season cause extensive leaching and result in nutrient-poor soils. Nevertheless, annual net production ranges between 800 and 8,000 kcal/m^2.

African savanna supports a rich diversity of grazing animals whose cycles of reproduction and migration are correlated with the rainy seasons. Tropical savanna has been greatly disturbed by human activity, particularly by agricultural development and overgrazing by domestic animals. For example, the large herds of native grazing animals that once covered the African savanna, such as wildbeast, zebra, and many species of antelope, are rapidly disappearing, as are many of the predators, including lions, leopards, cheetahs, and hunting dogs that prey on them.

Grassland

Grassland is found on the temperate plains areas of North America, Eurasia, and South America. Rainfall is variable; in many of these areas the rainfall could support tree growth, but the encroachment of trees is halted by fires and periodic drought. In North American prairies, along an east to west gradient of decreasing moisture, there is a transition from tall-grass to mid-grass, and finally to short-grass grassland.

Compared to forest ecosystems, grasslands exhibit little biomass accumulation, and much of the annual production dies each year. Thus, the turnover of biomass and nutrients is rapid. Furthermore, because of relatively low rainfall and high evaporation rates, leaching is not excessive, and nutrients tend to accumulate in the lower part of the soil. At the same time, organic matter accumulates in the upper part of the soil. As a result, grasslands have very fertile soil and they have been exploited world-wide for agriculture. Most of the native temperate grasslands of the world have been converted to agricultural usage.

Grasslands, like tropical savannas, can support large populations of herbivores. Insect herbivores, notably grasshoppers and ants, are very common. In North America, millions of bison once roamed the native grasslands. There were also large numbers of pronghorn antelope and deer; their numbers were kept under control by predators such as coyotes and wolves.

Today, however, most native grassland herbivores have been largely replaced by cattle, sheep, and goats, and in many places overgrazing has resulted in severe deterioration of the grassland ecosystem.

Desert

Desert forms a worldwide belt at about 30° N and 30°S latitude. Desert is associated with the rainshadows of mountains, coastal areas next to cold ocean currents, and the deep interiors of continents. It is characterized by rainfall of less than 25 cm per year, low humidity, a high evaporation rate, and high daytime temperatures and low nighttime temperatures. Desert plant life consists of widely scattered shrubs that have short, waxy, drought-resistant leaves. The soil is high in inorganic salts and low in organic matter. Because of lack of water and extreme temperatures, annual net production is very low (below 200 kcal/m^2).

Desert may be hot or cold. In North America, hot deserts, dominated by cactus and desert shrubs, are found in the southwestern United States and Mexico. Northern cold deserts are dominated by sagebrush.

Even deserts are subject to extensive human disturbance. Irrigation has made it possible to utilize deserts for agriculture. However, water for irrigation is often pumped to the surface from deep wells, and it is usually removed faster than it can be replaced by natural processes. Also, as a result of the evaporation of irrigation water, salts accumulate in surface soils. Finally, there are grave questions concerning the future availability of water for direct human use in areas such as the southwestern United States, where the human population is growing very rapidly.

Environment Concerns

The world has serious environmental problems, and many natural ecosystems are being threatened. Generally, most ecosystems can recover from minor disturbances, but there is a point beyond which an ecosystem loses its ability to return to normal. This usually occurs when it is subjected to a very large disturbance or to continuing disruptive influences over a long period of time, or when key species become extinct.

Since an ecosystem consists of a complex array of interconnected parts, a disruption of even one of its aspects eventually affects the entire system. Despite our sometimes arrogant attitudes, we humans are an integral part of ecological systems, and our activities very often have negative effects on the environment. A large percentage of the environmental problems to which we contribute arise simply because there are so many of us, and it looks as if the total human population of the world will continue to grow until at least well into the next century.

Pollution

To begin with, world population growth contributes to environmental pollution problems. *Pollution* is an undesirable change in the characteristics of an ecosystem, and pollution can be very harmful to all living things in the ecosystem, including humans.

7

Biological Communities

Trophic Complexity and Stability

Food chain length may influence the resilience of the community. Models of communities with different levels of trophic connectance show that complexity reduces resilience and stability. However, such studies should be interpreted with caution, as real communities may possess important attributes not found in the communities of null models. Stability also depends on environmental conditions—a fragile (complex or diverse) community may persist in a stable and predictable environment, while in a variable and unpredictable environment only simple and robust communities will survive.

The Community

The community is an assemblage of species populations that occur together in the same place at the same time. It is the biological part of an *ecosystem,* as distinct from its physical environment. The community has interesting and complex properties arising from the interaction between species. Competition, predation, parasitism, and mutualism occurring among individuals appear to lie behind many patterns in *community organization*. At the other extreme, the community can be viewed from the wider perspective in terms of species diversity and distribution, food-webs, energy flow and the interactions among **guilds** of similar species.

Early this century the community was perceived as a *'superorganism'* whose member species were somehow bound together such that the community was born, lived, died and evolved as a whole. This view has been largely rejected in favour of the *'individualistic'* concept which focuses on the community as a collection of individual species where community patterns can be explained by processes at the level of the individual.

Community Structure

A community can be characterized by its diversity which is a function of the number of different species it contains and their abundance. Diversity depends on *species richness* (the number of species) and on the *evenness* or equability of species abundance. *Diversity indices* such as the *Shannon Index* take into account both of these components. Two hypothetical communities comprising the same species can differ greatly in structure and in their diversity, depending on their relative abundance distributions (Fig.7.1). Communities in which the species are all more or less equal in abundance exhibit evenness, whereas communities with one or a few abundant species and many rare ones show *dominance*.

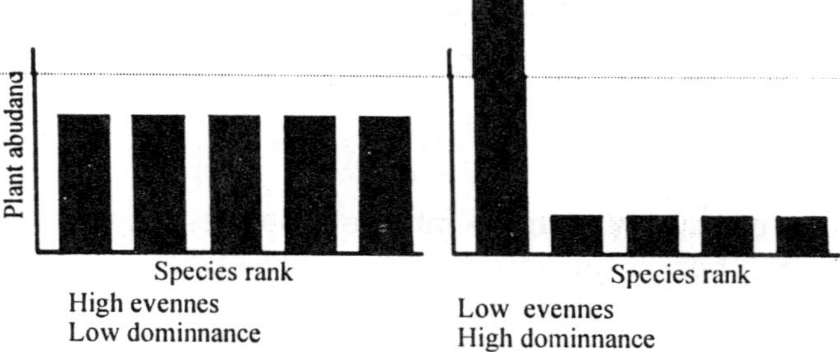

Fig. 7.1 Two hypothetical dominance-diversity distributions with the same species richness (5), but different levels of species evenness. The right-hand community is dominated by one species and has a lower overall diversity than the species-even community (left).

One way of demonstrating species richness and evenness is to plot the relative abundance of species against their *rank order;*

that is, to plot the largest value first followed by the next largest and so on. This yields either a straight line relationship or a characteristically-shaped curve. Figure 7.2 shows five of these relationships plotted on the same axes for the relative abundance of plant species in old fields in five stages of abandonment in southern Illinois, USA. For each line, the most abundant species appears on the left, the least abundant on the right. It is apparent that the species richness and the species evenness increase over time as more species colonize the site and a more complex community is established.

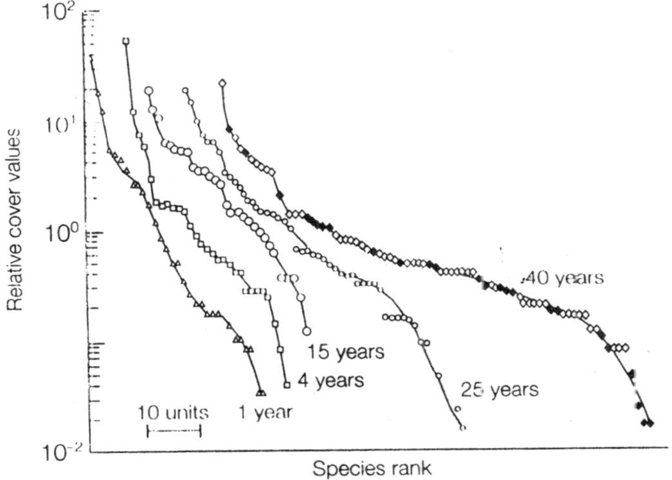

Fig. 7.2 Relative abundance data (in terms of cover) for plants species in old fields in 5 stages of abandonment in southern Illinois, USA.

It is useful to distinguish between diversity measurements over a range of *spatial scales*—a common convention is to divide these into three categories:

(a) *alpha (α)-diversity* is the diversity of species within a habitat or community, as described above;

(b) *beta (β)-diversity* measures the rate of change in species composition along a gradient from one community to another on a regional scale;

(c) *gamma (γ)-diversity* reflects the broadest geographical scale, representing species diversity across a range of communities in a region or a number of regions.

Community Boundaries

The vegetation of one area differs in characteristic ways from that of another area allowing communities to be distinguished subjectively, for example on the basis of dominance by particular tree species. Early plant ecologists sought to classify plant communities and draw boundaries around vegetation types. However, these boundaries are artificial as communities do to end abruptly but grade into each another. This can be explained by focusing on individual species distributions through *gradient analysis*. Each species has a unique distribution, determined by an environmental gradient such as latitude, which will overlap with those of other species (Fig.7.3). Since the limits of individual species are not sharp, but tail off gradually, the same must be true of community boundaries.

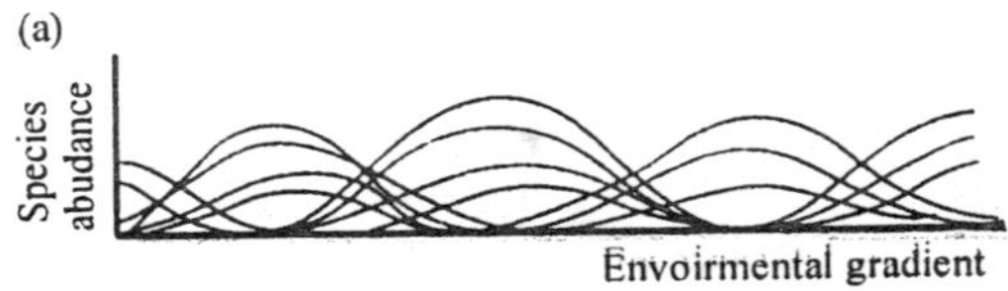

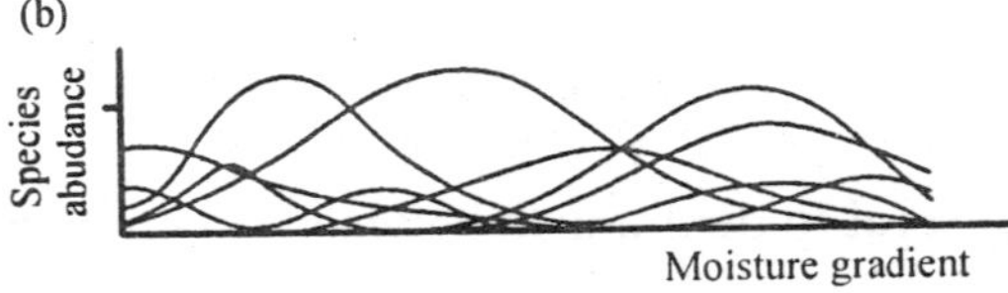

Fig. 7.3 Species distribution along an environmental gradient. **(a)** Hypothetical distributions which would result in distinct adjacent communities; **(b)** typical real distributions of plant species along a moisture gradient indicating the fuzziness of real community boundaries.

An objective method for classifying communities and superimposing boundaries on this continuum is to employ statistical methods which look for similarity between species distributions or the species composition of sites. *Classification* techniques separate different species or sites to yield distinct classes which can be taken as representative of different communities (Fig.7.4). In Britain, the National Vegetation Classification (NVC) has characterized and described vegetation types using a classification technique called TWINSPAN (two-way indicator species analysis). Species data from any community can be compared to the NVC database of community types and the community named and classified accordingly.

Ordination does not attempt to draw boundaries but group species or sites according to how different their distributions are. For example, species ordination used a number of theoretical axes along which species are assigned positions; plots of these axes reveal clumps of species that have similar requirements, occur together and belong to the same community as shown in Fig. 7.4. Questions can then be asked as to the underlying environmental factors determining the patterns revealed by ordination. Ordination and classification are often employed together as complementary techniques.

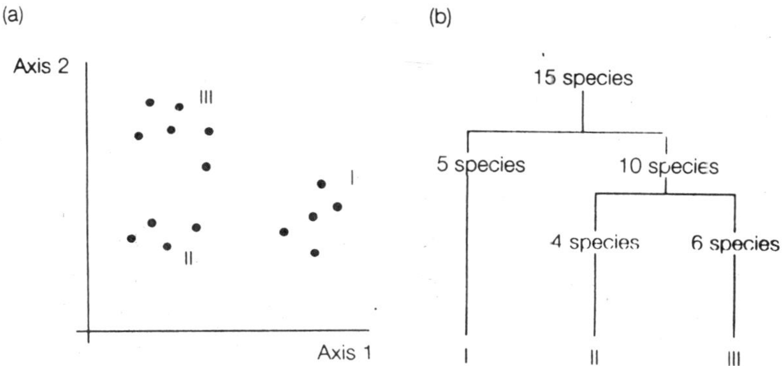

Fig. 7.4 Ordination and classification of species abundance data. **(a)** Ordination places species with similar distributions together to give three loose groups; **(b)** classification separates out the most different species (group 1) first and then divides those remaining along the same lines as the ordination.

Guilds

Species can be categorized into *guilds*, which are groupings of species that occupy similar niches; for example, insects feeding on broad-leaved trees form one guild. Ecological guilds derive their subsistence from common pools of resources. Where interspecific competition has been shown to occur it is usually among members of the same guild. Some studies have found constancy in the proportion of total species in certain guilds within a community, especially the ratio of predator species to prey species, which indicates that there may be certain common 'rules' governing community structure.

Community Complexity, Diversity and Stability

Complexity is a function of the number of interconnections among community elements. Complexity increases as the number of interacting species in the community increases. These interactions can be *horizontal* such as competitive interactions which tend to occur among species at the same trophic level, or *vertical*. *Trophic interactions*, such as plant-herbivore, prey-predator, host-parasite, are vertical interactions as they involve species at different trophic levels within the community.

Stability involves two components: *resilience*, the ability of a community to return to its original state quickly following displacement, and *resistance*, the ability to avoid displacement. One question that has exercised ecologists is. Does complexity enhance stability? Intuitively, complex communities are considered stable, because the impact of sudden population change in one species will be cushioned by the large number of interacting species and will not produce drastic effects in the community as a whole. For example, it has been suggested that such buffer mechanisms operate in tropical rain forest communities where insect outbreaks are unknown. This contrasts with simple cultivated communities where pest outbreaks are common.

The analysis of complexity has usually emphasized the trophic structure of communities, although in some studies competitive interactions have been considered also. Some studies involving theoretical communities have concluded that increased complexity leads to *instability*. Although these models have been criticized for being unrealistic, the consensus is that complex communities are not necessarily stable.

Experiments on natural communities yield a variety of results. Species—poor (simple) and species-rich (complex) Serengeti plant communities were perturbed by increasing buffalo grazing. This caused a *reduction* in diversity in the species-rich communities only. In this case, the more complex communities were less resistant. In contrast, species diverse grassland communities in Yellow-stone National Park were more resistant to summer drought than species-poor communities.

The Serengeti study demonstrates a decline in stability with increasing complexity. However, it also showed that *biomass stability* was greatest in the more complex communities. Another study in Minnesota, USA, found that species-poor grasslands took longer to recover their biomass following drought than diverse grasslands. Thus, in terms of energy, complex communities may be the more stable. As biomass stability suggests, there may be different kinds of stability for different properties of the community.

The flux of energy through a community also appears to have an important influence on its resilience. Models of contrasting communities show that the higher the energy input and turnover of the system, the less time the community takes to return to equilibrium after disturbance. Tundra communities with very low turnover appear to be the least resilient.

There is little theoretical support for the notion that a complex species network brings about stability. However, there is mounting evidence that competition and diversity are related. Intense competition leads to low diversity, while weak competition permits coexistence and hence higher diversity. Theories of stability have tended to treat horizontal (competitive) and vertical (trophic) interactions alike. Thus, weak competition reduces instability in theoretical communities.

Trophic Complexity and Stability

Trophic complexity increases with the number of trophic levels and vertical connections in the food web. A community can be considered to be only as resilient as its least resilient species. Consider a lake ecosystem where the density of phytoplankton has been reduced by manipulation or disturbance. This will reduce the density of species at higher trophic levels. The phytoplankton may be able to recover quickly, but the fish will recover more slowly. Zooplankton will be abundant and will keep the density of phytoplankton down. The resilience of this system is limited by that of the predatory fish. This indicates that food chain length may influence that overall resilience of the community.

Food-web connectance is a function of the number of vertical links in the food-web and is considered to influence community stability. In general, theoretical work has demonstrated that if food webs are randomly constructed, increasing the number of species in the food web, increasing the connectance and increasing the intensity of trophic interactions all result in a *decrease* in stability. This is in direct opposition to the holistic view of communities where a dynamic balance is thought to be maintained by the network of species present in the community. These findings should be interpreted with caution however, as some models incorporating more realistic community attributes have shown the reverse. Real communities are different from the randomly constructed communities of null models; it is conceivable that they are complex in a particular way which enhances stability.

In natural communities, stability appears to depend on the balance between the stability of the community and the variability of the environment. In a stable environment, a community that is complex and dynamically fragile may persist (e.g. tropical rain forests). In a variable environment, only dynamically robust, and simple communities will persist.

RECOGNITION OF COMMUNITIES

Community as an Organic Entity

Although the major community or ecosystem is the generally accepted unit of analysis in synecological studies, there is a difference of opinion as to whether the community constitutes a discrete organic entity. Two different points of view are incorporated in the organismic and individualistic concepts which are usually associated with the names F.E. Clements (1916) and H.A. Gleason (1926), respectively, and more recently Phillips (1934-1935), Tischler (1951), and Emerson (1952) on one side and Bondenheimer (1938), Whittaker (1951, 1952, 1956, 1957, 1967), Churtis (Brown and Curtis 1952), and McIntosh (1963) on the other. Ramensky (1926) stated the individualistic concept independently in Russia as early as 1924.

The *organismic concept* considers the community to be a supraorganism, a complex organism, or a social organism. As such, it is the highest stage in the organization of living matter: namely, cell, tissue, organ, organ system, organism, species population, community. There is emergent evolution, so to speak, at each higher stage in this hierarchy; the whole is more than merely the sum of its parts. Tissues have properties, characteristics, and functions over and above those of the individual cells involved; the organ, the organ system, or whole organism functions in a way not to be predicted from a knowledge of the parts of which each consists. The species population has inherent characteristics of density, rate of natality, rate of morality, and age distribution, while the total community has such unique functions as dominance, cooperation, trophic balance, competition, and succession which are beyond the characteristics of the individual organisms of which the community is composed. The community behaves as a unit in its competitional and successional relations with other communities, in its local and geographic distribution, in its seasonal activities and response to climate, and in its evolution. Although the community varies in its taxonomic composition and structure in different environmental situations, this variation is proportionally no greater than occurs in different cells of the same type or between individuals belonging to the same species.

The *individualistic concept* places emphasis on the species, rather than the community, as the essential unit for analysis of interrelations, activities distribution, and evolution. Each species responds independently to the integrated influence of the various facts of the physical environment and biotic coactions. The environment may be conceived as a pattern of *gradients* with the intensity of the various factors changing gradually in space from one extreme to the other (Fig. 7.5). The gradient may be a short one, as from the subterranean to the tree stratum in

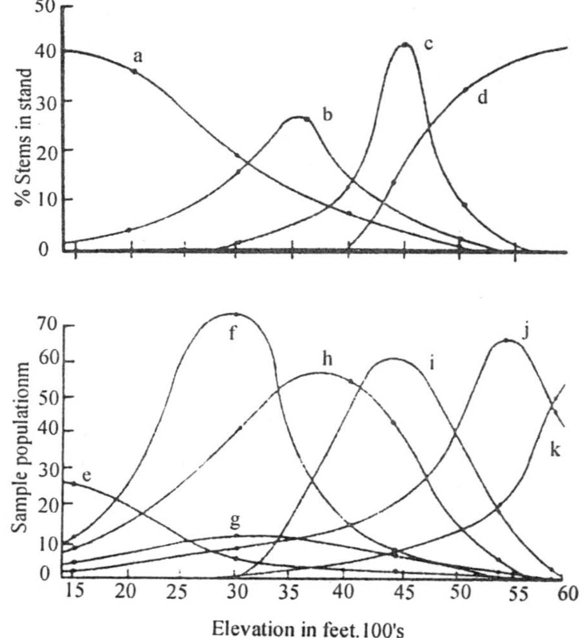

Fig. 7.5 Continuum of tree (a-d) and foliage insect (e-k) species in an elevation gradient in the Smoky Mountains, Tennessee: (a) Tsuga canadensis; (b) Helesia carolina; (c) Acer spicatum; (d) Fagus grandifolia; (e) Graphocephela coccinea; (f) Caecillius sp.; (g) Agalliopsis novella; (h) Polypsocus corruptus; (i) Anaspis rufa; (j) Cicadella flavoscuta; (k) Oncopsis sp.

a forest or from the open water of a pond to a nearby swamp or climax forest, or it may be longer, as from the bottom to the top of a mountain or even from the tropical to the arctic zones of a continent. The population density of each species is distributed in a form resembling a normal curve when plotted along the gradient of a given factor, and the curves of many species in relation to various environmental gradients overlap in a heterogeneous, and apparently random, manner. There is seldom agreement between the limits of the distribution curves of any two species; species are not in general bound together into groups of associates which must occur together. Furthermore, the vegetation and its associated animal life very often form a *continuum* of gradually changing composition and complexity from one extreme of the environmental gradient to the other. A vegetational continuum has no sharp boundaries between individual communities of different types, and the ecologists must choose the manner in which he distinguishes these units so as best to suit his interests and objectives. The statistical orientation of species or communities in a continuum along a gradient is called *ordination*.

These two points of view are not necessarily incompatible. There is no doubt that each species is distributed according to its own physiology, its own complex interrelations with other species, and its own tolerances, and that no two species are exactly alike in these various respects and in their responses to the environment. On the other hand, one can be convinced that the community and its habitat, collectively the ecosystem, is a functional system and that every species of necessity occurs in and as a part of such a system so that its distribution is importantly modified by these interactions and community relations.

Kinds of communities are usually recognized and identified by their most important organisms, the dominants and predominants. Subdominants and member species, however, are not usually dependent on the dominant species directly; rather, on the environmental conditions that the dominants establish. Different species of dominants in adjacent or related communities may react on the environment in a manner so nearly the same that subordinate species find suitable conditions for existence in each, although they are usually more characteristic of one than the other.

The *community-stand* is an actual aggregation of organisms occurring in a particular locality. In a sense, it is a collection of niches occupied by a particular set of interacting species populations, but it is something that one can see and study in the field. Because of the great variation in composition and character of community-stands in different habitats and parts of the world, they need to be evaluated and classified in some logical manner for reference purposes. *Community-types* are abstract groupings of individual community-stands which resemble one another and consequently must be defined rather arbitrarily. Different systems of community-types have been proposed, each designed to emphasize a particular point of view. We are, in this book, using the biome system, the various parts and concepts of which will unfold as we proceed. We will consider the community as being at least analogous to an organism in being a functional unit of interacting parts and having some degree of structural integrity. Although community-types are certainly not highly discrete and absolute units, recognition and naming of them is one way of indicating positions in the continua along environmental gradients that are occupied by particular aggregations of plant and animal species.

Physiognomy

The gross structure of a community or its physiognomy is an important basis for its recognition. In terrestrial communities, physiognomy is determined by the life-forms of the dominant plant species and their spacing. The life-forms that prevail in given area depend on the climate and the substrate or other special features of the habitat and give character to the landscape. The distribution of animal communities is closely correlated with the structure of the vegetation, hence these vegetation-types need to be recognized and defined:

Desert: Hot, arid habitats with scattered scrubby or thorny vegetation or, in extreme cases, none.

Thorn scrub: Tropical dry thorny deciduous tall shrubs, succulents, ephemerals.

Steppe, plains: Semi-arid grassland covered with short grasses.

Prairie : Semi-humid grassland covered with mid and tall grasses.

Temperate shrubland: Semi-arid areas covered with bushes and shrubs, either deciduous or broad-leaved evergreen (chaparral).

Savanna: Grassland with scattered trees or groves of trees or shrubs covering 10 to 25 per cent of the ground.

Woodland: Open stand of small deciduous or evergreen trees covering 25 to 60 per cent of the ground and with undergrowth of grassland or desert vegetation.

Forest-edge: Mixture of trees, shrubs, and open country, ordinarily occurring as a narrow belt on the margin of forests.

Forest: Closed stand of trees forming a continuous canopy over at least 60 per cent of the area.

Deciduous forest: Broad leaves fall during cold or dry seasons.

Two aggregations of species occurring naturally in different areas or in the same area at different times are to be considered as distinct communities when at least 50 per cent of the species of each aggregation are *exclusive* or *characteristic* to the aggregation. This we may call the 50 per cent rule. It is necessary to have quantitative information on the size of the populations to evaluate the importance of each species before community classification is attempted.

The distinctiveness of communities must work in both directions; that is to say, 50 per cent or more of this species of each aggregation must be different from the other aggregation. This means that the two aggregations are more different than they are alike. If the species composition does not exhibit the 50 per cent distinction, the two aggregations are considered as belonging to the same community. If the difference approaches but does not equal 50 per cent, it is often worthwhile to designate the two aggregations as *facies* of the same community if they are seral, or as *faciations* if they are climax. A number of other statistical procedures for distinguishing communities have been used by different investigators, but no method has yet been suggested that incorporates, as does the 50 per cent rule, not only the presence or absence of a species, but also its relative abundance and ecological significance.

Naming Communities

Since communities are distinguished by differences in life-form and taxonomic composition of the dominant or predominant organisms, these characteristics are usually used also in naming the community. Where the habitat is well defined but vegetation is largely or wholly lacking, as in many aquatic communities, habitat may be used in the terminology. Since names are largely a matter of convenience, they should be short and be derived from some easily recognized feature of the community or habitat. Very often the generic names of two, sometimes three, conspicuous dominants are used to name plant communities. The prevailing type of vegetation or habitat is commonly employed to name animal communities, although two or three predominant characteristic or exclusive animal species may also be used. In case of some large communities, geographic names are more convenient.

Large geographic units, differentiated on the basis of the climax type of vegetation or life-form of animals, are called *biomes*. They are specifically named by the characteristic form of vegetation or life-forms present; tundra biome, grassland biome, or pelecypod-annelid biome, for instance.

Secondary communities within the biome can be distinguished as climax or seral, respectively, by the suffixes*iation* and *-ies*. An *association* is a climax plant community identified by the combination of dominant species present; and *associes* is an equivalent seral plant community. Thus we may speak, for instance, of the *Fagus-Acer* association, which is a climax deciduous forest community, and of the

Calamogrostis-Andropogon associes, which is a grass stage in a sand sere (Celements and Shelford 1939).

Animal communities on land are related to different life-forms of plants or types of vegetation, but only seldom to plant communities distinguished by the taxonomic composition of the plant dominants. Thus animal communities must be analyzed and named independently of plant communities. A *biociation* is a climax animal or biotic community identified by the distinctiveness of the predominant animal species; a *biocies* is the seral equivalent. The North American deciduous forest *biociation* is to be contrasted, for instance, with the pond-marsh *biocies*.

When two plant or animal communities merge, either by intermingling of species in the same habitat or by juxtaposition of different communities in the same region, the resultant transitional state is called an *ecotone*. Ecotones occur between consecutive communities in seral development on an area as well as between adjacent existing local or geographic communities.

PRESERVATION OF ENVIRONMENTAL QUALITY

Up to the advent of civilized man, the biosphere of the world consisted of a complex pattern of biotic communities responsive to a variety of environments. The harmony prevailing between the biotic and physical realms was the result of hundreds of millions of years of evolution and adjustment of one organism to another and of each to the habitat in which it lived. The relationship was a resilient one. A sporadic outbreak of a species as the result of some temporary advantage, or a catastrophe to a population as the result of a change in the physical environment, soon became tempered and a balance restored. Primitive man occupied a niche and assumed a role comparable to other animals, and the ecosystem maintained itself in a vigorous healthy condition.

Modern man, however, has initiated a number of abrupt changes during the last few centuries beyond the capacity of the ecosystem to adjust. He has become the dominant organism in the biosphere, forcing changes to fit his needs and desires, and creating ecosystems different from any that have before existed.

Emergence of new dominants and evolution of new communities is a characteristic of the geosere, but the replacement of the biosphere by an 'anthroposphere' or 'noosphere' (from the Greek word, *noos*, mind) is unique in the speed and extreme changes produced. This acquisition of dominance is the result of the extraordinary evolution in man of intelligence, ingenuity, and use of tools. He has greatly modified or replaced the natural processes that control stability and balance within the ecosystem with new processes and manufactured products which the environment cannot absorb. The result has been a general deterioration of the environment, a population explosion of the human species, and a disharmony in the world ecosystem that will, unless resolved, lead quickly to catastrophe. Whether man will survive as the dominant organism, be reduced again to the role he occupied as a primitive, or become extinct may well be decided within the next few decades. There is conflict here between the demands of man for expanding population, productivity, power, and pleasures and the functioning of natural laws for maintaining stability and equilibrium between the living and physical environments.

Immediate problems of concern in the preservation of environmental quality are the elimination of air, water, and soil pollution; the disposal of solid wastes; and climate modification on a planetary scale. Perhaps noise pollution should also be added. *Pollution* is change in environment, caused by man's discharge of matter or energy into it, which alters the normal function of an ecosystem. Fundamental to the attainment of a stabilized and balanced worldwide ecosystem are the control and regulation of the human population at a suitable level and preserving the habitat and the basic community of plant and animal life upon which human life depends. Considerable technical knowledge already exists for solving these problems. The immediate need is concerted attention for implementation of the necessary remedial measures. Satisfactory resolution of the problems will involve not only science and technology but also law, sociology, politics, and economics. In the United States, an Environmental Protection Agency became established only in 1970. Emphasis on existing knowledge, however, should not obscure the fact that continuing basic research is required to elevate man's understanding of the environmental system. Scientists are just beginning to study ecosystems on the multidisciplinary basis that is clearly required. The various problems involved in the survival of man and in maintaining environmental quality will be identified and analyzed at appropriate places in the text that follows.

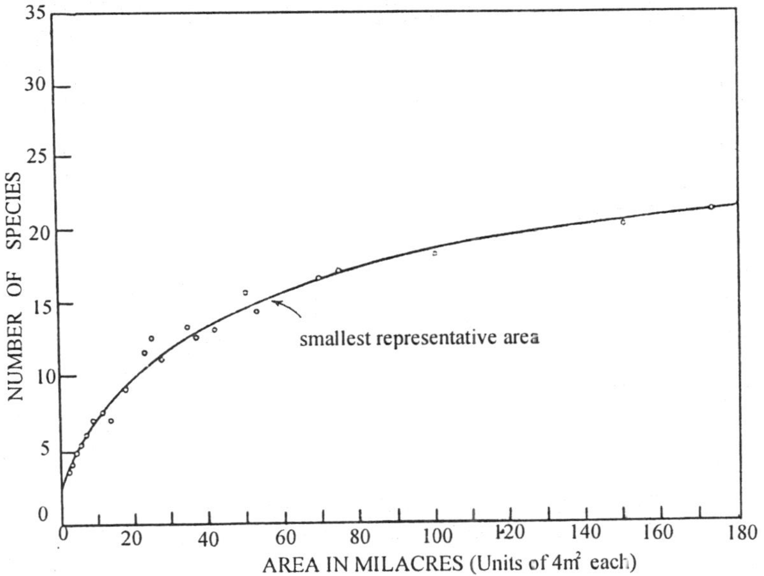

Fig. 7.6 Species-area curve for trees to illustrate the method of determining sampling area size adequate for analysis of community species composition. Plant ecologists commonly consider a minimum adequate sampling area as represented by a point on the curve where a 10 per cent increase in the sampling area increases the total number of species by only 10 per cent.

Organisms in a community are united in *dynamic relationships*, that is they influence one another's distributions and abundance. How they do so was the subject of the previous chapter. Organisms also are united by *coevolved relationship*, that is, they have influenced one another's evolution,. Coevolution occurs whenever the characteristics of one species influence the evolution of the traits of another species. Dynamic and *coevolved relationships* are not always closely correlated. For example, just because there is a coevolved relationship between a plant and its insect pollinator does not mean that the population size of either species strongly influences the population size of the other. The plants, for instance, may be limited by competition with other plants for suitable germination sites, whose abundance is independent of the abundance of the insect pollinator. The insect, in turn, may be regulated by overwinter survival of adults or pupae, factors unrelated to the abundance of the plant.

Macrodescriptors

To build up pictures of complex communities, we need a number of *macrodescriptors*, indices or terms that describe some aggregate feature of groups of organisms. A very familiar macrodescriptor is a non-biological one—the *gross national product*, which is the sum of the economic performance of a nation. All macrodescriptors lose precision just as do higher systematic categories, but they provide a simplified summary of many relationships that, if we have picked our macrodescriptors well, capture some important part of the process.

A useful ecological macrodescriptor is the *guild*: a group of organisms that share a common food resource. Two examples are seed-eaters, which constitute a guild dominated by rodents, ants, and birds, and foliage-gleaning birds of a region. The guild concept is useful because guild members have strong interactions with one another but weak interactions with members of other guilds. At the beginning of an ecological study, the strength of interactions is unknown, and one of the first tasks of the investigator is to make observations and perform experiments that will show if the guild initially selected is in fact dynamically united.

Another useful macrodescriptor is the *food subweb*: a group of organisms united as a result of being food for the same or similar sets of predators. A food subweb is, in effect, a guild extended upward through one or more trophic levels. Its reality can be established by removing a top predator and observing major changes in abundance of other members of the subweb but little effect on members of other subwebs. Alternatively, we can remove one or more members from the bottom trophic level of a subweb and observe which high trophic populations are affected most by this.

A third important macrodescriptor is *species richness*: The number of species of organisms living in a region. For some purposes it is useful to weight species according to their abundance or size, giving more weight to large and common species than to small and rare ones. When this is done we have a measure of *species diversity* rather than species richness.

Plant Growth Forms

Plants provide the pathway through which energy enters ecological communities. Because of their size and complexity, plants also form most of the structural environment for other organisms. Therefore, we begin our discussion of ecological communities by considering ecologically useful groupings of plants.

The most obvious way of grouping plants is by their growth form. The familiar term *trees, shrubs,* and *herbs* do just that, but we can be a bit more precise. The Danish ecologist C. Raunkaier proposed in 1905 that plants be classified according to the position of their overwintering buds. Raunkaier used the position of the overwintering buds because the unfavourable season in northern Europe is winter, but the idea works just as well when the unfavourable season is a dry one, as it is at low latitudes (See Table 7.1). These categories represent a rough indication of the cost to the plant of surviving the unfavourable season. A large tree maintains a great bulk of living tissues that respire throughout the unfavourable season (even though much of wood is dead), whereas an annual plant passes the unfavourable season as a seed with a very low metabolic rate. The position of the buds also influences the susceptibility of the plant to various environmental disturbances such as fire and grazing. Plants that die back to the ground during unfavourable periods, that have their growing points at or below the ground surface, and that have most of their stored energy below the growing points, are much more resistant to fire and grazing than are plans with their growing points and much of their energy reserves above ground.

Table 7.1 Raunkaier's six major life-forms of terrestrial plants

Category	*Characteristics*
Epiphytes	Air plants; no roots in the soil
Phanerophytes	Aerial Plants; renewal buds exposed on upright shoots; trees, shrubs, stem-succulents, vines
Chamaephytes	Surface plants; renewal bud at the surface of the ground
Hemicryptophytes	Tussock plants; bud in or just below soil surface
Cryptophytes (Geophytes)	Earth plants; bud below surface on a bulb or rhizome
Therophytes	Annuals; survive unfavourable season as seeds

Within these growth-form groups, plants differ markedly in their shapes and the forms and longevities of their leaves. Plants that die back to the ground in the unfavourable season inevitably lose all of their leaves. Trees and shrubs may retain them throughout the year or drop them for part of it. The cost of surviving the unfavourable period is greater if leaves are retained than if they are lost because leaves respire whether or not they photosynthesize.

The larger a tree, the more supporting structures must be fed by the leaves which are, for most plants, the only energy-producing structures. All other parts depend upon the leaves. A large tree has a less favourable ratio of leaves to supporting structures because the number of leaves increases little with size whereas the amount of wood and roots does. If it costs more for a large plant to survive unfavourable periods, why don't all plants die back to the ground? We shall discuss answers tu this question in the following section.

Plant Height

Several advantages may accrue to plants that grow tall. First, microclimates are most extreme at and very close to the soil surface. This is especially true in hot regions where surface temperatures at midday are extremely high (Fig. 7.7). At these temperatures, leaves respire faster than they photosynthesize. Just a short distance above ground, temperature variations are less extreme.

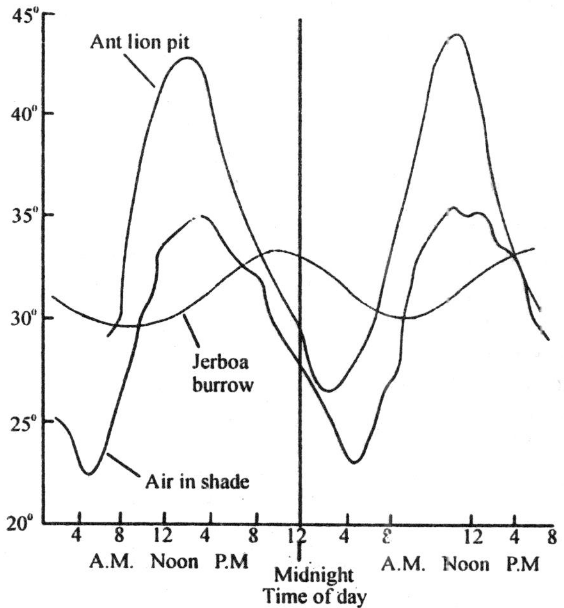

Fig. 7.7 Daily temperature cycle in a hot desert.

Tall plants are also above the reach of grazing and browsing animals. In those few remaining areas of the world where large mammals are still abundant, they exert profound influences on the growth forms of trees. Large browsing animals evolved in the Jurassic, providing continuous pressure favouring taller plants ever since then. Heavy grazing and browsing pressures also favour plants with their growing points close to the ground, below the reach of grazers. Most botanists believe that the great success of grasses lies in their ability to continue growing in the face of very strong grazing pressures.

For two reasons, the most important advantage of increased height is success in competing for light. First, a tall plant can position more leaves above the piece of ground that it occupies and, hence, fix more carbon by photosynthesis than a short plant. Second, by growing taller it can overtop and shade its competitors, increasing the area from which it can harvest light. There are two reasons why a tall plant can capture more sunlight per unit ground area than a short one. One is that sunlight usually arrives at an angle, so that a pyramidal shape intercepts much more light than a flat layer of leaves over the same amount of ground. This is most pronounced at higher latitudes, where the sun is always low in the sky; the most strikingly conical plants grow at high latitudes. This is not simply a taxonomic artifact because tropical conifers have much broader crowns than high-latitude ones.

Second, a tall plant can intercept more sunlight than a short plant because the sun, although it is approximately 15.53 x 10^7 kilometers from Earth, is nonetheless broad enough that it is not a point source of light. It is actually wide enough that on a clear day a shadow disappears behind a leaf at a distance 108 times its diameter (Fig. 7.8). On hazy days, sunlight is bounced off water vapour particles, arrives from many directions, and shadows are less distinct. On such days, shadows disappear much more quickly than they do on clear, dry days. The

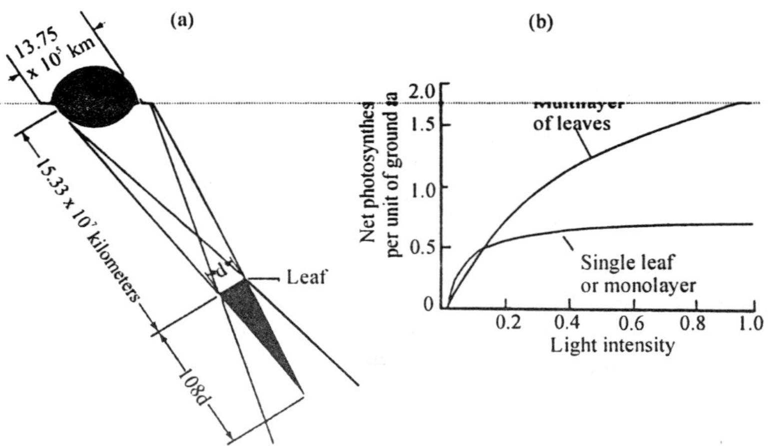

Fig. 7.8(ab) Shadows, photosynthesis and light intensity.

disappearance of shadows behind leaves is readily observed by holding a branch with leaves, preferably small ones, above a flat, bright object and moving the branch closer to and farther from the object. There is, of course, less sunlight behind it, but the light is generally reduced; there are no distinct patches of full sunlight and deep shade.

Another important fact is that the rate of photosynthesis of leaves of most trees does not increase linearly with light intensity but reaches its maximal value at approximately one-fourth to one-fifth of full sunlight. Therefore, if a loose layer of leaves removes approximately one-half the sunlight falling on it, a second layer 108 daimeters behind it will also photsynthesize at its full rate. Similarly, a third layer 108 diameters behind the second receives approximately 25 per cent of full sunlight, enough for it also to photosynthesize at the maximum rate. Therefore, a tree with leaves in a number of open layers, known as a *multilayer*, can achieve more photsynthesis per unit of ground area than can a tree with a single, dense layer of leaves, a *monolayer*. Notice that a multilayer does better than a monolayer only if a high intensity of light falls on the top of its canopy. If little light reaches a plant, it achieves more photosynthesis by arranging its leaves in a single, dense layer. This is why plants of forest understories are mostly monolayers and why, during the succession of plants on a disturbed area, the first invaders are fast-growing multilayers that are later replaced by monolayers better adapted for growth in the shade of the canopies of the tall multilayers.

Plant Dependencies and Mutualisms

Plants can also be grouped according to their dependency on other plants. A simple classification of plants by *dependency guilds* is given in Table 2, which stresses both physical and energetic support. Plants use one another extensively for physical support in the tropics. Up to one-half of the canopy of a mature forest may consist of the crowns of vines, and all branches and even the surfaces of the levels of most rain forest plants are covered with other plants. Vines, epiphytes, and epiphylls are much less prominent features of high latitude communities, but they are common in some temperate rain forests.

Plants may also be grouped according to their relationships with animals. Because all plants are eaten by animals, mutualistic relationships are a more interesting basis for establishing groupings (Table 3). Plants do not compete for the wind because it is not used up when it carries the spores, pollen, or seeds of any individual. However, plants do compete for animal transporters of pollen and seeds. Wind-pollinated plants are especially common at high latitudes and in arid areas, whereas animal-pollinated plants dominate low-latitude forest communities. Aquatic plants have fewer interactions with animals because most of them do not produce animal-transported pollen, seeds, or fruit.

Table 7.2 Dependency guilds among plants

Guild	*Characteristics*
Energy dependency	
Parasites	Obtain their energy directly from other living plants
Saprophytes	Obtain their energy from dead remains of other plants
Physical support	
Epiphytes	Grow on the trunks and branches of other plants; orchids, bromeliads
Epiphylls	Grow on the leaves of other plants
Vines	Depend on other plants for support but are rooted in the ground

Table 7.3 Mutualistic interrelationships of plants with animals

Category	*Examples*
Animal-dispersed pollen	Most composites, many flowering plants of deserts
Animal-dispersed seeds	Conifers such as junipers, yews, and *Podocarpus*
Both pollen and seeds animal-dispersed	Most tropical trees; shrubs at all latitudes
Neither pollen nor seeds animal-dispersed	Conifers such as pines, sprucess and firs; grasses, sedges

Finally, plants may be grouped into *defense guilds*: groups of species that interact with one another in their relationships with predators. For instance, insects often locate their host plants by chemical signals. If plants mimic one another or if the chemicals emanating from any one species become mixed up with those emanating from others, it is much more difficult for insects to locate their hosts.

Plants also mimic one another visually, as illustrated strikingly by similarities in the shapes of leaves of parasitic mistletoes and the eucalyptus trees upon which they grow. Relatively little is known about defense guilds because most of these interactions are chemical. Ecologists find it easier to discover and study visually based interactions than chemical ones.

Folivores

The tissues of plants—leaves, flowers, fleshy fruits, seeds, and wood—are very distinct. The groups of animals that utilize them overlap very little in membership. Therefore, these categories of plant tissue form appropriate units of community structure. Guilds based on food source are named by adding the suffix *-ivore* (eater of) to a word describing what is eaten. Thus, a folivore is an eater of foliage and a frugivore is an eater of fruits.

The leaves of plants are produced in great abundance, but they are, in general, a poor source of food. They are high in cellulose and, except when very young, low in proteins. Most of them are also defended chemically by substances that act in one of two principal ways. One group of defensive chemicals, the *acute toxins,* acts within the cells of the bodies of the folivores to interfere with some vital metabolic function. Some toxins, such as nicotine, interfere with transmission of nerve impulses to muscles. Others imitate insect hormones and interfere with insect metamorphosis. Some are unusual amino acids that become incorporated into the proteins of the folivore and interfere with their functioning. Acute toxins are widespread in plants and effective in their action because animals have tissues and organs not found in plants. Plant toxins can act on animals without being autotoxic to the plant producing them. It is more difficult for one animal to poison another without poisoning itself, particularly if predator and prey are closely related.

Digestibility-reducing substances are produced in larger quantities than acute toxins. They act in the guts of the folivores, which are, technically, outside their bodies because no cell membrane has been crossed. Digestibility-reducing substances usually combine with proteins—the materials in shortest supply for folivores—to form compounds that are extremely difficult to digest. The most common group of such substances are the tannins, which are present in the leaves of most species of woody plants in the world. They may be present in such large quantities that waters draining from areas dominated by tanniferous plants are darkly coloured like tea. The most famous of such "blackwater rivers" is the Rio Negro in Brazil.

Short-lived, fast-growing plants are less well defended than are longer-lived plants because the former are more likely to escape being eaten even if undefended. For the same reason, long-lived tissues on a plant are usually better defended that short-lived tissues. Wood is very heavily defended, mature leaves less so, young leaves even less so, and flowers least of all. Plants can also respond to being grazed upon by increasing the level of defenses in their leaves. The mountain birch of northern Finland is eaten by caterpillars of the moth *Oporinia autumnata* Caterpillars grow more slowly when reared on leaves from birch trees that were defoliated the previous year than they do on leaves from birches that received only light damage the previous year. The lengths of the larval periods of *Oporinia* and another moth *Brephos parthenias,* as well as those of two sawflies, are longer when larvae are fed on leaves growing close to leaves that are heavily damaged during the *same* growing season. Chemical defenses in leaves may increase within a few days of attack.

Many folivorous insects are specialists on a single host species of plant. Butterflies are the best-known group and they illustrate the pattern well. A community in Ecuador contains 53 species of nymphalid butterflies coexisting with 44 potential host plants. Of 27 species whose life histories have been determined, 22 are restricted to a single plant species and the other five use two to five hosts.

Mammalian grazers feed on many different plants each day and only a few, such as the koala bear of Australia or the sloths of tropical America, are extreme host specialists. The larger the individual the more food it must eat every day and the larger its mouth. As a consequence, large species are much less selective in their diets than are smaller species that can more readily pick plants with higher nutritive value. Communities of grazing mammals in East Africa are organized this way. Tall, coarse grasses are first grazed upon by elephants and buffalo that are able to obtain enough energy from eating forage with a high proportion of stems. Once these animals have reduced the standing crop of plants, zebras are better able to graze and they reduce the plant biomass even more. Their activities make foraging better for wildebeest, which, being still smaller, graze more selectively from among the more nutritious, low-growing plants that remain. Finally, the closely grazed areas are fed upon by gazelles, the smallest of all the plains grazers, that crop very selectively from among the resprouting grasses. These animals, all of which feed on the plants growing in the same area and are, therefore, potential competitors, actually have a mutualistic relationship, or at least a commensal one. The larger species may not benefit from having the smaller species in the environment with them, but the small ones depend on the larger ones.

Flower Visitors

Wind-pollinated plants do not produce attractive rewards, but their copious pollen is often fed upon by insects. In fact, pollen feeding by early insects probably led to the evolution of the use of animal vectors by flowers and the rich mutualistic systems found in terrestrial communities today. The flower-visitor guild is especially interesting because the exploiters exert such a strong evolutionary influence on the nature of their resource. Because flower visitors are attracted by a reward, plants compete for pollen transporters. If an insect is attracted to a flower offering a better reward, it may not visit a nearby one offering a poorer reward. At the same time, a pollinator is not "used up" in the usual sense of the term because it can visit many different flowers and carry pollen from all of them as it moves around. Competition for pollinators is most intense when there is a shortage of flower visitors relative to the number of flowers. Most field studies have shown that flower visitors typically drain the nectar from most of the flowers of an area and by the end of the day very little is left.

Even though they compete for pollinators, plants can overlap completely in their use of pollinators and still reproduce successfully. In western North America, many different species of sympatric plants (living together) are adapted for pollination by hummingbirds, and yet inmost regions there is only a single species of hummingbird present.

From the point of view of the flower visitors, pollen and nectar are resources that are depletable but that have an unusual pattern of renewal. Nectar, in particular, is rapidly renewed in many flowers. A new supply is available every day for the life of a particular flower, and a given plant may produce new flowers over a period as long as months. At one extreme are those plants of the understories of tropical forests that flower for extended periods but never have more than a few flowers in bloom at any one time. They are visited by large bees and hummingbirds that establish 'trap-lines' through the forests along which they travel, visiting each flower at intervals long enough to allow recovery of the nectar supply after its depletion. At the other extreme are *mass-flowering plants* in which all flowers open at about the same time. In the tropics, where many mass-flowering trees occur an individual tree may be in flower for as little as 3 or 4 days, during which time it may harbour a number of territories of hummingbirds and be visited by thousands of insects. Intermediate cases are plants with flowers on spikes, with the oldest flowers at the bottom and the youngest on the top. Over the total flowering season hundreds of flowers may be produced, but at any one time no more than approximately one dozen are present.

In most ecological communities there are many more species of flowering plants than species of pollinators, and most pollinators visit many different species of flowers. The greatest species richness of pollinators occurs in an unexpected place—warm temperate deserts. Six hundred and sixty eight species of bees occur in the deserts of California and Baja California, about one-third of which visit flowers of only a single plant family. They are active for only brief periods of time during the flowering season and then spend most of the year buried in the soil as pupae.

Because flowers are small and scattered, pollinators must visit many different flowers daily to make a living. Most pollinators are very active animals that spend but a short time at each flower and have excellent powers of flight. Unlike folivores, they are very difficult for predators to capture and many of them also have powerful stings. This may contribute to their ability to deplete floral rewards on a daily basis in most communities where they have been studied. Depletion is also possible because nectar is nutritious. Most nectar is devoid of toxins, and plants do not increase the defenses of their nectar when it is depleted. As a result, flower visitors actively compete with one another for access to floral resources. Hummingbirds are among the most aggressive of all birds. Individuals of most species establish territories from which they evict all other hummingbirds. Larger species normally dominate smaller ones. It is very difficult, however, for hummingbirds to defend their territories against bees and flies.

Frugivores

Fleshly fruits are also nutritious. Nonetheless, fruit/fruit eater systems differ in several ways from flower/flower visitor systems. The advantage of producing fruits is that the seeds contained within them are moved greater distances from the plant that produced them than if they simply fell or were carried by the wind. Many seed predators find their food by first locating the plant producing the seeds and then searching near it. Therefore, the chance that a seed survives increases with distance from its parent (Fig. 7.9). But unlike pollen, where the appropriate target is a stigma of another conspecific individual, the appropriate target for seeds are difficult to specify. The vicinity of another conspecific individual is not a good target.

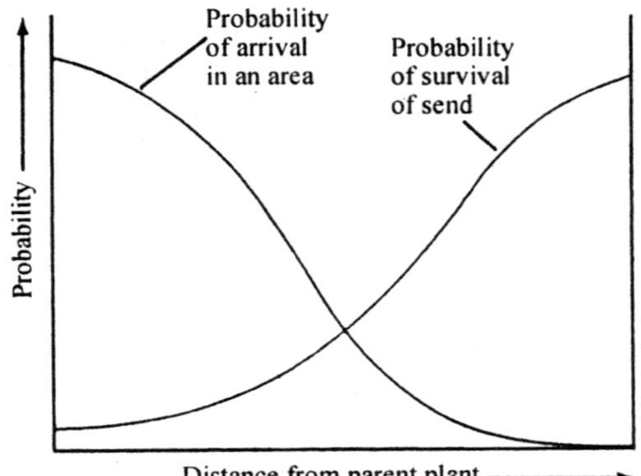

Fig. 7.9 Arrival and Survival of seeds

The area over which seeds from a particular plant can fall increases with the square of the distance from the plant. Therefore, the probability that a small area receives a seed from a plant decreases with distance from the plant, as shown in the black curve on this graph. However, seeds that land farther away from the parent plant are more likely to survive and produce seedlings because predators locate seeds more easily when they are close to the plant producing them. If survival probabilities increase with distance from parent plant (coloured curve), then the probability that a seedling will grow into a mature plant is maximal at an intermediate distance from the parent plant.

Places suitable for germination of the seed and establishment of the seedling are scattered in space and time. Moreover, there is no possibility of "payment on delivery," as exists for pollinators. Also, fruits are relatively large food items and are fed upon by larger organisms than those that visit flowers. These frugivores must maintain themselves throughout the year, whether or not fruits are being produced. Most of them are not obligate fruit eaters, but eat other types of food for at least part of the year.

The dominant frugivores in most ecological communities are birds, some of which are spectacular and rare. In the tropics many fruits are dispersed by bats. Surprisingly, many plants have their seeds dispersed by ants. These plants produce a nutritious body that is attached to the seed and that the ants carry back to their nests, seed attached. Ants do not move seeds long distances as do birds and bats, and they function primarily to move seeds to better microsites. For instance, violets initially disperse their seeds by an explosive action of the mature seed capsule and then the seeds are carried by ants to their nests. Seeds are actually moved farther on the average by capsule explosion than by ants, and ants are just as likely to carry a seed back towards the plant that produced it as to move it farther away. However, violets grow in the disturbed soil created when a tree falls over and its roots are torn from the ground. And nests are mostly located in those very sites and they carry the seeds to good germination sites.

It is unusual, however, for frugivores to carry seeds preferentially to suitable sites. A particular species of bird spends more time in certain habitats than in others and is more likely to deposit seeds there, but most seeds land in sites unsuitable for germination or seedling survival. Nonetheless, a few traits of fruits have evolved to influence the behaviour of frugivores. One of these is the evolution of mild toxicity, which discourages frugivores from perching in a tree full of fruit, eating their fill, and then sitting there while they digest the fruit and defecate the seeds. Mild toxicity forces frugivores to mix their diets and move to other species of plants. Toxicity is also a common feature of unripe fruits, where its function is to delay consumption until the seeds in the fruit are mature. Some temperate zone fruits, notably the conspicuous, white showberries, are mildly toxic when they are ripe, but they gradually become more palatable as they are frozen and thawed during the winter. They are eaten primarily late in the winter and their seeds are deposited on the ground not long before germination time. This trait is favoured if seeds are safer from seed predators or molds while they are within the fruit on the plant than when they are lying on the ground.

It is impossible for plants to produce fruits without also attracting predators who either eat both fruit and seeds or who eat the seeds without consuming the fruit. Among the frugivores that are also seed predators are many insects, parrots, finches, and most arboreal, fruit-eating mammals. These often eat many seeds while they are still on the tree. Some insects prey upon seeds before they are ripe by penetrating the fruit from the outside and sucking their juice. Others lay their eggs on or in the fruit, and the larvae burrow into the fruit and eat the seeds. The pea weevils (Bruchidae) that prey upon the seeds of legumes and some other plants are highly host specific. Of 95 species of pea weevils from 83 species of trees in northwestern Costa Rica, 78 are restricted to a single host tree and the maximum number of tree species utilized by a single species of weevil is six.

Granivores

Many plants, especially those of cool and dry regions, do not surround their seeds with fleshy rewards. All utilizers of these seeds are predators as well as seed dispersers. The seed-eating (granivorous) guild has been most intensively studied in hot deserts where it is dominated by ants, small rodents, and birds. Seed-eating rodents are adapted for carrying many seeds, often in their cheek pouches, and they gather large quantities during times of seed abundance and store them underground. Ants also store seeds. Their foraging columns travel long distances from their nests when recently fallen seeds are abundant. With rare exceptions birds do not store seeds and depend upon mobility to find patches of high seed densities. Virtually all of them feed upon seeds primarily in the autumn and winter and turn to insects in spring and summer. They consume a much smaller fraction of seeds in the environments so far studied than do ants and rodents and appear to exert little influence on the other two members of the guild.

Ants and rodents, however, both consume large quantities of seeds and strongly influence one another's populations. Experimental plots from which either ants or rodents were excluded were established in Arizona. Within 2 years ants had increased by 71 per cent on plots from which rodents had been excluded and rodents had increased 20 per cent numerically and 29 per cent in biomass on plots from which ants had been eliminated. The plant communities also changed because ants forage most intensively on smaller seeds whereas rodents prefer larger seeds. On plots with rodents and without ants, smaller-seeded species of plants increased in abundance relative to the larger-seeded species because their seeds survived better in the soil.

Seeds are a highly concentrated food source that is readily stored. Cereal grains have played a critical role in human societies since the beginning of agriculture, and even today these seeds constitute the most important component of international trade in food. World food reserves are always expressed in terms of quantities of storable seeds. Most major civilizations have been based upon the cultivation of one basic cereal grain. This was corn in tropical America, rice in Southeast Asia, millet in parts of India and Africa, and wheat over much of Europe and the Middle East. No other food is so readily stored and no other component of natural ecological communities relies as heavily on stored food as does the seed-eating guild. Long-term storage, however, depends upon the ability to keep the seeds relatively dry. Seed-storing animals are more abundant in drier places than they are in wetter ones.

If seed-eaters are really seed predators, why do so many plants have no means of seed dispersal other than being moved by their major predators? Several factors help explain this paradox. First, ants and rodents are sufficiently abundant in most arid and semiarid regions that even if seeds were transported by frugivores they would still land in places where they would be found rapidly by seed predators. Second, these animals do not eat the seeds immediately and many seed-stores are captured by predators before they have time to consume their seed reserves. Finally, it is difficult for plants to evolve seeds that are not good to eat. Energy reserves for the seedling must be concentrated in the smallest possible packages. Poisons deposited within the food stores in the seed interfere with the early development of the young plant.

Wood Eaters

Great quantities of wood accumulate as forests age. In mature forests, standing crops of up to 22,000 grams per square meter of wood are common in both temperate and tropical regions. As much as 1000 to 3000 metric tons of wood can be added per hectare per year in a rapidly growing coniferous forest in the favourable climates along the Pacific Coast of North America. Both the existence of forests and the rapid rate at which wood accumulates as forests age emphasize the fact that very little wood is eaten while the plants producing it are still alive.

Fungi begin to attack trees when they are weakened by cold and moisture stress and by competition with neighbours, but healthy trees are relatively immune to attack. Most animals that eat wood are not able to digest cellulose directly but depend upon the activities of microorganisms in their guts. Termites—

the most famous of all wood-eating animals—contain a rich community of microorganisms, which break down cellulose for their hosts. Young termites must acquire their microorganisms by eating feces of other termites directly; if they are prevented from doing so, they die quickly.

Food Subwebs

The eaters of plant tissues also have predators. Some of these eat primarily members of one guild, but others eat herbivores from several guilds. Carnivores specialize on members of a single guild when those species are sufficiently distinct from others that special hunting techniques are needed to find them. Sometimes members of a single guild are so different among themselves that distinct predators utilize each segment of the guild. This is the case for granivores. Rodents, ants, and granivorous birds have almost no predators in common. Except for nestlings, the birds are preyed upon mostly by hawks that specialize (they eat little else) on capturing birds. Rodents are preyed upon by a wide variety of vertebrates, especially snakes, owls, and moderate-sized mammals such as foxes and weasels. Ants are eaten primarily by specialized lizards and a few birds.

Folivorous insects are preyed upon primarily by other insects, spiders, and birds. Predatory insects and spiders are just as suitable prey for birds as the folivores themselves. In addition, by preying upon larvae that have been parasitized by wasps and flies, birds act as a major mortality agent and regulator of their populations, too. Interactions among folivores and their predators are complex and still very poorly understood. In grassland communities large grazing mammals are both competitors with, and predators on, many grass-eating insects.

The complexities of food subwebs are illustrated by the *Heliconius* butterflies of tropical American forests. These butterflies are unpalatable to birds and are involved in a number of Müllerian mimicry systems. Their larvae feed entirely on the leaves of passionflower vines, no more than 10 to 15 species of which grow in any one area. Moreover, each of these species is rare, and the individuals flower infrequently and asynchronously. The flowers are visited by generalized pollinators, and the *Heliconius* themselves play little part in that interaction. The species of *Heliconius* are associated with particular habitats and they participate in different mimicry complexes. The adult butterflies live for a long time and gather resources in part by mixing pollen with their saliva, thereby breaking down the proteins to amino acids and then using these amino acids for egg production. The flowers they use for this purpose are borne on vines of two different genera of the cucumber family—*Gurania* and *Anguria*—but the larvae of the butterflies do not eat the leaves of these vines. The abundance of adult butterflies and the amount of food they have for forming eggs does, however, depend on the abundance of these vines.

The passionflower vines have defenses against *Heliconius*. Some have poisonous hairs and all are full of toxic chemicals. In addition, most species have extrafloral nectaries that attract insects that also prey upon *Heliconius* eggs and larvae. The butterflies attempt to lay their eggs at the very growing tips of the vines where they are the most difficult for ants to find, but the plants also produce structures that look like butterfly eggs that fool butterflies into thinking that the plant is already occupied. The general picture of these complex interactions is given in Figure 7.10. This is already a complex picture, but it does not include other insects that eat passionflower vines, predators on the adult butterflies, pollinators of passionflowers,

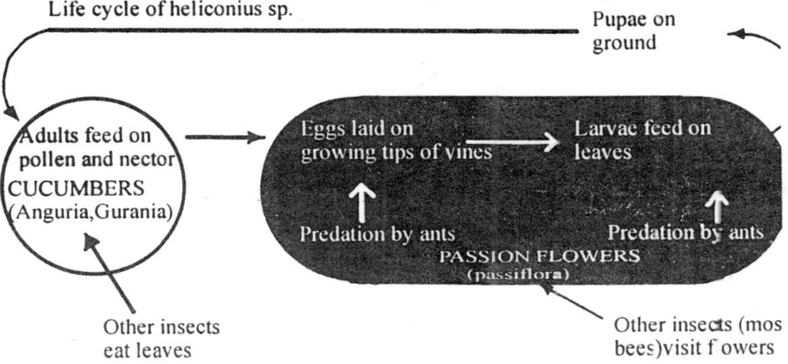

Fig. 7.10 The heliconius-passionflower-cucumber food web.

eaters of passion fruits and their seeds, or competitive interactions among passionflower vines and other plants. We do not know what determines the abundance of passionflowers. Do the butterflies significantly suppress reproductive output of the vines, or are the number of suitable germination sites and early competition with neighbours the key limiting factors? Are *Heliconius* butterflies controlled primarily by predation by ants on their eggs and larvae, by shortage of flowers at which the adults can gather pollen, or by difficulty in finding the scattered passionflower vines? Even with extensive knowledge, we cannot yet answer these important questions.

Aquatic Communities

Aquatic communities are built upon the same competitive predator-prey, and mutualistic interactions among their components species as the terrestrial communities, but the differences between water and air have caused different outcomes. Compared to air, water is a very dense medium that provides much more support for organisms living in it. At the same time, moving water creates much greater forces than moving air. There are no woody plants in aquatic communities, except at their edges. Submerged plants depend upon the water to support them and are flexible so that they move with water currents rather than resisting them. Another important difference is that movement through water is much more difficult than movement through air. Long-distance swimming is both slower and more costly than long-distance flying. As a consequence, aquatic plants have not evolved to use animals as dispersers of reproductive structures. Third, sunlight is attenuated quickly by passage through water but not through air. Communities at moderate depths in lakes and oceans receive little light whether or not there are other organisms above them casting shadows.

Correlated with these basic differences are a number of other community features. Aquatic plants are mostly very small and all of their tissues are eaten readily by animals. Plants provide relatively little physical structure to aquatic communities. Often a relatively small standing crop supports a large biomass of herbivores, the latter being sustained by high reproductive rates of the plants. Herbivores often determine the structure of aquatic communities. The removal of sea urchins, major grazers on marine algae, may cause a change from a community with small amounts of algae to one dominated by large algae that eliminate other competitors for space. Finally, water currents carry large amounts of food to organisms on the margins of oceans, leading to communities in which animals compete directly with one another and with plants for space. Nothing similar to this happens in terrestrial communities because air movement never delivers prey at such high densities. In most aquatic communities animals are the largest and longest-lived organisms, whereas in most terrestrial communities plants are.

Rivers and Lakes

Large quantities of solids are carried in suspension in turbulent river water, reducing the depth to which sunlight penetrates. In addition, only plants attached to the bottom can retain a fixed position. As a result there is little photosynthesis taking place in most streams and rivers. The organisms living there depend upon nutrients falling into the water from terrestrial communities and being carried downstream. There are attached plants in shallow water that support grazing animals, but most stream animals feed themselves with devices that capture food drifting downstream.

When water enters a lake, its velocity drops and its sediment load is deposited on the bottom, the larger particles settling out faster than the smaller ones. The increased clarity of the water increases the depth to which sunlight penetrates, and the lack of currents allows small plants to maintain themselves in the water column. Organisms suspended in the water column are called *plankton* and are divided into plants—called *phytoplankton*—and animals—called *zooplankton*. Animals, such as fish, that are able to swim against currents of water are called *neckton*. The plankton of lakes is dominated by green algae, filamentous blue-green bacteria, and small crustaceans.

Planktonic organisms are heavier than water and tend to sink in the water column, but many species are able to move enough to maintain themselves near the surface where the light is. Nutrients also tend to sink and are quickly exhausted unless replaced from below. This is in part accomplished by wind, which causes eddies in the water column, and by nutrients entering the lake from rivers, but most importantly by the annual overturn, which occurs in temperate lakes every spring and fall when the temperature of water column is uniformly 4°C. Most temperate lakes have a burst of productivity in the spring when the overturn of the lake coincides with longer days and warmer temperatures. There is usually much less biological activity in the autumn because the turnover occurs when days are very short and temperatures are dropping.

Animal communities in lakes are dominated by insects and fishes, the insects in particular occupying a wide variety of ecological niches. There are freshwater clams, cnidarians, various worms, and sponges, but they do not assume the prominence that their relatives achieve in the sea. Many groups of insects, such as dragonflies, damselflies, mayflies, stoneflies, caddisflies, and true flies, have aquatic larval stages and terrestrial adults. Their emergence and metamorphosis provides an important link between aquatic and terrestrial communities.

Marine Communities

Oceans cover approximately 71 per cent of Earth's surface, but most of that area is not very productive biologically. Attached macroscopic plants are limited to a very thin strip along the edges, and most of the photosynthesis of the ocean is carried out by tiny, single-celled organisms, particularly diatoms and dinoflagellates. These are most abundant in zones of coastal upwelling, where they are maintained in the surface waters by the upward-moving currents and where nutrient and light levels are high. The most important grazers of phytoplankton in the sea are crustaceans of species different from those that dominate fresh waters.

The shallow parts of the oceans situated over the continental shelves are collectively called the *neritic zone,* whereas the deeper ocean basins are known as the *oceanic zone*

(Fig. 7.11). At all depths, the bottom of the ocean is called the *benthic zone,* and the open water column is the *pelagic zone.* That part of the benthic zone below the level at which sunlight penetrates is called the *abyssal zone.* Animals living there depend entirely on the remains of organisms drifting down from the sunlit parts of the ocean for their energy. Energy is distributed very sparsely in the abyssal zone, and organisms living there have some remarkable adaptations enabling them to devour prey items that are very large compared to themselves.

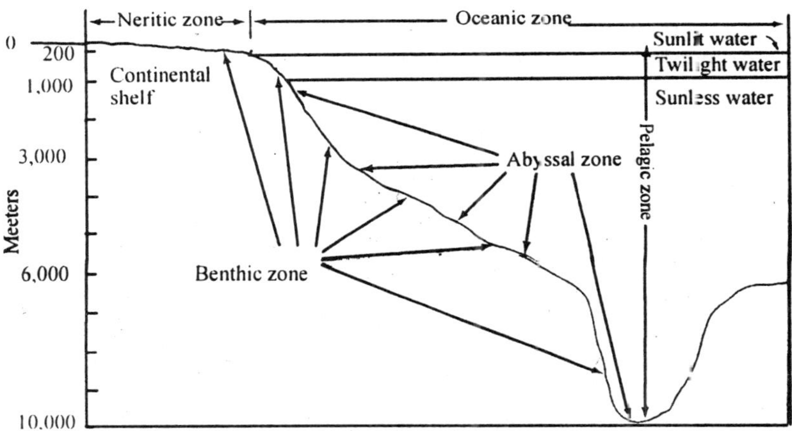

Fig. 7.11 Zones of the oceans.

Pelagic communities are relatively simple, consisting of zooplankton and phytoplankton, supplemented by larvae from benthic organisms during brief periods of the year. These small organisms are fed upon by planktivores ranging in size from small fishes to giant whales. Most pelagic fishes live in large schools, and they are streamlined in form for rapid movement though the open water.

Communities of the *littoral zone*—the coastal zone from the uppermost limits of tidal action down to the depth where the water is thoroughly stirred by wave action—are richer in species and more complex in structure than pelagic communities. Communities on and in mud and sand consist mostly of animals that bury themselves in the substrate. Some feed by filtering prey out of the water column above them, others set nets out on the surface of the mud, and still others burrow through the substrate, extracting out larger filter-feeding organisms.

Communities on rocky substrates are more obvious in their structure because most of the organisms are above the substrate, many of them glued to it. This is a zone of intense competition for space because waves carry to the littoral zone a supply of food that is undepletable by the thin strip of animals living there. Among competitors for space in a single small area may be algae from several different phylaplus animals from such diverse phyla as sponges, cnidarians, mollusks, annelids, bryophytes, echinoderms, and chordates.

The richest and most complex structurally of all marine communities are coral reefs, which fringe coasts of tropical continents and islands. The structural complexity of the reefs is due to corals, which produce distinctive skeletons—after which they are named (for example, staghorn corals, organ corals, brain corals, and fire corals). Like the trees of terrestrial communities, these corals provide the structures within which many different ways of life are possible. The richness of life in these environments, which may include over 100 species of corals alone, can only be hinted at in the best of illustrations. Fishes, in particular, assume a much wider variety of shapes than in the pelagic zone, in part because of the variety of prey on which they feed and in part because refuges from predators are always close at hand.

Continental shores are also important as breeding grounds for many pelagic species. All pelagic species that lay large eggs move to coastal waters, particularly sheltered bays and estuaries, to breed. The vitality of many populations of marine organisms, particularly fishes and arthropods, depends upon the existence of these spawning grounds. Unfortunately, these areas are often excellent ones for commercial and residential developments, and they also receive pollutants from the rivers that flow into them. Wise management of estuaries is a major environmental problem.

8
Ecological Succession

Ecological succession is defined as a continuous, unidirectional, sequential change in the species composition of a natural community. This sequence of community is termed a sere, and culminates in the climax community. Early successional stages are characterized by pioneer species, low biomass and often low nutrient levels. Community complexity increases as succession progresses, often peaking in the mid-successional stage. A mid-successional community is characterized by high biomass, high level of organic nutrients and high species diversity.

A typical community maintains itself more or less in equilibrium, but the members of the community are never in complete balance with each other or with the physical environment. Changes in the environment over a period of time are produced by variations in climatic and physiographic influences and also by the activities of the plant and animal inhabitants themselves. These modifications of the habitat may cause sufficient changes in the dominant species so that the existing community is replaced by a new community, or they may cause marked fluctuations in the abundance of certain species within the same community.

Progressive changes in communities take place from one geological epoch to another and also within much shorter periods of time. Detailed consideration of large-scale changes in the fauna and flora, such as those caused by the passage of an ice age or the uplifting of a moutain range, or those resulting from the evolution of new species, is beyond the scope of our present task; such alterations in the biota have great long-term significance, and they are discussed in treatments on paleontolgy, climatology, biogeography, and evolution. Here we shall concern ourselves primarily with the replacement of one community by another in particular areas and within the same general climatic conditions.

Observation has revealed the fact that in given biotopes certain communities tend to succeed one another. The occurrence of a relatively definite sequence of communities in an area is known as *ecological succession*. The change in the communities may be due in part to independent physiographic changes such as alteration of drainage, erosion, or deposition, but more especially it is caused by modifications produced by the action of each community on its own environment. The two types of causes are frequently operating together, as is seen, for example, in the replacement of a pond community by a marsh community. The filling of the pond is brought about by the deposition not only of a certain amount of inorganic silt, but also of a large amount of the organic remains of successive communities, and the accumulation of both types of deposit is enhanced by the presence of the roots and stems of the living community members.

The extent to which ecological succession is self-induced—as distinct from being caused by changes imposed from without—varies greatly in different situations. Similarly, the predictability of the course and speed of succession is variable. In many instances the presumed course of succession is based on inference derived from studies of surrounding areas so that "space is substituted for time"; but in other instances, some of which are described below, the nature of the succession is substantiated by actual records. Self-induced ecological succession is another outstanding example of the organism and the environment acting as a reciprocating system.

Living things modify their own habitat so as to cause one community to give way to another in a variety of ways. All species of animals and plants tend to increase in numbers and/ or in size. The conditions of the community consequently change because of the growth of the inhabitants even without any change in species composition. Consider a forest, for example. As the trees increase in size, they provide more shade, higher humidity, and different conditions of food and cover. New types of animals find suitable conditions here; old forms may be eliminated. Wildlife mangers have come to realize that the carrying capacity of a forest area for game changes with time because the availability of food and shelter in a stand of saplings is entirely different form that in a stand of mature trees.

When populations grow in respect to numbers or size of individuals, or both, the total weight of living material in the area tends to become larger. As predators and parasites increase in numbers, they tend to reduce the abundance of their prey, but as food becomes scarcer the consumers in turn are curtailed. At the same time the community causes changes in the physical nature of its biotope. Plants withdraw material from the soil as they grow and return it when they die, but the material returned to the soil is not in the same form. Humus accumulates, pH changes, moisture content is modified, and other alterations of the environment discussed in previous chapter are brought about.

The changed conditions caused by the varied activities of the inhabitants of the area may favour the growth of species other than those that have been dominating the scene. When this occurs, different species will soon get the upper hand. These may either be species already present in a subordinate capacity, or invaders from the outside. As one or more species take over the dominant position, a new community will be formed; its establishment constitutes a step in the ecological succession of the area.

The Mechanisms of Succession

There are three theories concerning the mechanisms which underlie and propel succession (Fig. 8.1). The classical explanation for the process of succession is known as the *facilitation theory*. In this hypothesis, species replacement occurs because existing species modify their habitat making it less suitable for themselves. The environment so modified favours other species which then outcompete and replace the existing plants. Thus succession proceeds in a predictable, orderly manner until the final climax community is reached.

The second theory is the *inhibition theory*. In this theory, which species succeed in establishing themselves depends entirely on which arrives first. No one species is competitively superior to another. Succession does not proceed in an orderly manner. Short-lived species are eventually replaced by longer-lived species.

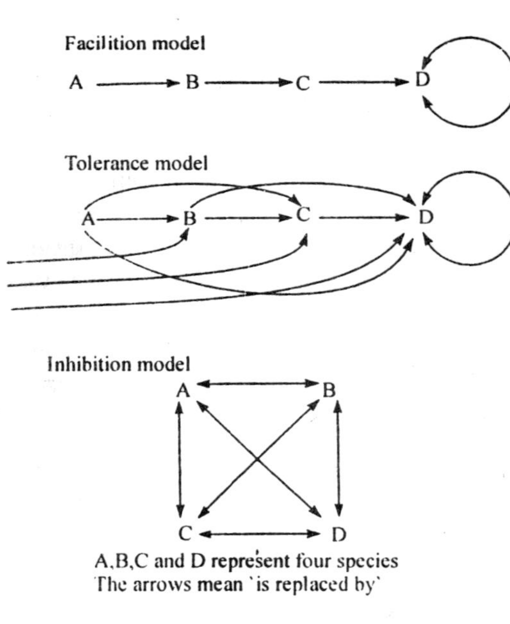

Fig. 8.1 Three models of succession

The final model is known as the *tolerance model* and may be viewed as intermediate between the other two. In this model, any species can colonise initially but some are competitively superior and come to dominate the mature community.

Characteristics of Plants of Early and Late Successional Stages

The pioneer plants are efficient colonisers which are able to utilise harsh physical habitats. They grow rapidly and produce large quantities of light seeds which are wind-dispersed. They are poor competitors and shade-intolerant.

Plant species characteristic of mature communities are adapted to competition. The average height of plant species tends to increase as succession proceeds because of competition for the light needed for photosynthesis. Species richness tends to increase with each seral stage, although it often falls off slightly in the climax community. Interspecific competition replaces the intraspecific competition characteristic of the earlier stages of succession.

Most studies of ecological succession concentrate on the plant communities present since their make-up determines to a great extent the composition of the animal community which they support.

The Concept of the Climax State

In the discussion to date, the climax community or state has been repeatedly referred to. How do we identify when this 'final' stage has been reached? The answer could be when the rate of change is so slow as to be imperceptible to man. However, care should be taken with this definition as the process of succession usually occurs over hundreds of years whilst the average life-span of humans is about 70 years.

The key here is the *persistence* of the community, which is self-perpetuating in its composition. However, gradual changes will still occur even in communities thought to be in the climax state as they follow climatic fluctuations. In addition, micro-successions will occur within communities, for example, when a tree falls to the ground, giving even a stable community a mosaic pattern. Therefore, the climax community in any absolute sense is probably regarded as a useful but essentially abstract concept.

It used to be thought that climate was the sole determinant of the composition of the climax community. Therefore, in a particular climatic zone, every sere would eventually end up as the climax state characteristic of that region. This was known as the *monoclimax theory*.

This theory has been superseded by the *polyclimax theory*. This states that whilst climate is the major determinant, there are a number of possible climax states depending on local conditions of topography, soil conditions and animal activity.

In some habitats, the expected climax community is never achieved due to the periodic interruption of some natural disturbance, for example, fire. The plant species able to withstand fire come to dominate the community because fire eliminates potential competitors. Examples of these so-called 'fire-maintained' communities are the pine forests of the southern and western United States. These should be regarded as climax communities even if they do appear as transitional stages elsewhere.

In some instances, natural succession is halted or perturbed by the selective grazing of herbivores. A striking example of this occurred in the 1950s in Great Britain when the introduction of the virus myxoma (a relative of smallpox) devastated the population of *Oryctolagus cuniculus* (the European rabbit). This resulted in a dramatic increase in the number of flowering plants, and shrub and tree seedlings.

Succession—the Classical Model

Ecological succession is the change in species composition and community structure and function over time. Succession is usually defined as a continuous, unidirectional, sequential change in the species composition of a natural community. The term succession was first used to describe the transition in abandoned old fields in eastern North America. After abandonment there appears to be a predictable sequence from grass and weedy herbaceous plants to shrubs such as sumac and hawthorn, eventually developing into a forest of maple, oak, cherry or pine:

Annual weeds ⇒ Herbaceous perennials ⇒ Shrubs ⇒ Early successional trees ⇒ Late successional trees

Such a sequence of communities is termed a *sere* and each of the distinct and recognizable successional stages is a seral stage. Each seral stage is a snapshot of a continuum that is changing over time, but each has its characteristic species composition. The sere is a generalization—some seral stages may be missed completely. Seral stages may persist for a few years or a few decades depending on the type of stage and the environmental conditions.

Eventually succession slows as the community reaches a steady equilibrium with the environment. This final seral stage is termed the *climax community*. In theory, at this point the community is stable and self-replicating. In its extreme form, the climax concept predicts that there is only one final community for a geographical region and that all succession will progress towards this *monoclimax*.

According to the classical model, structural complexity and organization of an ecosystem increase and mature over time as succession proceeds. Early successional stages are characterized by a few species known as *pioneer species*, low biomass and often low nutrient levels. Net community production is greater than respiration, resulting in an increase in biomass over time. Food chains are short and species diversity is low. The mature stage in succession is characterized by high biomass, high levels of organic nutrients, and gross production that about equals respiration. Food chains are complex and levels of competition are high. Accompanying the changes is an increase in species diversity. Species diversity often declines as the climax community is approached and the community becomes dominated by the most competitive species. Fig. 8.2 shows how environmental conditions can influence the accumulation of species. On mesic sites, forest diversity peaks at an intermediate stage in

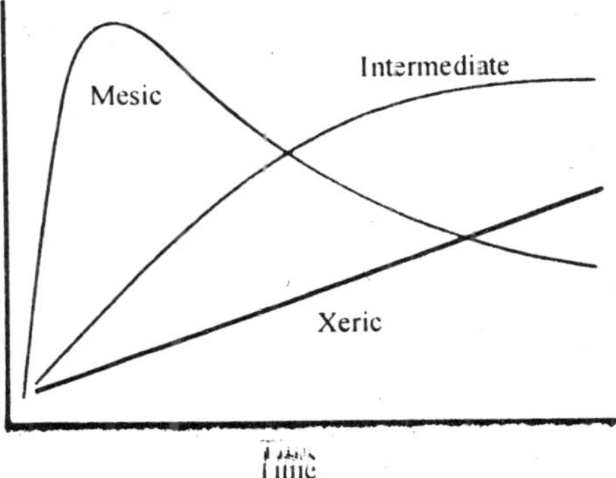

Fig. 8.2 Species diversity of forest sites over time for xeric, intermediate and mesic moisture conditions. Diversity increases over time as the climax community is approached on xeric and intermediate sites. On mesic sites diversity peaks in mid-succession.

the succession, but continues to increase on xeric (dry) and intermediate sites.

Autogenic Succession

Succession can be thought of as autogenic or self-driven. Changes in the environment occur as a result of the interaction between the organisms and that environment. Succession that begins on newly formed substrates not occupied by any organisms and where there is no organic material present, such as lava flows, newly exposed rock faces, alluvial deposits and glacial moraines, is termed *primary succession*. Succession where the vegetation cover has been disturbed by humans, animals or by fire, wind or floods, is called *secondary succession*.

In primary succession initial conditions are often severe. At first, the only organic matter will be wind-blown and no soil is present. On exposed boulder clay till deposited by retreating glaciers in Glacier Bay, Alaska, the nutrient poor clay is first colonized by mosses and shallow-rooted herbaceous species such as mountain avens (*Dryas* sp.). These early colonists alter the soil conditions allowing new species to colonize the habitat. Mountain avens has nitrogen fixing symbionts that increase the nutrient content of the substrate. Litter accumulation also leads to soil development, allowing colonization by shrub and eventually tree species such as cottonwood and hemlock. Alder (also a nitrogen fixer) has a strong acidifying effect, reducing the pH of the surface soil and creating conditions suitable for invasion by sitka spruce. The alder is eventually displaced by the spruce leading to a mature sitka spruce-hemlock forest.

Primary succession to woodland is a slow process, taking in the region of a hundred years or more. More rapid primary succession can be observed on intertidal boulders overturned by wave action. Bare substrate is rapidly colonized by the pioneer green alga *Ulva,* followed by several species of red algae. The red algae are slower to establish and can only invade where the *Ulva* has died, been grazed off or removed by some other disturbance. Selectively grazing by a species of crab particularly partial to *Ulva* accelerates succession to the tougher, longer-lived red or brown algae.

Old field succession is an example of secondary succession. Here the pioneer species are opportunistic annuals that can respond quickly to the appearance of a new habitat. The germination of *Ambrosia artemisiifolia* is triggered by disturbance, unfiltered light fluctuating temperature and reduced CO_2 concentration, all of which are associated with a newly cleared site. Summer annuals such as *Ambrosia* are replaced by winter annuals which establish earlier in the season, giving them a headstart in the competition for space, light and other resources. Late successional species are shade-tolerant and slow colonizers.

Degradative Succession

The term degradative succession describes a particular type of autogenic, primary succession: the colonization and subsequent decomposition of dead organic matter. Different species invade and disappear in turn, as the degradation of the organic matter uses up some resources and makes others available.

For example, pine needles fall in August and are first colonized by fungi which digest and soften them allowing other fungal species and mites to penetrate. After about 2 years in the A_0 layer the tightly compressed needle fragments are invaded by other soil microfauna which feed on the fungus as well as the needles. Bascidiomycetes attack the needle fragments, digesting cellulose and lignin. After about 7 years the needles have been completely decomposed, forming an acidic humus with almost no biological activity. All degradative successions terminate when the organic substrate is completely metabolized.

Allogenic Succession

Serial replacement of species can result from external environmental factors, such as geophysico-chemical changes. This is in contrast to autogenic succession (above) which is dependent on the biological action of the organisms on their environment.

Allogenic succession has occurred over most of North America and northern Europe in response to climate warming, following the retreat of the last pleistocene ice sheet about 10000 years ago. The ice sheet advanced and retreated three times in Europe causing a similar advance and retreat in the biota, each advance consisting of a somewhat different mix of species. As the climate ameliorated, easily dispersed, light-demanding trees such as pine and birch advanced northward to be replaced by the slower dispersing, shade-tolerant species such as oak and ash. The next glacial period favoured spruce, fir and eventually treeless tundra vegetation.

These changes have taken place over thousands of years and are evidenced by pollen cores taken from lake deposits. An example of a pollen map from Pennsylvania (Fig. 8.3) showing the transition over time (depth) from fir to pine and then to oak, hickory, beech and hemlock (with grass in forest openings) which occurred as the climate warmed.

Allogenic transition over shorter time scales occurs where sediment is accreting (e.g. on sand dunes or estuaries). Silt is accreting in the Fal estuary in Cornwall, England causing the salt marsh to extend seawards and the valley woodland to invade the landward limits of the marshland. The recent spread of the salt marsh grass, *Spartina anglica,* is also contributing to the extension of salt marsh. Species are able to colonize at a particular height above sea level according to their tolerance of the brackish mudflats, while slightly higher up tidal scrub and then tidal woodland have developed. This geographic transition mirrors the changes apparent from the woodland soil profile, and hence the historic origins of the woodland community.

Successional Processes

The example above and studies involving models of succession indicate that it is strongly influenced by three processes.

(i) *Facilitation*: changes in the abiotic environment that are imposed by the developing community. Pioneer species 'prepare the ground' allowing other species to invade, e.g. mountain avens and alder on boulder clay till.

(ii) *Inhibition*: species of one seral stage resist invasion by later successional species such that invasion is only possible following disturbance or by replacing individuals as they die, e.g. green algae (*Ulva*) on intertidal boulders.

(iii) *Tolerance*: late successional species invade because they are able to tolerate lower resources level and can

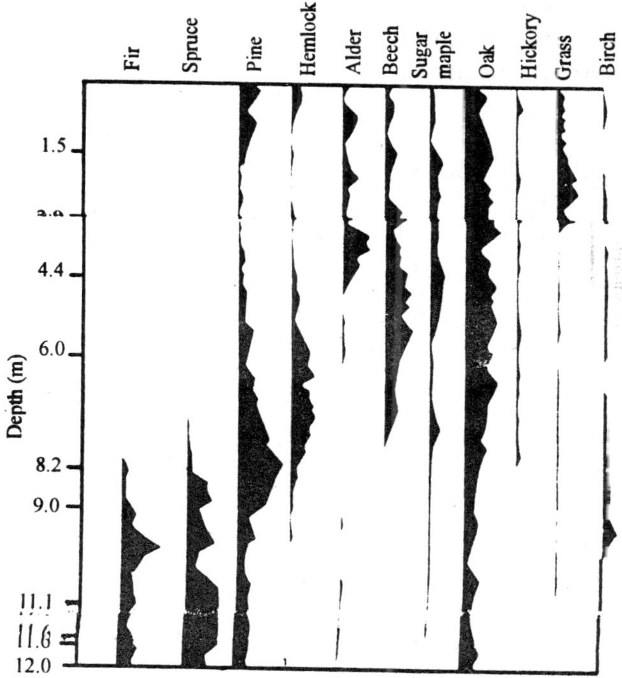

Fig. 8.3 Pollen diagram for Crystal Lake, Harstown Bog, Crawford County, Pennsylvania. From bottom to top: initially, spruce-fir forest with some oak invaded as the ice sheet retreated. This was replaced by pine, oak and birch as the climate warmed. The slower colonizing species, such as beech, hickory and hemlock were next to dominate.

outcompete early successional species for limited light and space, e.g. old field succession.

The tolerance model suggests a predictable sequence of species replacement based on species' strategies for exploiting resources. Tilman, in his *resource ratio hypothesis,* argues that species dominance is determined by the relative availability of two resources, light and nutrients. During succession, nutrient availability increases with litter accumulation and soil formation, while light levels decrease as a result of shading. Figure 8.4 shows how species A, with the highest requirement for light and the lowest requirement for nutrients is the first to colonize, followed by a predictable sequence of invaders until eventually the community is dominated by nutrient hungry, shade-tolerant species E.

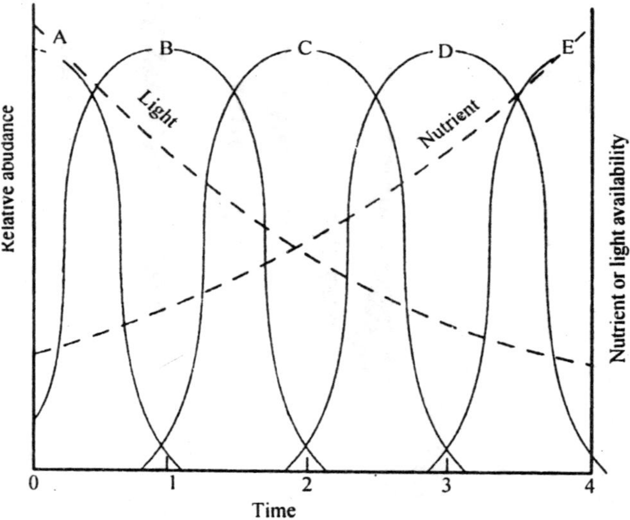

Fig. 8.4 The resource ratio hypothesis of succession.

Nobel and Slatyer also stress the importance of individual species properties in determining their place in succession. These properties are their method of recovery after disturbance and their ability to reproduce in the face of competition. Species unable to reproduce in competitive environments and able to recover rapidly after disturbance (e.g. by means of a seedling pulse from a seed-bank) occur early in the succession. Competitive species only able to invade slowly through seed dispersal into the habitat dominate later.

Typically, early successional species are *r*-selected, having developed efficient colonization mechanisms to escape from competition. Late successional species have evolved characteristics enabling them to persist longer in competitive situations and are thus responding to *K*-selection. In general, good colonizers are poor competitors and *vice versa*, i.e. there is a trade-off between competitive strength and colonization ability.

The three primary strategies proposed by Grime-competitive (C), stress-tolerant (S), and ruderal R—allow any species of plant to adopt a combination of strategies, trading off performance in one type of environment with performance in another. Using this model, succession in a temperate forest clearing involves a transition from ruderal (pioneer) species to fast growing trees and shrubs which have a competitive strategy (Fig. 8.5). At later stages the course of succession begins to deflect downwards towards the stress-tolerance corner of the

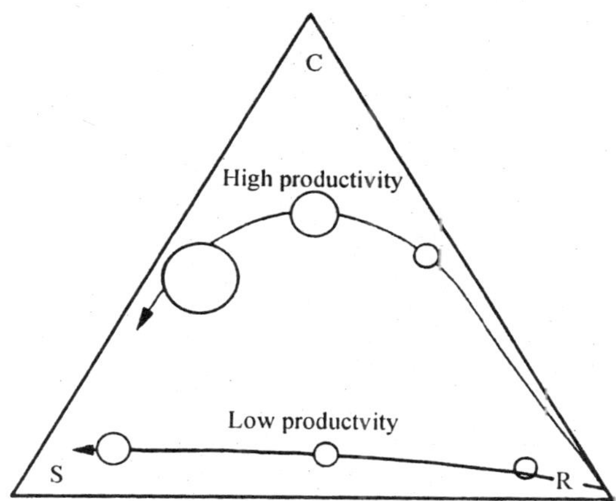

Fig. 8.5 Successional trajectories in relation to Grime's triangle of competitive, stress-tolerant and ruderal (CSR) strategies.

triangle. Secondary succession on sites with low soil fertility has a more shallow parabola.

Fluctuations and the Climax Community

Studies have shown that the model of the *single* climax community is inaccurate. The climax community will vary depending on local variations in climatic and edaphic conditions and in response to other factors such as the disturbance regime and the influence of animals, particularly herbivores. There is no single climax but continuum of climax types varying along environment gradients.

The *stable* climax community also proves elusive in practice. No community is stable for long because of natural disturbances such as storms and fire. A forest climax may take 300 years to develop but the probability of disturbance may be so high that succession never reaches this stage. Nonsuccessional, short term, reversible changes in the floristic composition of a community, or *fluctuations*, are also common. These changes are the result of environmental stochasticity and range from changes in species dominance to more subtle variations such as changes in the age structure of a species.

Succession was considered a predictable unidirectional process, but it is now known that this is not always the case. For example, *cyclic succession* occurs on a small scale in most communities. Where a forest tree dies and falls, pioneer species germinate in the gap to be replaced by mid- and late-successional species until the gap is again filled by trees. Thus, even old, seemingly stable communities are in a state of flux and are a shifting mosaic rather than a steady state. In disturbed habitats cyclic succession will prevent the development of a climax community and arrest seral development at an early stage. Although succession is undoubtedly an important and widespread process, the physical and biological structure of many communities may be dominated by other influences.

Types of succession

Primary Succession: Ecological succession that begins on a bare area where no life has existed, or where the previous fauna and flora have been completely destroyed, is known as *primary succession*. Habitats that become available for initial colonization include: new islands, sand bars, deltas, or glacial moraines; recently formed ponds; fresh alluvial shore, or volcanic deposits; and various types of substrata exposed by erosion. These diverse areas may be classified as xeric, mesic, or hydric according to whether the initial moisture conditions are dry, intermediate, or wet. Seres starting from these types of situations represent *xerarch, mesarch,* and *hydrarch* succession, respectively.

A striking example of primary succession, and a classical one, is the hydrarch succession in which a pond and its community is converted to dry land with an entirely different community. The vegetation rooted along the margins of the pond is able to push out from shallow water into deeper water in a variety of ways. As the vegetation invades the open water, the margin of the pond is reduced. At the same time the growth of the plankton and of other aquatic organisms adds organic matter, and much of this is deposited on the bottom. Beavers, muskrats, and other animals carry material into the pond, deciduous vegetation blows in from the shore, and silt is carried in from surrounding land. Rafts of vegetation from the pond margin drift offshore, strand, and take root, thus establishing islets that grow in size until they meet and also joint the shore. At the same time the area available for completely aquatic plants, such as the water lilies, becomes reduced. As the free water is changed to swampy land, the water lilies and similar species give way to sedges and rushes, and these are subsequently replaced by heaths and shrubs. As succession continues, the soil is further built up, so that it becomes drier and is also changed chemically. In time certain smaller species of trees invade the area, taking the place of the shrubs, and eventually full-sized forest trees will dominate the scene.

The existence of zonation in a community does not necessarily mean that succession is going on since distinct horizontal subdivisions may occur in a static community, as described in the previous chapter. Furthermore, the change in the vegetation is not always self-induced but may be caused by outside influences. Sometimes the conversion of swamp to dry land is brought about primarily by a lowering of the water table caused by a physiographic change. But in other instances, as in the situation described in the preceding paragraph, the vegetation itself is chiefly responsible for building up the land as ecological succession goes forward.

At the same time that the vegetation is undergoing these profound changes in the hydrarch succession, the animal life of the community is correspondingly altered. Fish, beavers, and muskrats will gradually be excluded and excluded and terrestrail vertebrates will enter. Less conspicuous but just as significant will be the manifestations of succession among the invertebrates and the microorganisms. These trends in the animal members of the community are indicated schematically in Figure 8.6 for a hydrosere in Illinois. The changes in bird species associated with a hydrarch succession are shown in Table 8.1. The changes in the insect members of seral communities leading to a red oak-maple climax are discussed by Smith (1928).

Secondary Succession: When a natural area is disturbed so as to destroy the community inhabiting it and so set back the course of succession, the new series of communities tending again toward the climax constitutes *secondary succession*. This situation arises when the principal species of the community have been destroyed by fire, disease, tornado, flood, or by human activities such as farming or lumbering. In some instances of secondary succession a community is established that is essentially the same as a stage in the previous primary succession, but in other instances a quite different community is brought into being by the special conditions resulting from the disturbance. However, later stages tend toward the type of community found in the primary succession, and ordinarily the same climax is eventually reached.

An admirable illustration of secondary succession may be observed today in old fields of central New England, and the phenomenon occurred on a large scale during the last century

Primary Succession

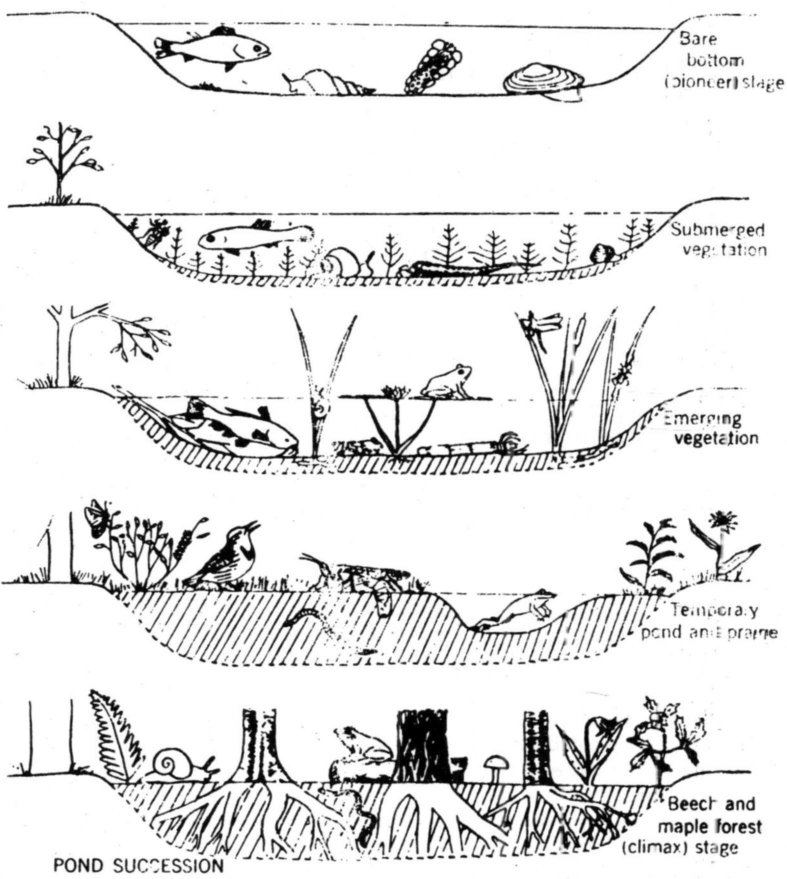

Fig. 8.6 Schematic representation of some of the changes in animal life during successional stages from pond to forest in Illinois.

when wholesale farm abandonment took place. The early settlers of this region cut down the forests to build farms and homesteads. Where the land was intensively cultivated, the stumps and roots of the forest trees were completely removed from the soil, and the cleared areas were planted to crops or used as pastures. Beginning shortly after 1830 the opening of rich farmlands in the west, the building of railroads, and the growth of industrial centers, brought about the exodus of the farmer from the rocky hillsides of New England that were so

Table 8.1 Representative species of birds present in various stages of a typical hydrosere in northeastern Ohio to show the change in bird life that accompanies the change in vegetation

Stage	1	2	3	4	5
Bird species	*Water Lily*	*Loose-strife-Cattail*	*Button-bush-Alder*	*Maple-Elm-Ash*	*Beech-Maple*
Pied-billed grebe	S	S			
Common mallard	X	S			
Virginia rail		S	X		
Long-billed marsh wren		S	X		
Eastern red-wing		S	S		
Eastern swamp sparrow		S	S		
Eastern kingbird			S		
Alder flycatcher			S		
Eastern yellow warbler			S		
Catbird			S	X	
Eastern goldfinch			S	X	
Northern yellow-throat			S	S	
Mississippi song sparrow			S	S	
Northern blue jay				P	
Eastern hairy woodpecker				P	P
Northern downy woodpecker				P	P
Eastern wood pewee				S	S
Eastern white-breastednuthatch				P	P
Black-capped chickadee				P	P
Tufted titmouse				P	P
Red-eyed vireo				S	S
Eastern oven-bird				S	S

X = present at times; *P* = permanent resident; *S*= seasonal.

difficult to work. The abandoned fields were covered by a thick sod of grass. In this turf the seeds of white pine trees blown from neighbouring forested areas found a favourable environment for germination, but other forest trees were not able to establish themselves effectively in the abandoned fields. The result was that the white pines grew into a dense, uniform stand, and eventually produced a forest in which all the dominant trees were of this species. A few hardwood saplings later became established as an understory throughout the pine forest, but their further growth was suppressed by the pines.

In the course of fifty years or so, the pines had grown to a size that made a valuable timber harvest possible. Great areas were lumbered off; in regions where lumbering was not carried out the even-aged trees grew old and finally fell. In both situations the trees which then grew into dominant position were not white pines but were hardwood trees that had existed in subordinate position almost unnoticed in the pine forest. Pines will not sprout from stumps, and, being an intolerant species, the pine seedlings are unable to develop in the shade of the mature trees. The root sprouts and seedlings of beech, maple, and other tolerant species that existed as minor members of the community grew into mature trees when they were "released" by the removal of the pines. White pine stands formed in abandoned fields of New England were therefore not selfperpetuating. Once the mature pines were gone, the species composition of the forests changed completely, and gave way to a community of mixed hardwoods. Thus, secondary succession originating in old fields of this region produces a transient community of pines but eventually results in the restoration of the typical climax forest.

Convergence: Since the bare areas in which primary succession can be initiated are extremely diverse, it is not surprising to find a correspondingly great variety in pioneer communities. As succession proceeds, however, the different biotopes tend to be modified towards a more nearly uniform condition that will support similar communities. This convergence is particularly clear in relation to the moisture factor—seral communities cause hydric habitats to become drier, and xeric habitats to become moister, so that a trend toward a more nearly mesic condition is followed from either extreme. As a general principle we recognize a convergence in succession such that seres originating in diverse habitats within a region of similar climate tend to progress toward similar climax communities.

An examination of succession in the deciduous forest region of Indiana (Table 8.2) revealed a tendency to converge that could be traced in seres originating from five kinds of biotopes. Pioneer communities established on a sand ridge, clay bank,

Table 8.2 Convergence in Succession in Indiana

SAND RIDGE		**CLAY BANK**
1. Cottonwood *Cicindela lepida*		1. Bare ground *Cicindela limbalis*
2. Jack pine *C. formosa generosa*		2. Shadbush *Polygyra monodon*
3. Black oak *Cryptoleon nebulosum*		3. Cottonwood *Ploygyra monodon*
4. White oak-Black oak-Red oak *Hyaliodes vitripennis*		4. Hop-hornbeam *Fontaria corrugatus*
5. Red oak-White oak *Cicindela sexguttata*		5. Red oak-Hickory *Cicindela sexguttata*
	BEECH SUGAR MAPLE *Plethodon cinereus*	
5. Hickory-red oak *Cicindela sexguttata*	5. Soft maple-Tulip *Plethodon cinereus*	5. Birch-Soft maple *Plethodon cinereus*
4. Elm *Panorpa venosa*	4. White elm-White oak *Anguispira striatella*	4. Tamarack *Hyla crucifer*
3. River maple *Helodrilus caliginosus*	3. Buttonbush *Asellus communis*	3. Poison sumac *Hyla versicolor*
2. Willow *Succinea ovalis*	2. Cattail-Bulrush *Chauliodes rastricornis*	2. Cattail-Bulrush *Sistrurus catenatus*
1. Ragweed *Tetragnatha laboriosa*	1. Water-lily *Musculium partumeium*	1. Water-lily *Musculium partumeium*
FLOOD PLAIN	SHALLOW POND	DEEP POND

flood plain, shallow pond, or deep pond initiated the successions indicated by the serial numbers with the dominant plants named. The exclusive presence, or relative abundance, of a dominant plant served as a *seral index* to the stage of succession in each sere. An animal species characteristic of each stage could also be recognized, and the names of these animal seral indices appear in the diagram. Succession in each of the widely different original areas tended to bring about the eventual establishment of the same type of climax community in which beech and sugar maple trees and the salamander, *Plethodon cinereus,* are the indices.

Succession in Special Habitats: The instances of succession discussed above are typical of broad areas in temperate regions. Other manifestations of succession occur in special habitats—either large or small. A classical example of succession in a microhabitat is the sequence of bacteria, protozoans, and other microorganisms that follow one another in a hay infusion formed by allowing hay to rot in a quantity of water. The seasonal succession of different kinds of phytoplankton and of zooplankton in natural waters is in part caused by temporary modification of the medium by the organisms themselves and in part by seasonal changes in the physical environment,. The special circumstances of succession in arctic areas in which the substratum is repeatedly disturbed by forest action are described by Hopkins and Sigafoos (1951).

Succession on a small scale, but of a complex nature, takes place on and in the trunks of fallen trees. The sound wood, cambium and bark are first attacked by group of boring insects, saprophytic fungi, and various kinds of microorganisms. This pioneer assemblage is followed by a series of more elaborate communities that cause and accompany the further disintegration and decomposition of the tree trunk. The presence of these organisms attracts predators and scavengers until a very large number of species may be represented in the dead-tree-trunk biotope. Mosses, lichens, ferns, and, later, higher plants find the rotting log a favourable substratum for growth. With the establishment of autotrophic plants a new cycle of constructive growth begins, in which the new pioneers take root literally as well as figuratively from the remains of the previous cycle.

The building up of coast lines is another activity in which the course of succession and the physiographic changes are mutually interdependent. The ecological steps involved in and following the formation of sand dunes along the southern shore of Lake Michigan as one community succeeds another have been studied by a number of ecologists. In simplest terms the course of succession was found to involve the capture of moving sand by grass, and the development of communities dominated successively by cottonwoods, pines, and oaks, leading to the beech-maple forest. Subsequent investigation has shown that the succession is far more elaborate and also more flexible than was at first supposed and that several interlocking channels for advancement or recession may be followed by the various series of communities involved (Fig. 8.7).

Along marine coast lines in tropical regions a well-known land-building succession takes place (Fig. 8.8). Here red mangroves (*Rhizophora mangle*) work out from the shore by dropping viviparous seedlings that will root only in water more than about 25 cm deep. By means of spreading stilt-like roots this species of mangrove can maintain itself in and slightly below the tidal zone in spite of wave action. Mud that collects around the dense jumble of roots builds up the bottom, causing the gradual elimination of the red mangrove, and prepares the way for the establishment of the black mangrove (*Avicennia nitida*), the seedlings of which will grow only in water shallower than about 25 cm. As the bottom is raised and the soil becomes drier, buttonwoods replace the mangroves, and these are subsequently succeeded by the climax palm community.

In the situations described thus far the controlling organisms are plants. On land, plants are usually the dominant members of the community, and vegetation development is chiefly responsible for causing succession, but in some communities animals are found to exert control. For example, birds nesting in colonies are sometimes so abundant that their

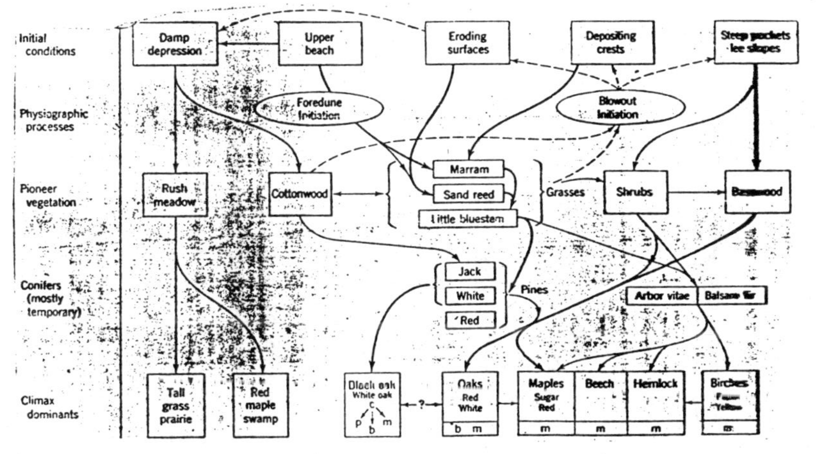

Fig. 8.7 Generalized diagram of ecological succession in the sand dunes region of Lake Michigan. Heavy arrows indicate major trends; light arrows indicate certain alternative courses of succession. Mixed or transitional communities also occur between stages. Depending on fire history and microclimate, contrasting types of undergrowth may develop under the same forest dominants: p = prairie grasses and forbs, b = blueberry-sedge type, c choke-cherry type, m = mesophytic herbs and shrubs. Broken arrows indicate destruction of pioneer communities by blowout initiation, after which new lines of succession may begin; once started, blowouts may undermine or smother communities at any stage.

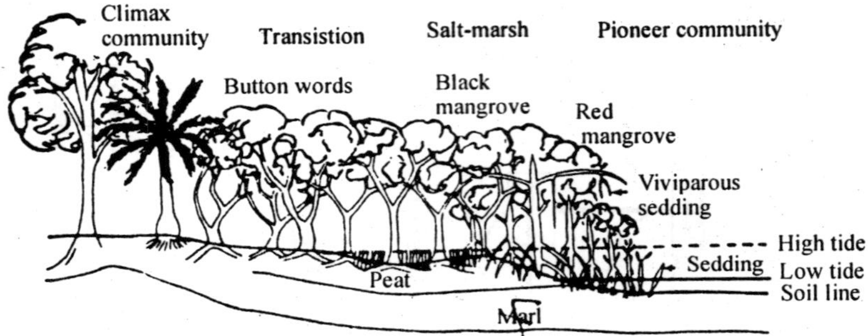

Fig. 8.9 Diagram of succession along the margin of tropical shores as seen in southern Florida.

activities and their droppings cause a change in the vegetation. In this way a rookery of herring gulls on Kent Island off the coast of New Brunswick caused the elimination of a grove of spruce trees and its replacement by grass.

Control of the environment by animals is more commonly found in aquatic habitats where sessile forms are prominent. Oysters became established on the sea bottom in this biotope, but mussels soon began attaching to the oysters' shells. Gradually the growth of the mussels smothered the oysters, and the latter were replaced by an almost continuous carpet of mussels. Subsequently, barnacles became attached to the shells of the mussels in sufficient numbers to kill them. After the death of the mussels their shells broke loose from the bottom and the barnacle population was swept away by wave action so that no enduring change was brought about in this instance.

Frequently both plants and animals are involved in well-defined succession in the littoral zone as described by Dexter (1947) for mud-bottom communities; similar relationships occur on hard bottoms and on submerged surfaces subject to attack by fouling organisms. When a jetty or pier is built, or a boat without anti-fouling paint is left at her moorings, various plants and animals attach in recognizable sequences. Bacteria are the first to establish themselves, and they commonly form a film in which benthic diatoms and various filamentous algae find a foothold. The subsequent course of succession varies according to circumstances, but in some situations Bryozoa appear next, followed by mussels, whereas in other situations barnacles, tube worms, or tunicates attached at an early stage. The varied surfaces of the attached organisms, and particularly the crevices between them, provide abodes for a host of mobile forms that join the community in these later stages and that could not inhabit the biotope if it were not for the presence of the earlier arrivals. Thus in these small habitats, as well as in the larger areas considered, the early inhabitants change the conditions so as to cause their own displacement and the establishment of successive new communities.

9

Population Dynamics

Introduction

A *population* may be defined as a group of individuals belonging to the same species which live in a given area at a given time. In studying populations, we are concerned not only with their size but with the way in which populations change over time. The study of population change is known as *population dynamics*. Environmental scientists are interested in this subject because it is when the combined effects of individual responses to change become manifest at the level of the *population* that the environment implications of change become significant.

In this chapter, the growth of populations and the factors which serve to regulate population growth, including competition and predation, are examined in detail. This discussion includes a number of classic experiments which have contributed greatly to our understanding of the mechanisms which underlie population growth and regulation.

The study of plant and animal populations has very important ecological applications, for example, in the conversion of threatened wildlife and in the biological control of pests. Many of the concepts which apply to such populations apply equally to *human* populations. However, there are some features of human populations which are unique, for example, our ability to largely control our own environment. An understanding of human population dynamics is extremely important when considering the impact of various human activities upon the environment. Therefore, the dynamics of human populations are discussed, as a special case, in the concluding section of this chapter.

POPULATION SIZE

One of the important features of any population is its size. Population size has a direct bearing on the ability of a given population to survive. Very small populations are the ones most likely to become extinct. Random events or natural disturbances are more likely to endanger a population if it contains only a few individuals. Inbreeding can also be a negative factor in the survival of a small population. Besides the lowering of vigour by direct genetic effects, inbreeding can also reduce the level of variability, which is likely to detract from the population's ability to adjust to changing conditions. If an entire species consists of only one or a few small populations, that species is likely to become extinct, especially if it occurs in areas that have been or are undergoing radical changes.

The size of the population is also affected by the reproductivity. The rate at which a species reproduces and the frequency of the population *turnover* can affect the speed with which it occupies new areas, becomes adapted to new niches, or evolves into new races. In order to analyze the population dynamics of a species, it is necessary to know its life history. This involves the stages in its life cycle, morality rates of each stage, longevity, sex and age ratios, age at which individuals become sexually mature, fecundity, factors causing mortality, and so forth. The proportion of different ages and sexes gives the population a definite structure.

FECUNDITY

Species vary greatly in the characteristic number of generations, broods, or litters produced per year, and in the size of them. Protozoans often divide so rapidly that they produce a new generation every few hours. Plankton organisms, less fecund, may produce a new generation every few days. Many vertebrates breed but once a year, some large animals only once every two or three years. Several species of small birds and mammals have two or more broods per year. The female woodland white-footed mouse in Michigan may produce three litters between early April and early June, and two more between middle August and early October. Under favourable environmental conditions, rodents may continue to breed throughout the winter, so that their reproductive potential is enormous.

Innate Capacity

The maximum size of a litter is determined by the physiological and morphological characteristics of the species. With mammals, which produce viviparous young, the size of the uterus and body cavity as well as the number of mammary glands for suckling the young after birth are limiting factors. With birds there is a limit on the number of eggs that one individual can cover and successfully incubate. In species that do not take care of their eggs after laying, the number produced may be limited only by the energy resources of the parent. This is indicated in part by the inverse relation between number of eggs produced and their average size.

Parental Care

The number of eggs or young produced per litter is correlated inversely with the amount of attention that they require. When parental care is altogether lacking, invertebrates may lay 1,000 to 500,000,000 eggs at one maturation; where there is some protection afforded by brood pouches, 100 to 1,000 eggs may be laid; with a high degree of brood protection, 1 to 10 or more eggs may be laid. Mammals seldom have more than a dozen young in a single litter and, in larger species, usually only one. Characteristic clutch size among birds varies from 1 to 15; rarely, 20.

There is a limit on the size of the brood or litter that adult warm-blooded animals can successfully feed and raise to maturity. There is no advantage, for instance, for starlings to have broods larger than five (Table 9.1). In larger broods, each individual receives less food, and hence has less vigour and weight on leaving the nest. Mortality increases either before fledging or in immediately subsequent months. In those species that feed their young in the nest, the clutch size has evolved through natural selection to the greatest number that can be hatched and raised successfully through efforts of the adults. In those species whose young leave the nest and feed themselves at hatching, the clutch size depends in large part on the capability of the female to mobilize energy in her body to produce eggs of a particular size. The variability in clutch and litter size for most species allow them to take advantage of temporarily improved conditions.

Table 9.1 Reproductivity in the starling in relation to brood size.

Brood size	*Number banded*	*Per cent recovered after 3 months*	*Brood size × per cent recovered*
1	65	0	0
2	328	1.83	3.7
3	1278	2.03	6.1
4	3956	2.07	8.3
5	6175	2.07	10.4
6	3156	1.68	10.1
7	651	1.54	10.2
8	120	0.83	
9	18	0	0
10	10	0	0

Weather

Clutches laid by birds during periods of hot weather are usually small than those laid when temperature is moderate. Clutches laid by related species in temperate latitudes tend to be larger than those laid in the tropics. The fecundity of white-tailed deer is higher with good forage than with poor forage. Reproduction is generally more successful after periods of high mortality than during years of abundance.

The mobilization of energy, usually within a definite period of time, is a limiting factor in warm-blooded organisms. The house wren, for example, lays 5, 6, or 7 eggs per clutch, the total weight of which is 7.0, 8.4, or 9.8 g respectively; yet the adult female herself weights only 11.5g. It is estimated that under average conditions about one-third of the daily energy intake of the bird, above its needs for existence, is deposited in the eggs being produced. Any appreciable change in temperature or rate of feeding thus affects the size of the egg, the number laid, or whether laying is undertaken at all.

Among invertebrates, clutch size also varies under different conditions and in different localities. The copepod *Eudiaptomus gracilis* commonly carries 11 eggs in April, 3 in early August, 9 in early November, and 5 or 6 over the winter. There is a decrease between spring and summer in the number of eggs carried in its brood pouch by the cladoceran *Daphnia*. *Diaptomus Siciloides* carries but 4 eggs in mountain lakes of California, as many as 18 in the Illinois River. These variations appear to be correlated with differences in temperature and food supply.

Death Rate

Death rates vary among species and are correlated with rates of reproduction (Table 9.2). The death rate of a species is influenced by a number of factors, but of fundamental importance is the number of young that are born in relation to the carrying capacity of the habitat. When more young are born than the habitat can support, the surplus must either die or leave the area. When populations are stabilized at a constant level, the death rate must fluctuate with the birth rate. Evolutionary adaptation tends to lower the frequency at which the population replaces itself and to raise reproduction to the highest rate compatible with the energy resources both of the species and of the habitat.

SURVIVAL OF YOUNG

Success in raising young depends not only on the ability of the adults to care for the young, but also on the vitality of the embryo and on the chance destruction of nests, eggs, or young by storms, wind, floods, predators, accidents, and desertion of

Table 9.2 Relation of reproductive to mortality rates per year

Local differences in same or related species

Species and locality	*Young produced per pair*	*Adult mortality rate (%)*
Starling		
Switzerland	5.8	63
England	4.7	52
Blue tit		
Britain	11.6	73
Spain and Portugal	6	41
Canary Isles	4.3	36
Wall lizard		
Italian mainland	24	40
Italian islands	11	20
California fence lizard		
Plains	8.5	80
Mountains	3.3	30

Species differences in same locality

Species and locality	*Young produced per pair*	*Average further life of half-grown young (days)*
White-footed mice in California		
Peromyscus californicus	6.2	275
Peromyscus truei	11.7	190
Peromyscus maniculatus	20.0	152

the parents. Considerable data are available in this connection with birds (Fig. 9.1).

Nest failures in birds are most frequent early in the nesting cycle and decrease progressively as nesting proceeds: 2.4 per cent per day during nest-building, 2.2 per cent per day during egg-laying, 1.2 per cent per day during incubation, and 0.5 per cent per day while the young are in the nest. Location of the nest is a factor in the successful raising of young (Table 9.3).

The relatively low percentage of nests that produce young successfully in many species is not a true index of annual

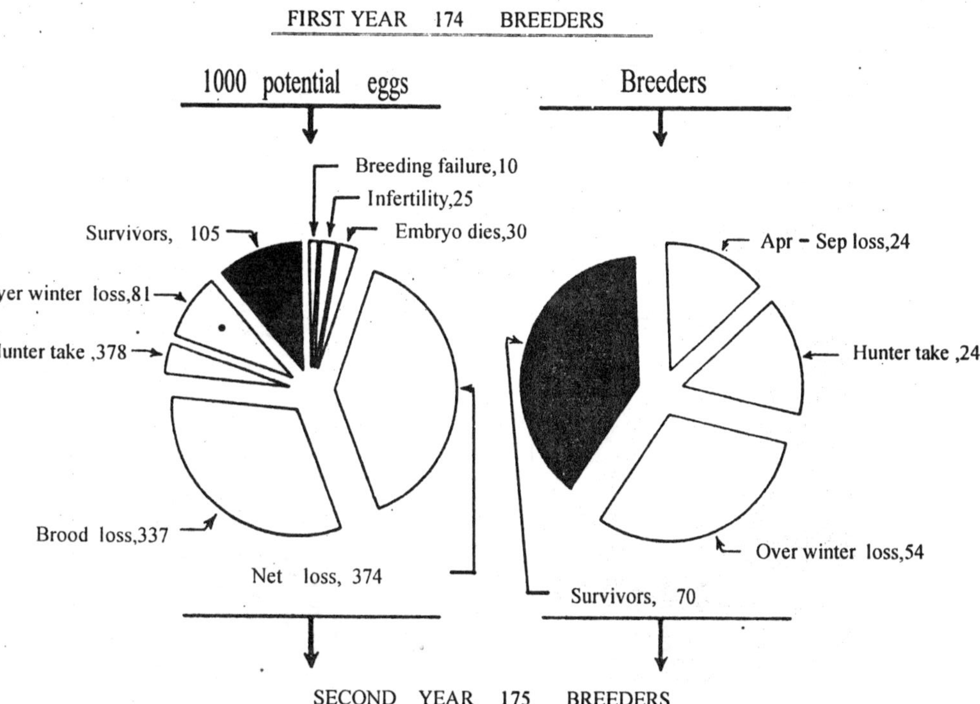

Fig. 9.1 Average life equation of a stabilized ruffed grouse population in New York State.

Table 9.3 Correlation between type of nest or nest location in birds and percentage of fledglings raised from eggs laid

Category	*Number of studies*	*Per cent successful*
Precocial gallinaceous species nesting on the ground	17	44
Open nests of altricial species	27	46
Waterfowl in aquatic habitats	22	60
Hole-nesting altricial species	32	66

reproductivity, since birds commonly make a second or even a third attempt if earlier nesting were failures. The ring-necked pheasant in Iowa has maximum nestings success averages of only 41 per cent; yet, before the season is over, by making repeated efforts, between 70 and 80 per cent of the hens are successful in raising broods. Full reproductive success is not assured, however, in raising the young to the stage of leaving the nest. In the study of the ring-necked pheasant, the average number of young hatched in successful nets was 8.7; after 1 to 3 weeks the average size of the brood was reduced to 6.7; after 4 to 5 weeks to 5.9; after 6 to 7 weeks to 5.3; and after 8 to 10 weeks to only 4.9.

MEASUREMENT OF POPULATIONS

Recognition of communities, calculation of species diversity, measurement of productivity, evaluation of species influence, following seasonal and yearly fluctuations, and investigation of a number of other community characteristics and functions depend on a knowledge of species density or biomass. In spite of its fundamental importance, available methods for measuring population size are only moderately satisfactory and are in need of improvement. In most types of ecological research, the aim should be to determine absolute abundance or the actual number or biomass of a species in an area of known size. This is no more difficult than in correcting relative indices of abundance for all the variables that are involved.

Since it is seldom possible to count all the individuals present in a large area, it becomes necessary to take samples over small areas where accurate counting of individuals is practical. *Strip censuses* involve counting all individuals of conspicuous species seen within a prescribed distance on each side of a line of travel over a measured distance. *Sample plots* may be distributed at random over an area but care must be taken that they are of proper size, shape, and number adequately to determine both species composition and densities.

Mammals

Small mammals usually need to be trapped because of their inconspicuousness and nocturnal habits. Killer traps, live traps, and pitfalls may be used, and all animals are captured over areas of known size. Considerable study is underway to standardize procedures. Live trapping has the advantage of allowing the determination of home ranges.

Birds

The spot-map method of censusing (Fig. 9.2) is commonly used for small birds, Enemar, but a variety of procedures are used in making inventories of waterfowl and upland game species.

Foliage Arthropods

In order accurately to determine the insect and spider composition is the foliage, net sweepings, light traps, bait traps, adhesive snares, and the like are necessary. A series of 48 strokes with a sweep net through the herb and low shrub strata furnishes an approximation of the number of arthropods per square meter. Variations in population estimates obtained by sweep-net samples are not usually significant unless differences of the order of two- or three-fold are obtained. Sampling of arthropods in tree foliage is more difficult, but special techniques make this possible.

Ground Animals

Hand sorting through the ground litter over measured areas permits the obtaining of good estimates of the larger millipedes,

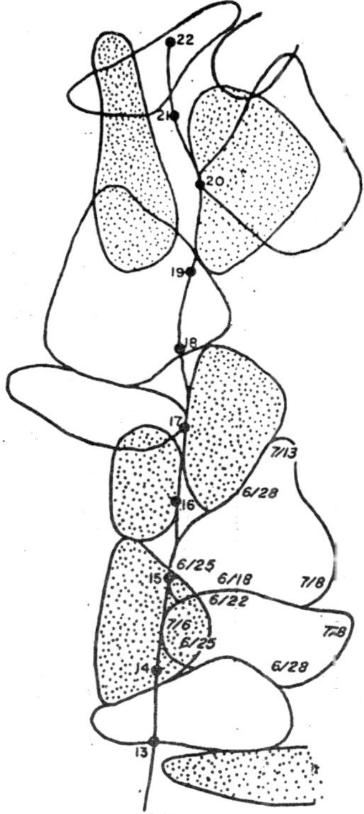

Fig. 9.2 Spot-map trail census of birds showing individual territories of two competing species, wood pewee (strippled) and least flycatcher. Two of the territories show the dates and locations of observations of particular individual birds. The numbered points on the trail are about 50 m (162 ft) apart. Note that territories of individuals of the same species do not overlap, and that territories of the two species are largely but not entirely exclusive.

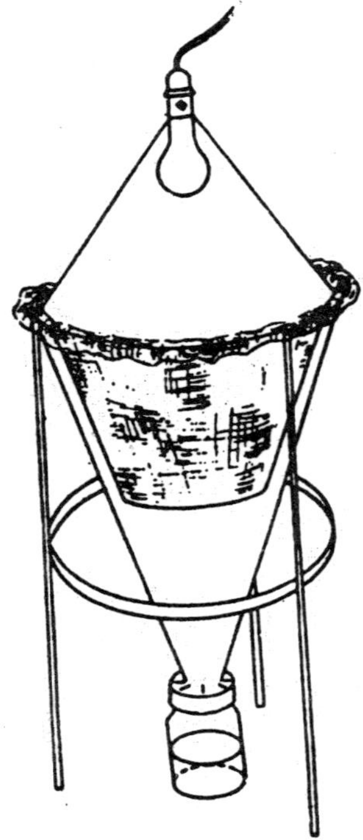

Fig. 9.3 Tullgren modification of a Berlese funnel for quantitative sampling of soil animals.

centipedes, snails, amphibians, and small reptiles. For the smaller invertebrates, especially insects, spiders, and mites, a common procedure is the use of the *Berlese funnel* (Fig. 9.3). A ground sample of soil and litter, 0.1 m^2 in size or smaller, is placed bottom side up in an open mesh basket inside a funnel.

A light bulb above the sample causes the animals to retreat downward and to fall into a bottle of preservative at the bottom. With the *flotation method,* the litter or soil is placed in a pan and covered with warm water, sugar solution, or water with chemicals added, such as magnesium sulfate, to increase the specific gravity. The material is agitated and the animals come to the surface, where they may be collected. For the microfauna or for other taxonomic groups, such as the annelids, a variety of special methods are available.

Fish

Seines may be used to sample fish populations in shallow water and trammel, gill, or Fyke nets in deeper water. Small bodies of water or representative areas of larger bodies can be blocked off by nets and the fish collected by the use of poison, such as rotenone. The fish are killed by this method, however, as are most zooplankton and some kinds of larger invertebrates. A less drastic method, that of shocking, employs two electrodes inserted into the water a short distance apart. The electric charge temporarily stuns the fish so that they float to the top, where they can be captured.

Plankton

Plankton nets are commonly made of silk bolting cloth arranged as a conical bag and attached to a wire frame. The net may be towed behind a boat either at the surface or submerged to any depth by appropriate weights attached to the tow line. Since the depth at which plankton occurs varies with the time of day, vertical hauls with a Wisconsin plankton net may be preferred. The Kemmerer sampler is used extensively for bringing up known volumes of water from measured depths for plankton and for chemical analyses. Various other types of closing nets or traps are available for sampling at different depths. Net plankton may be counted with the use of a Sedgwick-Rafter cell. For estimating the smaller nannoplankton, filtering or centrifuging is required.

Bottom Organisms

Dip-nets are commonly used in shallow water for obtaining macroscopic bottom organisms and those attached to submerged vegetation. Collections may be made quantitative if scoops are taken out of a bottomless cylinder covering a known area. The Surber swift-water net is standard equipment for sampling rocky stream bottoms (Fig. 9.4). A frame marks out 0.1

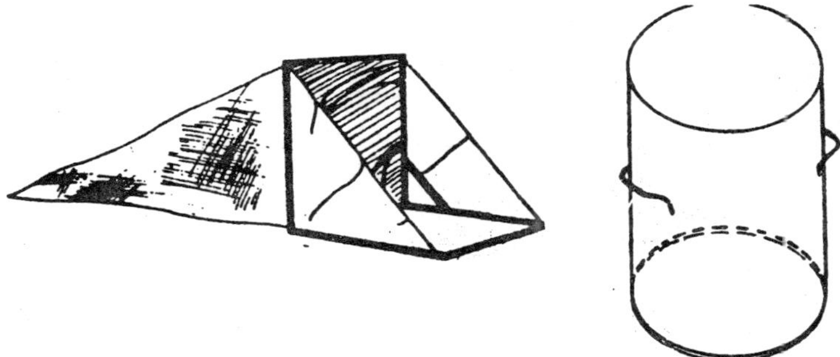

Fig. 9.4 Apparatus for collecting quantitative samples of bottom organisms in streams: Above, swift-water net, covers. 0.1 m^2; Right, sampling cylinder for use in pools, covers 0.2 m^2, has shar pened lower edge.

m^2 and a net downstream catches organisms dislodged as the rocks are agitated or removed for closer examination. Dredges of various shapes and sizes may be pulled along the bottom for measured distances to get organisms in deep water, but quantitative determinations obtained in this way are generally too low. Much more reliable are the Ekman bottom sampler on soft bottoms and the heavier Petersen sampler, used also on sand and harder bottoms (Fig. 9.5). Bottom samples must ordinarily be washed through sieves to remove the debris and separate the animals for identification and counting. Mesh of different sizes is used in the sieves to permit sorting out animals of different sizes.

Capture-Recapture Method

K.Dahl in 1917, using methods of marking and recapturing fish developed in 1894 by C.G.J. Petersen of the Danish Biological Station; F. C. Lincoln, of the U.S. Fish and Wildlife Service in 1930, trying to estimate the number of ducks on the North American continent; and C.H.N. Jackson (1933), working with tsetse flies in Africa, all independently derived a formula for determining the population size of various species of animals, much used in recent years. The method depends first on capturing a fair sample of individuals in a unit area, marking them in a distinctive manner, releasing them for uniform rediffusion over the area, then, after a short interval, retrapping the area. The ratio of marked individuals recaptured to the total number marked should theoretically be the same as the total marked and unmarked animals captured during the second trapping is to the total population:

$$\text{total population} = \frac{\text{total marked}}{\text{marked recaptured}} \times \text{total captured}$$

Other formulas make use of accumulating totals of marked and unmarked individuals during successive periods of trapping. There are several possible errors in arriving at accurate estimates of the total population for which compensation must be made.

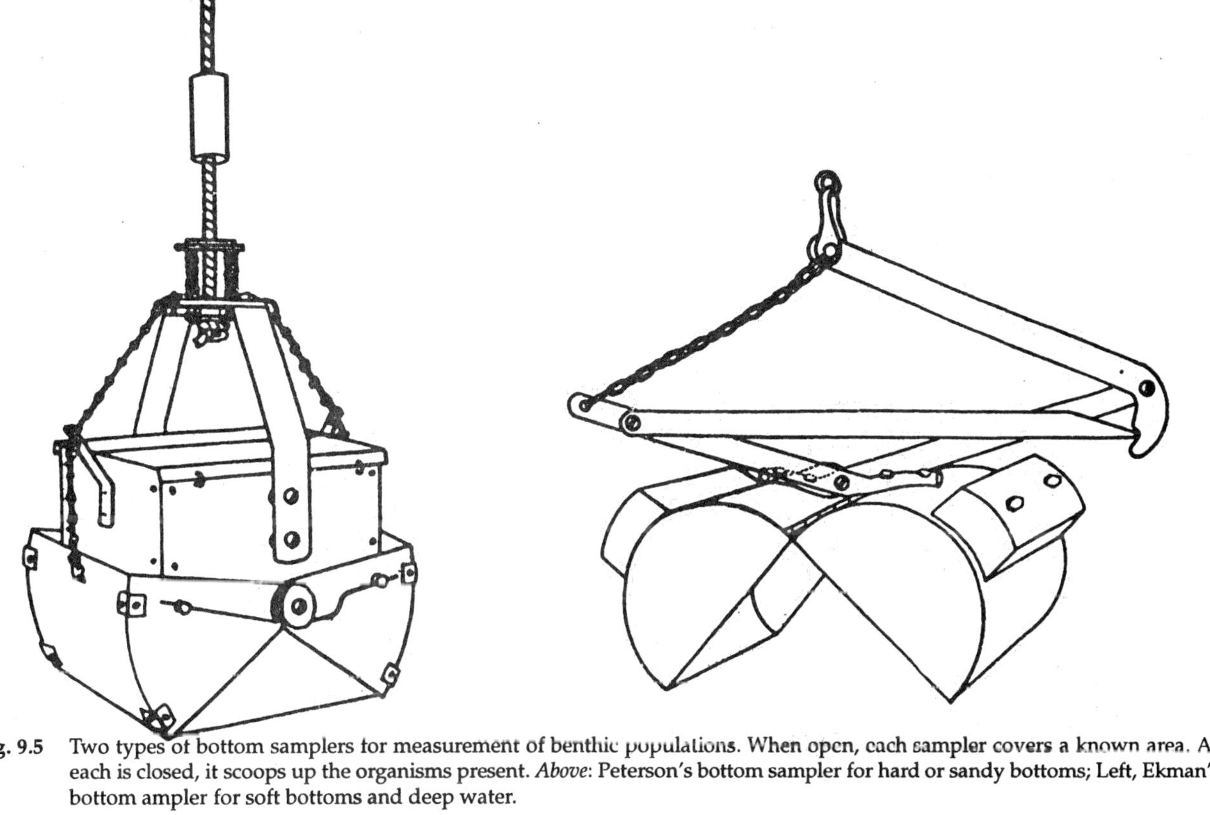

Fig. 9.5 Two types of bottom samplers for measurement of benthic populations. When open, each sampler covers a known area. As each is closed, it scoops up the organisms present. *Above*: Peterson's bottom sampler for hard or sandy bottoms; Left, Ekman's bottom ampler for soft bottoms and deep water.

Marking of small mammals, birds, lizards, and amphibians is commonly done by toe clipping, ear or wing notching, tattooing, dyes, or tags. Birds may be individually identified by means of numbered bands placed around their legs; such a program is sponsored by the U.S. Bureau of Sport Fisheries and Wildlife. Snakes are marked by removing scales from conspicuous locations on the body, frogs may be identified by punctures in the web between the toes, a fin may be clipped in a characteristic manner with fish, and so on. For determining home ranges, a different procedure is labeling individuals with radioactive material and following their movements by means of Geiger counters. Still another method becoming increasingly popular involves attaching miniature radio transmitters to animals and monitoring their movements by means of telemetry.

Capture Per Unit of Effort

In a closed or stabilized population, when the same time, traps, and effort are employed to capture or count individuals in the same area at different times and there is no loss or increment in the original population, and weather and other conditions remain the same, the number of new individuals captured or discovered with each subsequent effort becomes less and less, and should eventually reach zero (Fig. 9.6). When the number of new individuals captured per unit of effort is plotted against the cumulative number of animals captured, a straight line results. A line thus derived from a few catches may be extended to zero, and the total population of animals in the area determined. A variation of this method is to use the increasing percentage of marked animals in the total number captured at successive intervals of time, as the increase in these percentages follows a definite trend that would eventually include the total population.

POPULATION GROWTH

A second key characteristic of any population is its capacity to grow. Most populations tend to remain a relatively constant size, regardless of how many offspring are produced. Darwin

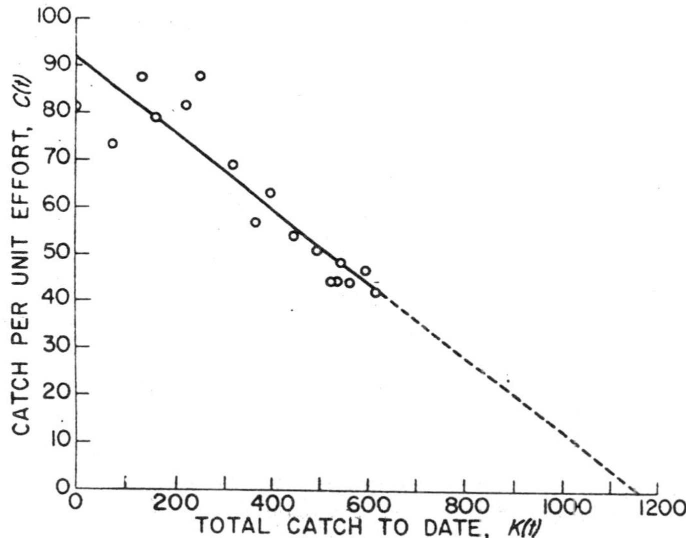

Fig. 9.6 Total population (K=1170) calculated by extension of a straight line through data on successive catches of new individuals per unit effor, $C(t)$, plotted against the accumulating total catch, K (t) (from DeLury 1947).

partly based his theory of natural selection on this seeming contradiction. Many calculations have shown that houseflies, bacteria, or even elephants would soon cover the world if their reproduction were unchecked. Under certain circumstances, a population size can increase rapidly. We must consider what these circumstances are and what factors operate in nature to limit population growth.

Biotic Potential

When discussing these matters, it is useful to consider the intrinsic rate of natural increase, or *biotic potential*, of a population. This term refers to the rate at which a population of a given species will increase when there are no limits on its rate of growth. In mathematical terms, this is defined by the following formula:

$$dN/dt = r_i N$$

where N is the number of individuals in the population, dN/dt is the rate of change of its numbers over time, and r_1 is the intrinsic rate of natural increase for that population its innate capacity for growth.

In practice, the biotic potential of a given population is difficult to calculate, since the assumption of unrestricted growth is unrealistic for most organisms. However, for varying sets of conditions, r, the actual rate of population increase, defined as the difference between the birthrate and the death rate per given number of individuals per unit of time, can be calculated. The actual growth rate may also be affected by net *emigration* (movement out of the area) or net immigration (movement into the area). For example, the increase in human population of the United States during the closing decades of the twentieth century is mostly due to immigrants. Less than half of the increase comes from the reproduction of the people already living there.

The innate capacity of growth for any population is exponential and can be expressed by a curve. The *rate* of increase remains constant, but actual increase in the number of individuals accelerates rapidly as the size of the population grows. This sort of growth pattern is similar to that obtained by compounding interest on an investment. In practice, such patterns of growth occur only for short periods, usually when an organism reaches a new habitat with abundant resources. Natural examples of this include dandelions reaching the fields, lawns, and meadows of North America from Europe for the first time; algae colonizing a newly formed pond; or the first terrestrial immigrants that arrive on an island recently thrust up from the sea.

Carrying Capacity

No matter how rapidly populations grow under such circumstances, they eventually reach some environmental limit imposed by shortages of an important factor such as space, light, water, or nutrients. A population ultimately stabilizes at a certain size, called the *carrying capacity* of the particular place where it lives. The carrying capacity is the number of individuals that can be supported at that place indefinitely, a measure that is dynamic rather than static as the characteristics of the place change. In practice, the number of individuals oscillates around a mean.

The growth curve of a specific population, which is always limited by one or more factors in the environment, can be approximated by the following equation:

$$dN/dt = rN(K - N/K)$$

In other words, the growth rate of the population under consideration (*dN/dt*) equals its rate of increase (*r*) multiplied by *N*, the number of individuals present at any one time, and then multiplied by an expression equal to *K*, the carrying capacity of the environment, minus *N* divided by *K*. As *N* increases (the population grows in size), the fraction by which *r* is multiplied becomes smaller and smaller, and the rate of increase of the population declines. In other words, as *N* approaches *K*, the rate of population growth (*dN/dt*) begins to slow, until it reaches 0 when *N* = *K*. In practical terms, factors such as increasing competition among more individuals for a given set of resources, the buildup of waste, or an increased rate of predation causes the decline. Graphically, this relationship is the S-shaped sigmoid growth curve characteristic of biological populations (Fig. 9.7). The curve is called *sigmoid* because it resembles the Greek letter sigma (s). As the size of a population stabilizes, its rate of growth slows down, and eventually it does not increase further.

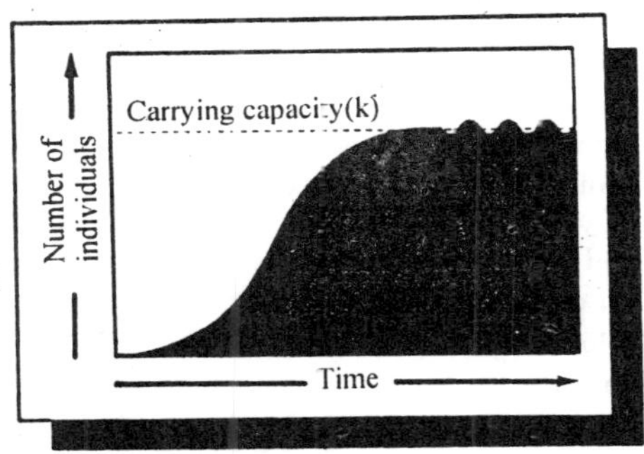

Fig. 9.7 The sigmoid growth curve.

Processes such as competition for resources, emigration, and accumulation of toxic waste products all tend to increase as a population approaches its carrying capacity for a particular habitat. The resources for which the members of the population are competing may be food, shelter, light, mating sites, mates, or any other factor necessary for the species to carry out its life cycle and reproduce itself. Among animals, territoriality, which is the result of competition for food and other resources between members of a species, limits the size of the population.

Population Strategies

The terms r (the intrinsic rate of increase) and K (the carrying capacity of the environment) have already been introduced in this chapter in the equations for population growth. These terms may also be applied to different species depending on the 'strategies' they adopt in order to maximise their chances of long-term survival.

At one extreme of the continuum of population strategies are the *r-species* (also known as *r-strategists*) (Table 9.4). The *r*-species are so called because they reproduce very rapidly and therefore have a high r value. They generally have a survival

Table 9.4 Some characteristics of *r*- and *K*-species

r-species	*K-species*
Rapid reproduction	Slow reproduction
Many offspring	Few offspring
No parental care	Parental care
Short generation time	Long generation time
High r value	Low r value
Reproduction rate not sensitive to population density	Reproduction rate sensitive to population density
Population size may exceed K and then crash	Population size tends to stay close to K
Good dispersal	Poor dispersal
Poor competitors	Good competitors
Exploit temporary habitats	Found in stable habitats
Small size	Large size
Short-lived	Long-lived
Examples	
Annual plants	Trees
Flour beetles	Albatrosses
Bacteria	Man

r = intrinsic rate of increase.
K = carrying capacity of the environment

curve approximating to Type III. The *r*-species include the pioneer species which exploit new or disturbed habitats and are typical of the early stages of ecological succession.

The *r*-species quickly exhaust the resources of the habitats they colonise and can exceed the carrying capacity of the environment (*K*) if they remain too long. When local conditions start to deteriorate, *r*-strategists usually disperse to colonise other unexploited habitats. They are poor competitors. Bacteria, annual weeds and aphids are all good examples of *r*-species.

K-species (also referred to as *K-strategists*) are good competitors. They occupy more stable habitats than those favoured by *r*-strategists. As their name suggests, they exist at levels close to the carrying capacity of the environment (*K*). They have low values of *r* and their reproductive rate is sensitive to population density. Their survival curves are generally Type I. In ecological succession, *K*-species are typical of the later, more mature communities. *Diomedea exulans* (the wandering albatross) is a good example of a *K*-strategist. It takes several years to reach maturity and, when it does, only lays a single egg every two years.

In reality, most species exhibit a mixture of *r* and *K* characteristics and therefore occupy intermediate positions on the *r*-*K* continuum. Nonetheless, this concept is a useful one in population and community ecology.

Population Regulation

Natural populations do not increase in size indefinitely but encounter in due course, environmental resistance, which places a ceiling on population size. There are several factors which can produce environmental resistance and these will be examined in turn. Only the first, intraspecific competition, originates from within the population itself. It is known therefore as an intrinsic factor and acts in a density-dependent fashion. All other factors originate from outside and are termed extrinsic factors. These may involve interactions with other populations, i.e. interspecific competition or predation or may be physical effects of climate.

It should be noted at this early stage that the factors responsible for population regulation rarely operate singly. Usually a combination of factors are involved. However, amongst these, there is often one factor deemed to be the most influential. This is known as the *key factor* and in most cases it exerts its effect through increased mortality, rather than decreased natality.

Intraspecific Competition

Intraspecific competition occurs between individuals belonging to the same species. In single-species laboratory experiments, intraspecific competition over dwindling food supplies and the build-up of toxic waste products are responsible for the eventual levelling out of population size. These intrinsic factors act in a density-dependent fashion. This means that as the population size increases, so too does the pressure on food supply and the deleterious production of toxic wastes.

In natural populations, intraspecific competition over space can regulate population size on a local scale. This may result from either territorial behaviour or overcrowding

Territorial behaviour occurs in a wide range of animals. *A territory* is an area of suitable habitat which is defended by the occupant(s) against intruders of the same species. Territories may be occupied permanently. For example, breeding pairs of *Strix aluco* (tawny owls) stay within a fixed exclusive area for the whole of their adulthood. In other territorial animals, territories are only occupied on a temporary basis. This is common in birds, for example *Parus major* (great tits), when territories are established for the duration of the breeding season only. In yet other animals, territorial behaviour occurs sporadically and appears to be correlated with the availability of food, for example in *Vesitaria coccinea* (the iiwi, a species of Hawaiian honeycreeper). To individuals fitness must outweigh the costs of defending it. The benefits are usually ones of adequate food supply and/or attracting mates. The size of occupied breeding territories must be just large enough to provide sufficient food for the rearing of the offspring. The costs are mainly ones involving defence.

In a particular habitat, for example, a mixed woodland, the territorial behaviour of a particular species will result in a mosaic pattern of defended patches. In a number of studies, nearest neighbour analysis has shown that individuals are not distributed in a random fashion but are spaced out by their territorial behaviour, for example in great tits (Fig. 9.8).

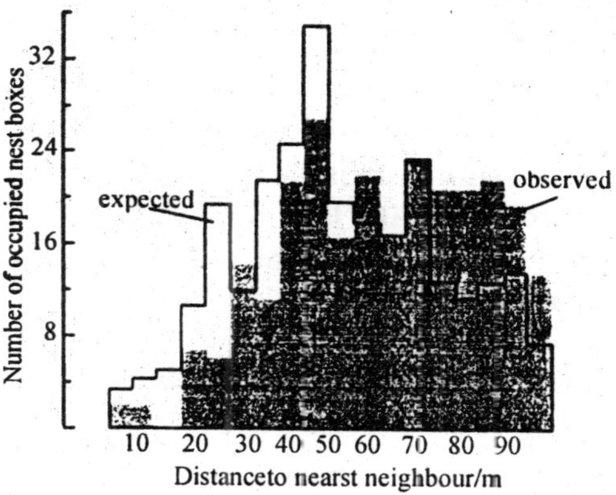

Fig. 9.8 Distribution of *Parus major* (great tits) in Wytham Woods, Oxfordshire.

Such behaviour serves to limit the population density within a given area. This has been demonstrated in a number of studies, where territories have become available through the natural death or experimental removal of their occupants. New individuals quickly establish themselves in the unoccupied territories. In the case of *Lagopus lagopus scoticus* (red grouse), replacements were non-territorial grouse which otherwise would have failed to breed. The possession of a territory therefore conveys a selective advantage to the occupant. From the population dynamics point of view, territoriality regulates population density on a local scale. It is important to emphasise that this is a product, and not a function, of territorial behaviour.

Overcrowding within populations can also result in population regulation through a variety of mechanisms. In laboratory conditions, *Rattus norvegicus* (brown rats) show a very marked decline in fecundity when the population reaches a certain critical density. This occurs despite the presence of adequate food supplies. Hormonal changes adversely affect the reproductive behaviour of the rats. They may fail to breed at all, or if they do they may eat their young or abandon them prematurely. These changes are accompanied by an increase in aggression. This type of regulation may prevail in natural populations under similar conditions of overcrowding, for example in *Microtus* spp. (voles).

Interspecific Competition

Interspecific competition occurs when individuals from two or more *different* species compete over limited resources, for example of food and water.

The classical model for two-species competition was put forward, independently, by Lotka (1925) and Volterra (1926). They assumed that each species on its own would increase in accordance with the model already represented. However, together they would interact in accordance with the following equations:

$$\frac{dX}{dt} = \frac{r_x X(K_x - X - \alpha Y)}{K_x}$$

for species *X*, and

$$\frac{dY}{dt} = \frac{r_y Y(K_y - Y - \beta X)}{K_y}$$

for species *Y*, where α and β are the *competition coefficients*.

In Equation, the rate of change dX / dt is zero when $Xt = K$. In the above equations, dX / dt is zero in species *X* when $X + \alpha Y = K_x$ and in species *Y* when $Y + \beta X = Ky$.

The Lotka-Volterra model for two-species competition predicts four possible outcomes (Fig. 9.9). The points at which there is no change in number of individuals in population *X* of *Y* (i.e. where $dX / dt = 0$ and $dY / dt = 0$ respectively) are represented on the graphs by the *zero growth isoclines*. It is the relative position of these lines and whether or not they intersect which determines the outcome of the model.

If the isoclines do not intersect, the two species cannot reach an equilibrium. The species whose isocline is underneath the other becomes extinct (Fig. 9.9 **a&b**). If the isoclines do intersect, then two populations can coexist at that point. The equilibria reached may be stable or unstable (Fig. 9.9 **c&d**).

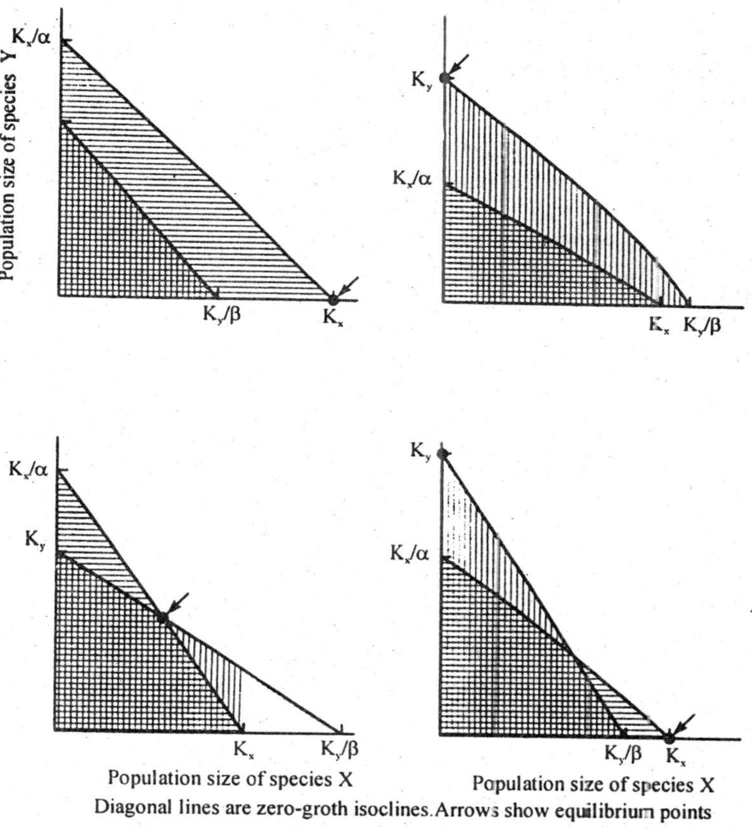

Fig. 9.9 Four types of outcome for the Lotka-Volterra two-species competition model.

Theoretically, models of two-species competition predict that either one species become extinct or that the two species may be able to coexist. Laboratory experiments have used a range of organisms, namely protozoans, yeasts, insects and plant, to examine the effects of interspecific competition between pairs of similar species. In most experiments, one of the species became extinct.

In the late 1940s, Park performed a series of competition experiments using the flour beetles *Tribolium confusum* and *Tribolium castaneum*. At a temperature of 29.5°C, *T. confusum* became extinct (Fig. 9.10(a). However, in later experiments, Park was able to alter the outcome of the competition by varying external physical factors of temperature and humidity. For example, at temperatures above 29°C, *T. confusum* was driven to extinction, whereas below 29°C, *T, castaneum* died out.

A further complication which affected the outcome of these experiments was the presence or absence of the parasite *Adelina*, a protozoan which attacks flour beetles. *T. castaneum* was more severely affected by the parasite. Park was therefore able to reverse the outcome of the experiment performed at 29.5°C. In the presence of *Adelina, T. castaneum* usually became extinct.

Park continued his investigations using different genetic strains of the two *Tribolium* species and found this factor too could influence the outcome of competition experiments. It should be noted here that the outcome of all these experiments was not absolute; rather the probability of one species outcompeting the other was higher depending on the experimental conditions.

In some laboratory experiments, coexistence between two species has been reported. Such an outcome seems to hinge on there being some slight difference between the requirements of the two species concerned. For example, Gause (1935)[3] found that coexistence between the protozoans *Paramecium bursaria* and *Paramecium aurelia* in a tube of yeast was possible because of a difference in their feeding habits: *P bursaria* utilised the bottom yeast layers whilst *P. aurelia* fed on the yeast suspension at the top of the tube.

In laboratory experiments, two-species competition often leads to the extinction of the 'inferior competitor' by the 'superior competitor. However, in the field situation, instances of the coexistence of similar species have been recorded, for example the many herbaceous species found living in intimate proximity in meadows. If resources are not limited in nature, then a lack of competition between the species could explain their coexistence. Alternatively, the explanation could be that intense competition in the past has caused slight differences between species to evolve, thus accounting for their current coexistence.

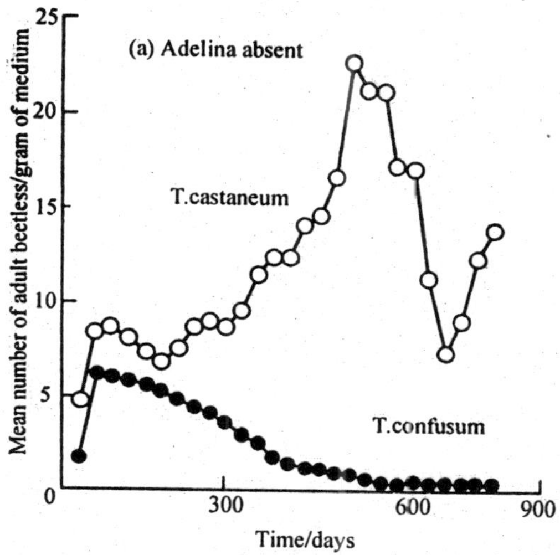

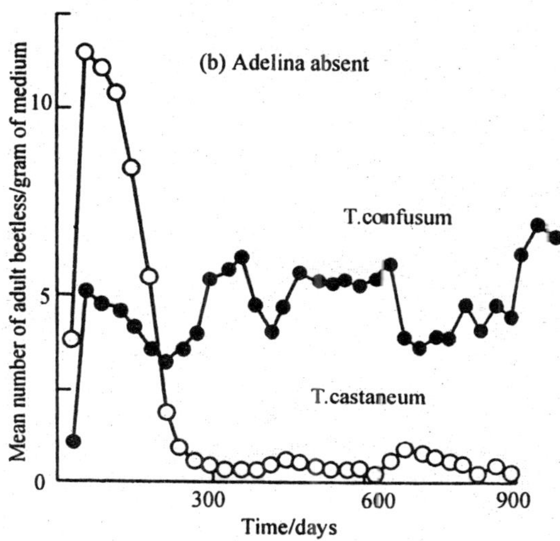

Fig. 9.10 Competition between two species of flour beetle, *Tribolium confusum* and *T. castaneum*, at 29.5°C.

In some natural populations, interspecific competition has been demonstrated to occur. One example is the competition between two species of barnacle, *Chthamalus stellatus* and *Balanus balanoides*.

Dispersion

Individuals in a population are dispersed (spread out) across the space they occupy. We can identify several different *dispersion* patterns.

Clumped distribution is the most common type of dispersion observed among organisms. Clumping may result because individuals of a species tend to form social groups. Animals that illustrate this pattern include bison, elk, caribou, wildebeests, and zebras, which live in herds of varying sizes. Clumping also results because few environments are uniformly suitable, and organisms occupy only the patches of favourable habitat. For instance, colonially nesting birds, such as gulls and swallows, congregate on suitable nesting sites. Plants also exhibit clumped distribution, especially plants such as strawberries and day lilies that reproduce vegetatively.

Even within clumped distribution, however, some spacing between individuals is maintained. Most animals seem to have a minimal distance within which they do not tolerate other individuals. Plants, as well as animals that remain attached to a substrate, require minimal distances for proper growth and development. For example, when trees grow very close together their growth may be stunted or some of them may be crowded out by more vigorous individuals.

In a *uniform* distribution pattern, which is less common than clumping, individuals of a population are more or less evenly spaced. Animals that defend a territory tend to be uniformly spaced because they divide the available habitat among members of the species. In desert communities and in dense stands of trees, plants tend towards uniform spacing, in many cases because older plants produce toxins that inhibit the establishment of seedlings near them. For example, the roots of desert creosote bushes produce substances that inhibit germination of seeds in the vicinity of growing plants.

Random distributions are rarely found in nature. In a random distribution, the location of one individual has no influence on the location of any other, and there is no tendency for individuals to attract or repel other individuals. Furthermore, the habitat is so uniform that it plays no part in determining the location of individuals. Some invertebrates of the forest floor and certain forest trees randomly distributed.

Population Patterns in Time

Populations also vary with time. Let us use populations of organisms in a woodland in the northern United States as an example. The populations of birds in the woodland in summer are somewhat different from those found there in winter. Robins, wood thrushes, and vireos are present in the summer, but as winter approaches they leave for warmer places. During the winter, populations of other birds, including pine siskins and evening grosbeaks, inhabit the woods.

Since animals are mobile, the seasonality of some animal species, such as our woodland birds, is due to local and long-range migratory movements. However, other species, including various insects, are active in summer but spend the winter in a state of dormancy.

Plants populations also change seasonally in the woodland. In early spring, plants such as large-flowered trillium, yellow violet, and spring beauty grow and flower. In the fall, however, they seem to have disappeared, and instead we see white wood aster, gray-stemmed goldenrod, and white snakeroot in flower. As winter approaches, plants enter a state of dormancy that allows them to survive the harsh season ahead.

Population Density

Another important dispersion characteristic of a population is its *density* (Fig. 9.11). Density is the number of organisms per unit area being studied. If we refer to 25 zebras per square kilometer or 150 mice per hectare, we are making a statement of *crude density*. Although crude density statements give us some information about a population, they do not describe the dispersion of the population *within* the area. Are the mice in the example scattered over the entire hectare, or are they clumped only within certain smaller portions of the area? Because organisms are generally confined to suitable habitat patches within any given area, we can also express density in terms of *ecological density*, the number of organisms per unit of suitable habitat. Thus, crude densities can be very misleading unless information is available about the dispersion of the population and the amount of suitable habitat available.

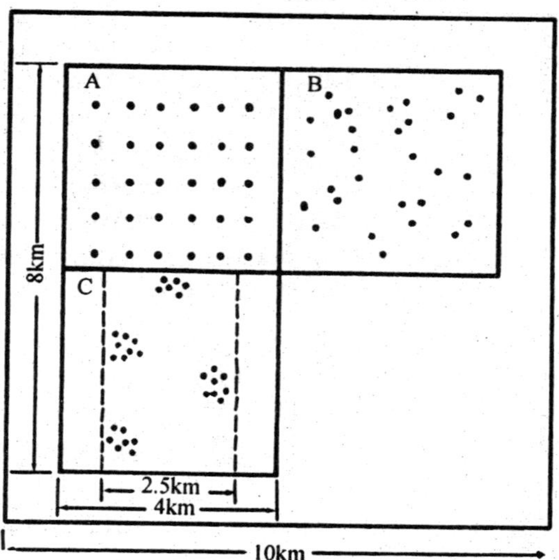

Fig. 9.11 Population distribution and density.

POPULATION FLUCTUATIONS, CYCLES AND CHAOS

Most real populations are not at a fixed equilibrium density for very long, if at all, but are dynamic and changing. Populations may be expanding or contracting because of changes to environmental conditions or because of predation or interspecific competition.

Population Fluctuation

Populations may fluctuate around their carrying capacity for a number of reasons:

(i) a time lag between a change in density and its effects on the population size, or *delayed* density dependence;

(ii) *overcompensating* density dependence

(iii) environmental stochasticity.

Delayed Density Dependence

Fluctuations can be easily induced in theoretical populations by introducing a delay between a change in density and density having an impact on the birth or death rate. The population can overshoot the carrying capacity and then show gradually diminishing, damped oscillations before eventually stabilizing at equilibrium.

In some cases the cycles can persist. The larch bud moth occurs in forests in Switzerland. The larvae emerge in the spring simultaneously with the flushing of the larch. Feeding has an effect on the physiology of the larch, reducing the needle size and their food quality *in the following year* (Fig. 9.12). High larval density results in poor host quality for consecutive years, causing the larch but moth population to crash. The low larval numbers allow the larch to recover which in turn leads to an increase in the number of larvae in response to improving food quality.

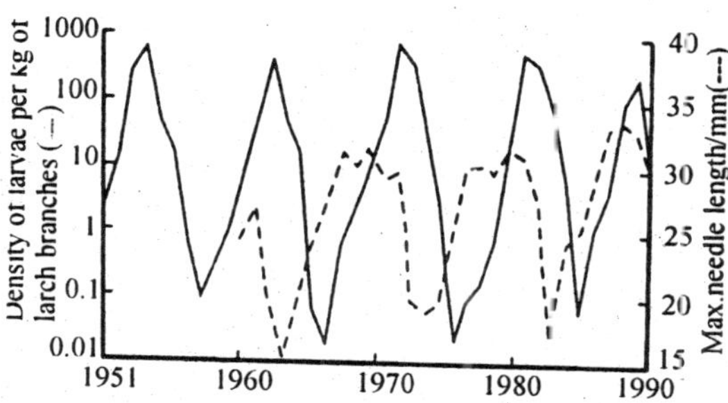

Fig. 9.12 Cycles of abundance of the larch bud moth in response to larch quality (needle length).

The delayed density dependence also produces cycles in predator and prey abundance (Fig. 9.13). A classic example of a predator-prey cycle is illustrated by the Canadian lynx (*Lynx canadensis*) and the snowshoe hare (*Lepus americanus*) (Fig. 9.14). In this case, high lynx numbers depress the snowshoe hare population. This in turn causes a reduction in the number of lynx in subsequent years, allowing the hare population to rise again, resulting in a roughly 10-year cycle. However, as with the larch bud moth, the plants the hares eat also influence this cycle. As the hare numbers increase, the food quality of the plant leaf tissue decreases, which decreases the hare's reproductive potential. Thus, snowshoe hare-lynx population cycling is best thought of as the result of three interacting components: plants, hares and lynx.

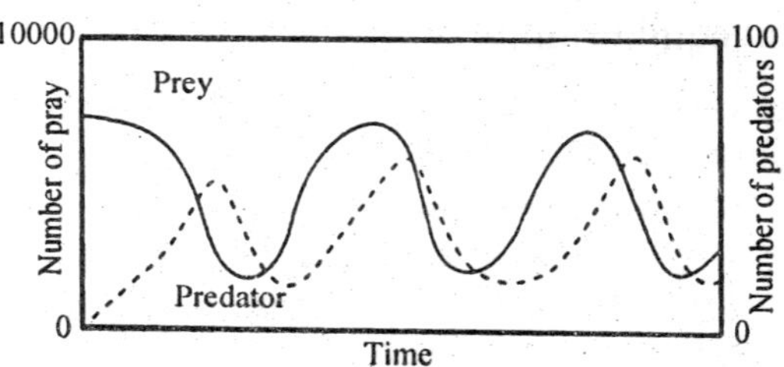

Fig. 9.13 Theoretical predator-prey oscillations

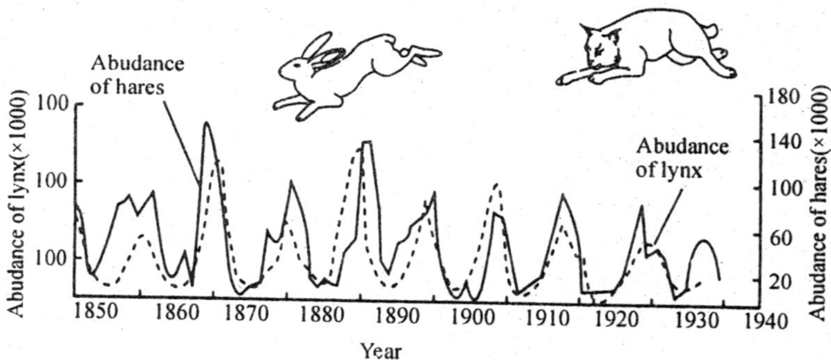

Fig. 9.14 Cycles in the numbers of a predator (Canadian lynx) and its prey (snowshoe hare) over 90 years.

Overcompensating Density Dependence

Density dependence is only stabilizing under certain conditions. Where there is no overcompensating density dependence the population will approach the carrying capacity with no oscillations. As density dependence becomes more overcompensating, damped oscillations and then population cycles occur (Fig. 9.15 **a&b**). These *stable limit cycles* have a fixed interval between each cycle and do not diminish in amplitude over time. Extreme overcompensation in combination with high reproductive rates can lead to chaotic fluctuations with no fixed interval or amplitude (Fig. 9.15c). Chaotic dynamics appears to be random but is actually deterministic in that its occurrence can be predicted.

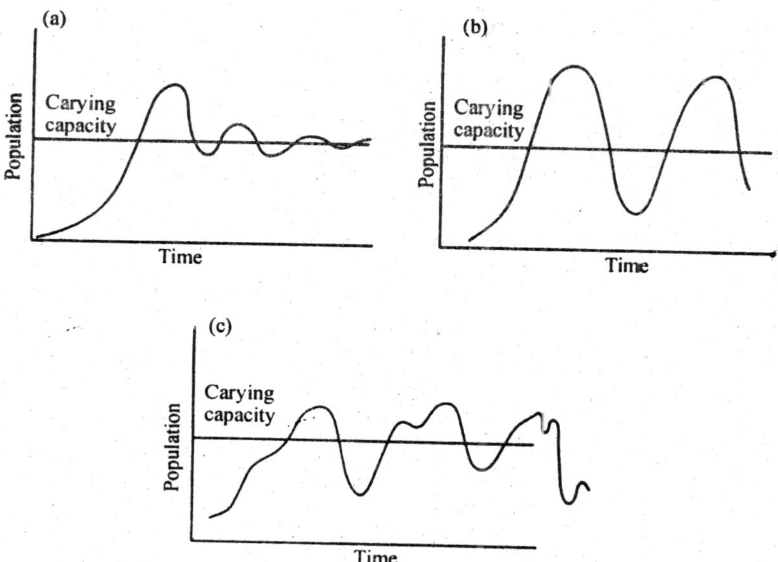

Fig. 9.15 **(a)** Damped oscillations; **(b)** stable limit cycles; **(c)** chaotic dynamics.

Environmental Stochasticity

The carrying capacity of the environment can fluctuate in response to environmental conditions such as the weather. As a result, many real populations experience unpredictable fluctuations in numbers in response to good years and bad years. These populations are said to experience environmental stochasticity, are regulated by environmental conditions and are not predictable, deterministic density-dependent processes.

Small, short-lived organisms are more likely to show dramatic fluctuations in numbers than large, long-lived organisms which are more tolerant of environmental variation. Algae are small, short-lived and able to reproduce rapidly making them sensitive to environmental changes. The algal population fluctuations (Fig. 9.16) are largely driven by temperature variation and consequent changes in nutrient availability.

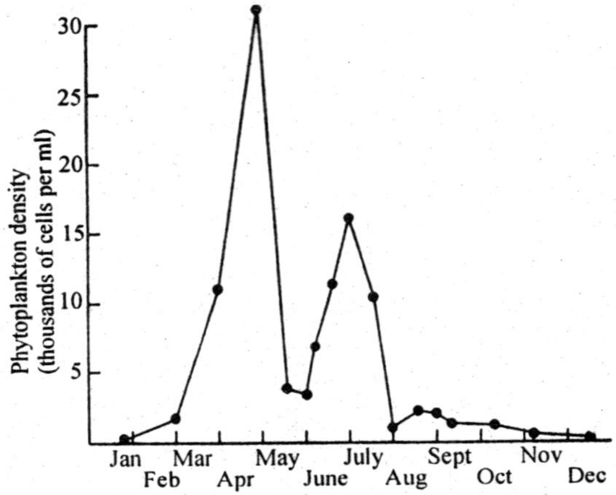

Fig. 9.16 Variation in the number of algae in Green Bay, Wisconsin in response to changes in the environment.

Chaos

Even in an environment which is wholly predictable and constant, population parameters can interact in such a way so as to result in chaotic dynamics. Chaotic systems are not the same as random systems— chaos occurs within fixed limits, so the population is regulated to some extent. However, the outcome of chaos is unpredictable. Two systems may eventually reach two very different equilibrium points because of minor differences in their starting conditions.

There has been a great deal of recent interest in how chaos may help ecologists to understand complex natural dynamics. Detailed mathematical analysis of large datasets of the incidence of measles in New York, USA and Liverpool, England suggests that measles outbreaks are chaotic. The abundance of voles and lemmings fluctuates dramatically and demonstrates some characteristics of chaotic dynamics. However, the prevalence of chaos in natural populations is still unclear, and the mathematical techniques to discern chaos require longer time series than most ecology studies are able to provide. It will not be possible to make long-term predictions about the behaviour of populations until the importance and prevalence of chaos is better understood.

Predation

Predation is another way in which individuals of one species interact with those of other species. We can define predation as one organism (the predator) feeding on another (the prey), such as a cougar feeding on a deer or a robin feeding on an earthworm.

Lotka and Volterra also examined the interactions between predator and prey populations. The mathematical models they developed predict that as a predator population increases, the population of its prey declines. Eventually, the prey population decreases to such an extent that the predator population also declines sharply due to lack of food. At this point the trend is reversed and the prey population increases, with the result that predators also begin to increase. Prey increase the fastest when both predator and prey populations are at their lowest. Predators increase most rapidly when the prey population is at its highest. These interactions produce oscillations, or cycles, in both populations, with the predator population lagging behind the prey population.(Fig. 9.17)

G.F. Gause set out to test these mathematical models experimentally. In the laboratory he raised together populations of two species of protozoans, a predator *Didinium* and its prey *Paramecium*. The predators soon eliminated the prey organism in these cultures, and without a food source, all of the predators subsequently died of starvation. Thus, Gause did not observe the population oscillations predicted by Lotka and Volterra. Only by periodically introducing new individuals of both species into such a culture was he able to generate such oscillations. Gause also discovered that if he provided a 'refuge'

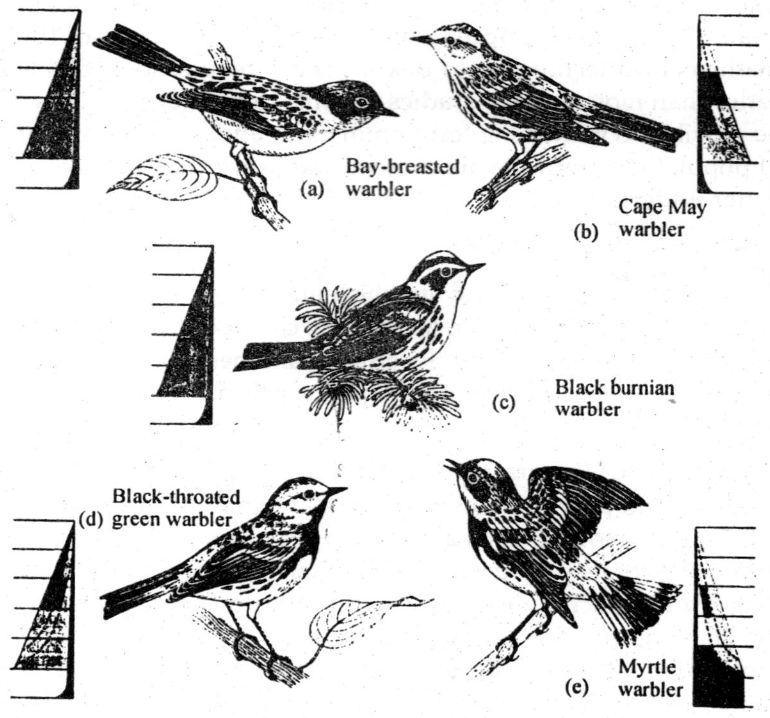

Fig. 9.17 Resource partitioning among five species of coexisting warblers. The time each species spent in various portions of spruce trees was determined. Each species spent more than half its time in the dark green zones.

for the prey, in this case sediment on the bottom of the culture tubes, part of the prey population was able to survive. Under these conditions the predators died of starvation, after which the remaining prey repopulated the medium.

In a series of experiments, Carl Huffaker modified a laboratory environment in an attempt to identify the factors that would prevent a predator from eliminating its prey. First he introduced two species of mites, one of which preys on the other, onto trays containing an array of oranges. The oranges constituted food for the prey species. In such a simple environment, the predators quickly ate all of the prey. Huffaker then designed a more complex environment in which he placed a number of similar-sized rubber balls among the oranges. By changing the number of oranges and rubber balls, covering parts of oranges with paper and sealing wax, and creating barriers of petroleum jelly to restrict the mites' movements, Huffaker could manipulate the complexity of the environment. He found that when food for the prey was not well dispersed (so that the prey population was highly clumped), the predators eliminated the prey. However, when the food was widely dispersed across the tray (so that prey organisms were also widely dispersed), both predator and prey populations fluctuated as the Lotka-Volterra model had predicted. However, even under the best conditions that he could design, Huffaker found that after several oscillations both predator and prey populations died out. Like Gause, Huffaker discovered that a predator population could not be maintained for a great length of time without adding prey from time to time.

If this were not the case, prey would be hunted to extinction or predators would be eliminated by starvation. Recall that when two species interact so closely that evolutionary changes in one influence the direction of evolutionary changes in the other, we call the process *coevolution*.

In natural environments, the hunting abilities of predators are constantly tested. Individual predators that are slow or that lack strength and agility are eliminated from the population (usually through starvation). Through natural selection, predators have acquired such 'tools' as claws, talons, sharp beaks, and shearing teeth that enable them to capture, hold, kill, and process their prey. Some predators have improved their hunting success by acquiring coloration that permits them to blend into their surroundings. Other predator species show *disruptive colouration*. Lines and marks tend to break up the outline of the animal, thus making it less visible to prey.

Colour patterns are also important in many prey species, enabling them to escape detection by predators. Some prey species closely resemble backgound objects. For example, some katydids look much like green leaves, walking sticks resemble twigs, and some moths may look like dead leaves or tree bark.

Avoiding Predation

Some prey species avoid predation by being noxious to predators. An example is the monarch butterfly, which lays its eggs on the leaves of the milkweed, a plant rich in cardiac glycosides. Cardiac glycosides are powerful poisons that are used in minute quantities to treat heart disease in humans. They also activate nerve centers that cause vomiting in many animals. The green caterpillar of the monarch consumes milkweed leaves and is able to incorporate the glycosides into its own tissues, where they remain through the pupal stage and into the adult stage. Thus, from the moment it emerges from its pupal case, the adult monarch butterfly has a built-in chemical defense that causes it to be vomited and rejected by predators. Of course, during the time that predators learn to avoid monarchs some individuals will be killed, but the monarch butterfly population as a whole will escape significant predation.

Unpalatable prey are often highly coloured; such bright colouration warns potential predators that these organisms possess noxious qualities and should be avoided. Once a predator has experienced such prey, it associates the bitter taste or unpleasant experience with the colour pattern and from then on avoids that prey. This is advantageous to both predator and prey. The prey is attacked less frequently, and predators do not waste time pursuing unpalatable prey.

Mimicry

Another means of escaping predation is close resemblance to a species that is protected by some defense mechanism. Such mimics, known as *Batesian mimics*, gain protection by looking and behaving like their models. The viceroy butterfly, for example, is a palatable species that feeds on willows and other nonpoisonous plants. The viceroy closely resembles the unpalpable monarch butterfly, and as a result, predators tend to avoid both model and mimic.

In another type of mimicry, called *Müllerian mimicry*, two or more protected species mimic each other. Such mimicry gives survival advantage to each of the species because the predator soon learns to associate a particular pattern with distastefulness or danger without sampling all of the species involved. Müllerian mimicry is found among a variety of species including wasps with yellow and black bands, many of which have potent stings.

Just as prey animals have developed defenses against their predators, plants have evolved defences against herbivores (animals that eat plants). Many plants possess certain chemical substances that negatively affect the herbivores that eat them. Tannin is one such chemical. When it is combined with plant protein, the protien cannot be digested by herbivores. Tannin also imparts a bitter taste to leaves and twigs and in this way discourages herbivores from consuming plants that contain it. Some plants contain toxic chemicals, such as nicotine, that disrupt the normal function of animal nervous systems. Spines or thorns discourage grazing animals from eating certain plants. Cattle avoid thistles growing in their pastures, and desert herbivores generally are put off by the spiny defenses of cactuses.

Can Predators Regulate Prey?

One of the questions ecologists ask about predator-prey relationships in nature is whether there are natural situations in which predators actually regulate the size of prey populations. Specific examples are hard to find because predation is only one of the factors that regulate populations. Other factors include such things as food supply, cover, and competition.

Human Populations

The size of human populations, like those of other organisms, is controlled by their environment. Throughout history humans have expanded the carrying capacity of the habitats in which they lived because of their ability to develop technical innovations. Earlier in our history, our populations were regulated by both density-dependent and density-independent effects, including food supply, disease, and predators. There was ample room on earth for migration to new areas to relieve overcrowding in specific regions. Unusual disturbances, including floods, extreme temperatures, and droughts also affected the pattern of human population growth.

Gradually, changes in technology have given humans more control over their food supply and enabled them to develop superior weapons to ward off predators, as well as the means to cure diseases. To a lesser extent, migration still plays a role in the adjustment of human populations to particular areas and improvements in transportation and housing have increased the efficiency of migration. At the same time, improvements in shelter and storage capabilities have made humans less vulnerable to climatic uncertainties.

As a result of our ability to manipulate these factors, the human population has grown explosively to its present level of nearly 6 billion people. It is continuing to grow at the rate of approximately 1.7% per year, so that nearly 90 million people are added to the world population annually. At this rate, the human population will double in 40 years.

MORTALITY AND SURVIVORSHIP

A population's intrinsic rate of increase depends on the ages of the organisms in it and the reproductive performance of the individuals in the various age groups. When a population lives in a constant environment for a few generations, its *age distribution*—the proportion of individuals in different age categories—tends to become stable. This distribution differs greatly from species to species and even, to some extent, from place to place within a given species. Depending on the mating system of the species, sex distribution can also have an important effect on population growth statistics. Another important factor in determining a population's rate of growth is *generation time*—the length of time that separates successive generations.

One way to express the characteristics of populations with respect to age distribution is the survivorship curve. *Survivorship* is defined as the percentage of an original population that survives to a given age (Fig. 9.18). In hydra, individuals are equally likely to die at any age, as indicated by the straight survivorship curve. Oysters, on the other hand, produce vast numbers of offspring, only a few of which live to reproduce. However, once they become established and grow into reproductive individuals, their rate of death, or *mortality*, is extremely low. Even though human babies are susceptible to death at relatively high rates, the highest mortality in people occurs later in life, in their postreproductive years.

Many animal and protist populations in nature probably have survivorship curves that lie somewhere between those characteristic of type II and type III. Plant populations, with high mortality at the seed and seedling stages, are probably closer to type III.

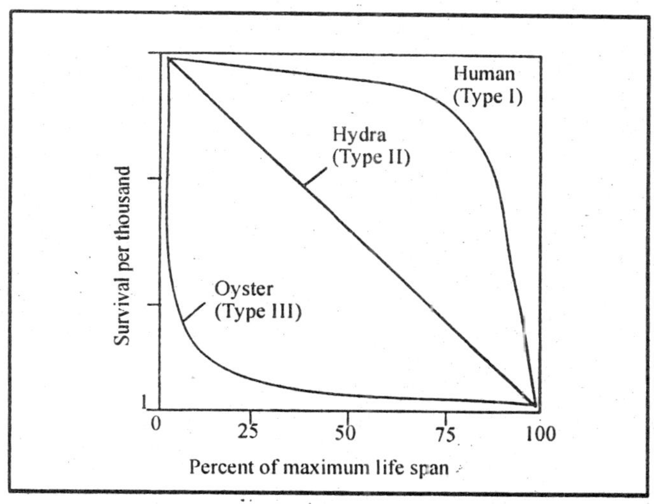

Fig. 9.18 Survivorship curves for the oyster, for a microscopic freshwater animal called the hydra, and for humans. Each population is assumed to have started with 1000 individuals. Humans have a type I life cycle, the hydra type II, and oysters type III. The shapes of the respective curves are determined by the percentages of individuals in populations that die at different ages.

DEMOGRAPHY

Demography is the statistical study of populations. The term comes from two Greek words, *demos*, 'the people' (the same root we see in the word 'democracy') and *graphos*, 'measurement.' It therefore means measurement of people, or by extension, of the characteristics of populations. Demography is the science that helps predict the ways in which the size of populations will alter in the future. It takes into account the age distribution of the population and its changing size through time.

A population whose size remains the same through time is called a *stable population*. In such a population, births plus immigration must exactly balance deaths plus emigration. Not only does the size remain constant, but so does the age structure.

The characteristics of a population can be illustrated graphically by means of human *population pyramid*—a bar graph using single-year or 5-year age categories. In a population pyramid, males are conventionally shown to the left of the vertical age axis, and females to the right. A population pyramid shows the composition of a population by age and sex. By viewing such a pyramid, one can see the historical trends of demographic events such as births and deaths.

In the population pyramid for the United States, the cohort (group of individuals) 55-59 years old is smaller in size than in those preceding (those people born during the Depression) and in following years. The cohorts 25-29, 30-34, 35-39, and 40-44 years old represent the "baby boom." In most human population pyramids, the number of females will be disproportionately large as compared with the number of males. Females in most regions have a longer life expectancy than males.

10
Biomes

The major boimes, which are terrestrial communities that occur over wide areas, are easily recognized by their overall appearance and characteristic climates. These biomes will be reviewed here, as will comparable assemblages that occur in the seas. Each biome, classified primarily by the general features of the vegetation, is similar in its structure and appearance wherever it occurs on earth and differs significantly from other kinds of biomes. Biomes could be classified in a number of different ways; those that we use here are chosen merely as a convenient means for discussing the properties of life on earth from an ecological perspective.

THE GENERAL CIRCULATION OF THE ATMOSPHERE

The distribution of biomes results from the interaction of the features of the earth itself, such as different soil types or the occurrence of mountains and valleys, with two key physical factors: (1) the amount of the sun's heat that reaches different parts of the earth and seasonal variations in that heat; and (2) global atmospheric circulation and the resulting patterns of oceanic circulation. Together these factors determine local climate, including the amounts and distribution of precipitation.

The Sun and Atmospheric Circulation

The earth receives an enormous quantity of heat from the sun in the form of shortwave radiation, and it radiates an equal amount of heat back to space in the form of long wave radiation. About 10^{24} calories arrive at the upper surface of the earth's atmosphere each year, or about 1.94 calories per square centimeter per minute. About half of this energy reaches the earth's surface. The wavelengths that reach the earth's surface are not identical to those that reach the outer atmosphere.

The world contains a great diversity of biomes because its climate varies so much from place to place. On a given day, Miami, Florida, and Bangor, Maine, often have very different weather. There is no mystery about this. The tropics are warmer than temperate regions because the sun's rays arrive almost perpendicular to regions near the equator. Near the poles the angle of incidence of the sun's rays spreads them out over a much greater area, providing less energy per unit area (Fig. 10.1a). Because the earth is a sphere, some parts of it receive more energy from the sun than others. This is responsible for

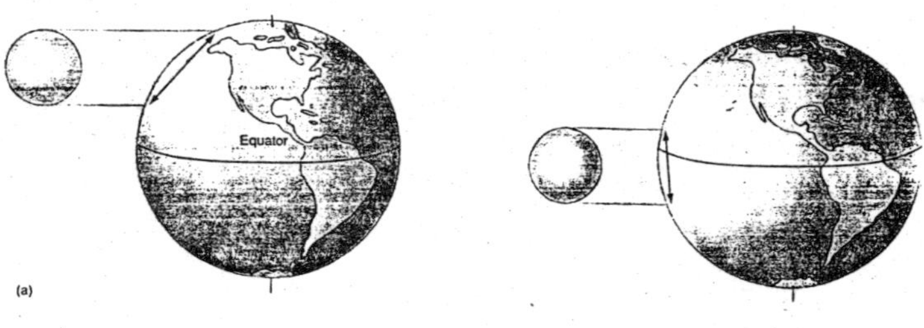

Fig. 10.1(a) Relationships between the earth and the sun are critical in determining the nature and distribution of life on earth. ***(a)*** A beam of solar energy striking the earth in middle latitudes is spread over a wider area of the earth's surface than a similar beam striking the earth near the equator.

many of the major climatic differences that occur over the earth's surface, and, indirectly, for much of the diversity of biomes.

The earth's annual orbit around the sun and its daily rotation on its own axis are both important in determining world climate (Fig. 10.1b). Because of the annual cycle, and the inclination of the earth's axis at approximately 23.5° from its plane of revolution around the sun, there is a progression of seasons in all parts of the earth away from the equator. One pole or the other is tilted closer to the sun at all times except during the spring and autumn equinoxes.

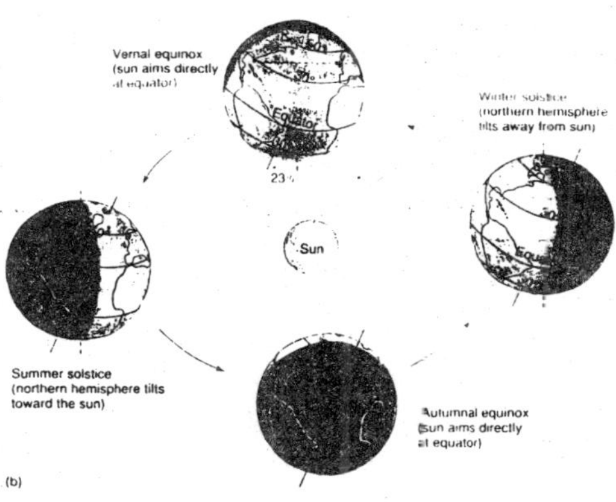

Fig. 10.1(b) The rotation of the earth around the sun has a profound effect on climate. In the northern and southern hemispheres, temperatures change in an annual cycle because the earth is slightly tilted on its axis in relation to its pathway around the sun.

Major Atmospheric Circulation Patterns

The moisture-holding capacity of air increases when it is warmed and decreases when it is cooled. High temperatures near the equator evaporate large amounts of water held in the warm air. As this warm moist air rises and flows towards the poles, it cools and loses most of its moisture (Fig. 10.2ab). Consequently, the greatest amounts of precipitation on earth fall near the equator. This equatorial region of rising air is one of low pressure, called the *doldrums,* which draws air from both north and south of the equator. When the air masses that have risen reach about 30° north and south latitude, the dry air, now cooler, sinks and becomes reheated. As the air reheates it absorbs moisture from the land, creating a zone of decreased precipitation. The air, still warmer than it is in the polar regions, continues to flow toward the poles. It rises again at about 60° north and south latitude producing another zone of high preeipitation at this latitude there is another low-pressure area, the polar front. Part of this rising air flows back to the equator and part continues north and south descending near the poles and producing another zone of low precipitation before it returns to the equator.

Related to these bands of north-south circulation are three major air currents, generated mainly by the interaction of the earth's rotation with the patterns of worldwide heat gain. Between about 30° north latitude and 30° south latitude, the *trade winds* blow, from the east-southeast in the southern hemisphere. The trade winds blow all year long and are the steadiest winds found anywhere on earth. They are stronger in winter and weaker in summer. Between 30° and 60° north and south latitude, the *prevailing westerlies,* often strong winds, blow from west to east and tend to dominate climatic patterns in these latitudes, particularly in lands that lie along the western edges of the continents. Zones of weaker winds, blowing from east to west, occur farther north and south in their respective hemispheres.

Atmospheric Circulation, Precipitation, and Climate

As discussed above, precipitation is generally low near 30° north and south latitude, where air is falling and being warmed, and relatively high near 60° north and south latitude, where it is rising and being cooled. Partly as a result of these factors, all

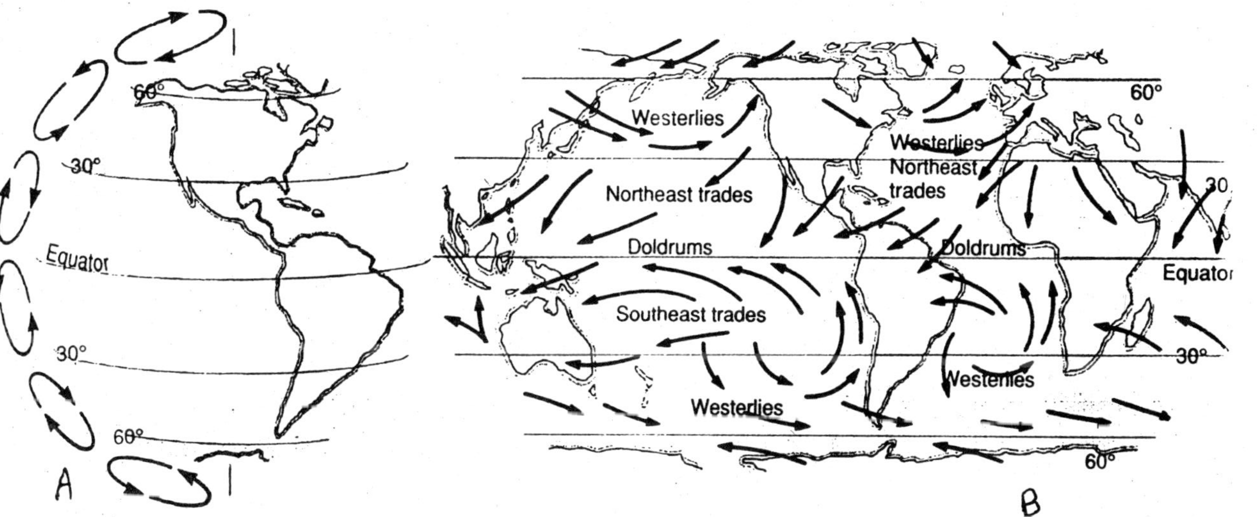

Fig. 10.2: General patterns of atmospheric circulation. **(a)** The pattern of air movement out from and back to the earth's surface. **(b)** The major wind currents across the face of the earth.

the great deserts of the world lie near 30° north or 30° south latitude. Other major deserts are formed in the interiors of large continents. These areas have limited precipitation because of their distance from the sea, the ultimate source of most precipitation. Other deserts occur because mountain ranges intercept moisture-laden winds from the sea (Fig. 10.3). When this occurs the air rises and the moisture-holding capacity of the air decreases, resulting in increased precipitation on the windward side of the mountains—the side from which the wind is blowing. As the air descends the other side of the mountains, the leeward side, it is warmed, and its moisture-holding capacity increases, tending to block precipitation. In California, for example, the eastern sides of the Sierra Nevada Mountains are much drier than their western sides, and the vegetation is often very different; this phenomenon is called the *rain shadow effect*.

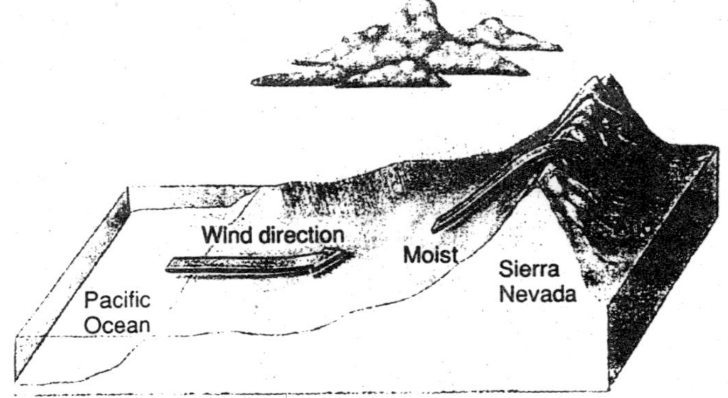

Fig. 10.3 The rain shadow effect. Moisture-laden winds from the Pacific Ocean rise and are cooled when they encounter the Sierra Nevada Mountains. As their moisture-holding capacity decreases, precipitation occurs, making the middle elevation of the range one of the snowiest regions on earth; it supports tall forests, including those that include the famous big trees (*sequoiadendron giganteum*). As the air descends on the east side of the range, its moisture-holding capacity increases again, and the air picks up moisture from its surroundings rather than releasing it. As a result, desert conditions prevail on the east side of the mountains.

Four relatively small areas, each located on a different continent, share a climate that resembles that of the

Mediterranean region. Such a climate is found in portions of Baja California, California and Oregon; in central Chile; in southwestern Australia; and in the cape region of South Africa. In all of these areas the prevailing westerlies blow during the summer from a cool ocean onto warm land. As a result, the air's moisture-holding capacity is increased, so precipitation is limited or completely blocked during the summer. Such climates are unusual on a world scale. In the five regions where they occur, many unique kinds of plants and animals, often local in distribution, have evolved. Because of the prevailing westerlies, the great deserts of the world (other than those in the interior of continents) and the areas of mediterranean climate lie on the western sides of the continents.

Another kind of major regional climate occurs in southern Asia. The monsoon climatic conditions that are characteristic of India and southern Asia occur during the summer months. During the winter the trade winds blow from the east-northeast off the cool land onto the warm sea. From June to October, though, when the land is heated, the direction of the air flow is reversed, and the winds blowing from the east-southeast south of India veer around to blow onto the Indian subcontinent and adjacent areas from the southwest. The duration and strength of the monsoon winds spell the difference between food sufficiency and starvation for hundreds of millions of people in this region each year. Under normal conditions, the monsoons bring heavy rains to the entire area, and the year's crop succeeds; if the rains do not come or are inadequate, the crop fails.

Patterns of Circulation in the Ocean

Patterns of ocean circulation are determined by the patterns of atmospheric circulation just discussed, but they are modified by the location of land masses. Oceanic circulation is dominated by huge surface gyrals (Fig. 10.4) which move around the subtropical zones of high pressure between approximately 30° north and 30° south latitude. These gyrals move clockwise in the northern hemisphere and counterclockwise in the southern hemisphere. The ways in which they redistribute heat profoundly affects life not only in the oceans but also on coastal lands. For example, the Gulf Stream, in the North Atlantic, swings away from North America near Cape Hatteras, North Carolina, and reaches Europe near the southern British Isles. Because of the Gulf Stream, western Europe is much warmer and more temperate than eastern North America at similar latitudes. As a general principle, western sides of continents in temperate zones of the Northern Hemisphere are warmer than their easter sides; the opposite is true of the Southern Hemisphere. In addition, and as we have just seen, winds passing over cold water onto warm land increase their moisture-holding capacity, limiting precipitation in such areas.

In South America the Humboldt Current carries phosphorus-rich cold water northward up the west coast. Phosphorus is brought up from the ocean depths by the upwelling of cool water that occurs as a result of the frequent offshore wind from the mountainous slopes that border the Pacific Ocean. This nutrient-rich current helps make possible the

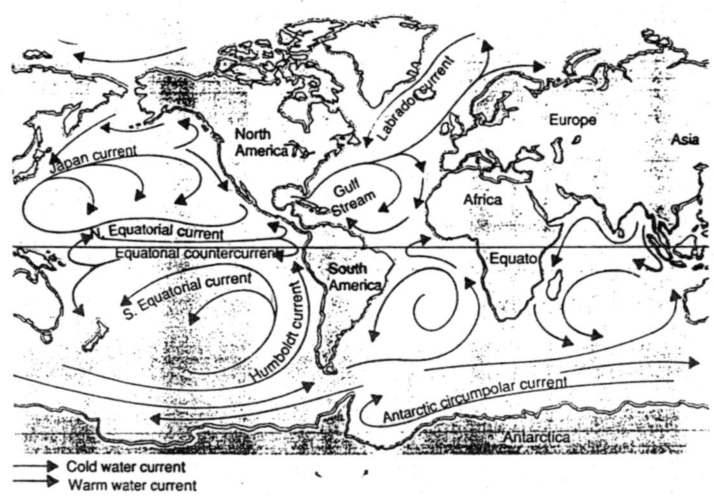

Fig. 10.4 Ocean circulation. The circulation in the oceans moves in great surface spiral patterns called gyrals; it profoundly affects the climate on adjacent lands.

abundance of marine life that supports the fisheries of Peru and northern Chile. Marine birds which feed on these organisms, are responsible for the commercially important, phosphorus rich, guano deposits of these countries. When the coastal waters are not as cool as usual, a devastating phenomenon called 'El Nino' occurs, which influences climatic patterns globally.

THE OCEANS

Nearly three-quarters of the earth's surface is covered by ocean. Oceans have an *average* depth of more than 3 kilometers, and they are, for the most part, cold and dark. Heterotrophic organisms are found at the greatest ocean depths, which reach nearly 11 kilometers in the Marianas Trench of the western Pacific Ocean. Photosynthetic organisms are confined to the upper few hundred meters of water. Organisms that live below this level obtain almost all of their food indirectly, as a result of photosynthetic activities that occur above.

The supply of oxygen can often be critical in the ocean and as water temperatures become warmer, oxygen becomes less soluble. In warmer marine regions of the globe, the amount of available oxygen is an important limiting factor for organisms. Carbon dioxide, in contrast, is almost never limited in the oceans. The distribution of minerals is much more uniform in the ocean than it is on land, where individual soils reflect faithfully the composition of the parent rocks from which they have weathered.

Despite the many new forms of animals, protists, and bacteria still being discovered in the sea and the huge biomass that occurs there, many fewer species live in the sea than on land. Probably more than 90% of all species of organisms occur on land. Each of the largest groups of organisms, including insects, mites, nematodes, fungi, and plants has marine representatives, but they comprise only a very small fraction of the total number of species. On land, barriers between habitats are sharper, and variations in elevation, parent rock, degree of exposure, and other factors have been crucial to the evolution of the millions of species of terrestrial organisms. In other words, there is a greater variety of available niches on land. In terms of overall diversity, the pattern is quite different. Of the major groups of organisms—phyla—most originated in the sea, and almost every one has representatives in the sea. Only a few phyla have been successful on land or in freshwater habitats, but these have given rise to an extraordinarily large number of species.

The marine environment consists of three major kinds of habitats: (1) the *neritic zone,* the zone of shallow waters along the coasts of continents; (2) the *surface zone,* surface layers of the open sea; and (3) the *abyssal zone,* the deep-water areas of the oceans.

The Neritic Zone

The neritic zone of the ocean is the area along the coasts of continents and islands, and less than 300 meters below the surface. The zone is small in area, but it is inhabited by large numbers of species as compared with other parts of the ocean. The intense and sometimes violent interaction between sea and land in this zone gives a selective advantage to well-secured organisms that can withstand being wished away by the continual beating of the waves. Part of this zone, the intertidal, or littoral, region, is exposed to the air whenever the tides recede.

Because of the accessibility to land, the intertidal zone must have been home for the ancestors of the first organisms that colonized terrestrial habitats. Perhaps the greater complexity needed to anchor and fasten the animals and plants that dwell in this zone constituted a kind of preadaptation of life on land, where the environmental stresses are even more extensive. The organisms that live successfully in habitats that are regularly exposed to air must have had some kind of waterproof covering or habits that protect them from the drying action of the air when the tide is out. Such adaptations are of central importance for terrestrial organisms.

The world's great fisheries occur in shallow waters over continental shelves, either near the continents themselves or in the open ocean, where huge banks come near the surface.

Nutrients, derived from land, are much more abundant in coastal and other shallow regions, where upwelling from the depths occurs, than in the open ocean. This accounts for the great productivity of the continental shelf fisheries. The preservation of these fisheries, which are a source of high-quality protein exploited throughout the world, has become a matter of growing concern. In Chesapeake Bay, where complex systems of rivers enter the ocean from heavily populated areas, environmental stresses have become so serious that they not only threaten the continued existence of formerly highly productive fisheries, but also diminish the quality of human life in these regions. For example, increased amounts of runoff from farms and sewage effluent in areas like Chesapeake Bay add large amounts of nutrients to the water. Increased nutrient supply allows for an increase in the numbers of some marine organisms. The increased populations use up more and more of the oxygen in the water and thus may disturb established populations of organisms such as oysters. Such effects may be enhanced by climatic shifts, and large numbers of marine animals may die suddenly as a result.

Many of the oceans of the world are highly unproductive biologically, resembling deserts in this respect. About three-fourths of the surface area of the world's oceans are located in the tropics. In these waters, where the water temperature remains at about 21°C, coral reefs can occur. These are highly productive ecosystems that can successfully concentrate nutrients, even from the relatively nutrient-poor waters that are characteristic of the tropics. Coral animals, which provide most of the structure of reefs, together with photosynthetic coralline algae and many other kinds of organisms that live in and around the reefs, make up one of the most complex and fascinating living systems on earth. Unicellular organisms known as dinoflagellates that live symbiotically within the coral animal contribute a great deal of the reefs productivity, as do other algae that grow among the coral animals.

The Surface Zone

Drifting freely in the upper, well-illuminated waters of the ocean, a diverse biological community exists, primarily consisting of microscopic organisms called *plankton*. Fish and other larger organisms that swim in these waters constitute the *nekton*, whose members feed on plankton and one another. Together, the organisms that make up the plankton and the nekton provide all the food for those that live below. Some members of the plankton, including algae and some bacteria, are photosynthetic. Collectively, these organisms account for about 40% of all photosynthesis that takes place on earth. Most plankton occur in the top 100 meters of the sea, the zone into which light from the surface penetrates freely. Perhaps half of the total photosynthesis in this zone is carried out by organisms less than 10 micrometers in diameter—at the lower limits of size for organisms—including cyanobacteria and the smallest algae. Such organisms are so small that their abundance and ecological importance have been unappreciated until relatively recently.

Many heterotrophic protists and animals live in the plankton and feed directly on photosynthetic organisms and on one another. Gelatinous animals, especially jellyfish and ctenophores, are abundant in the plankton but are relatively poorly known because their fragility makes them difficult to collect and study. Their bodies may consist of as much as 95% water. The largest animals that have ever existed on earth, whales, graze on plankton and nekton as their only source of food. They are heavier than the largest dinosaurs known. A number of heterotrophic, free-swimming organisms, such as fishes and crustaceans, move up into the plankton at times—in some instances on a regular cycle—to feed on the organisms there.

Populations of organisms that make up plankton are able to increase rapidly, and the turnover of nutrients in the sea is so great that the amount of productivity in these systems, although it is still low, has been seriously underestimated in the past. Even though nitrogen and phosphorus are often present in only small amounts and organisms may be relatively scarce, this productivity reflects rapid use and recycling rather than abundance of these nutrients. The smallest organisms turn over phosphorus much more rapidly than do larger ones, and their role in these complex and productive ecosystems is just starting to be understood properly.

The Abyssal Zone

The area of sea floor at depths below 1000 meters is about twice that of all the land on earth. The sea floor itself is a thick blanket of mud, consisting of fine particles that settle from the overlying water and have accumulated over millions of years. Because of high pressures (an additional atmosphere of pressure for every 10 meters of depth), cold temperatures (2° to 3°C), darkness, and lack of food, the first deep-sea biologists thought that nothing could live on the sea floor. In fact, although the number of individual animals decreases with increasing depth, the number of species that live at great depth is now known to be quite high. Most of these animals are only a few millimeters in their largest dimension, although larger ones also occur in these regions. Some of the larger ones are bioluminescent and thus are able to communicate with one another or attract their prey.

Many deep-sea animals are known from only a few samples. Today, they are sampled quantitatively by removing cubes of mud from the sea floor. Such samples have revealed a great diversity of animal species, indicating that the deep sea represents a major reservoir of largely unknown biological diversity.

Most organic matter in the plankton is recycled in the surface layers of the ocean. Animals on the sea bottom depend on the meager leftovers from organisms living kilometers overhead. The low densities and small size of most deep-sea bottom animals is in part a consequence of this low food supply. In 1977, oceanographers diving in a research submarine were surprised to find dense clusters of large animals living on geothermal energy at a depth of 2500 meters. These deep-sea oases were located as a result of a discovery of equal importance to our understanding of the chemistry of the oceans. Seawater circulates through porous rock at sites where molten material from beneath the earth's crust comes close to the rocky surface of the Mid-Ocean Ridge—the worldwide feature where basalt erupts through the ocean floor, causing the phenomenon of plate tectonics.

This water is heated to temperatures in excess of 350° C and, in the process, becomes rich in reduced compounds. These compounds, such as hydrogen sulfide, provide energy for bacterial primary production through chemosynthesis instead of photosynthesis. Mussels, clams, and large red-plumed worms in a phylum unrelated to any shallow-water invertebrates cluster around the vents. Bacteria live symbiotically within the tissues of these animals. The animal supplies a place for the bacteria to live and transports CO^2, H^2S, and O^2 to them for their growth. The bacteria supply the animal with organic compounds to use as food. Polychaete worms, anemones, and limpets live on free-living chemosynthetic bacteria. Crabs act as scavengers and predators, and some of the fish are predators. This is one of the few ecosystems on earth not dependent on the sun's energy.

FRESH WATER

Freshwater habitats are distinct from both marine and terrestrial ones, but they are limited in area. Inland lakes cover about 1.8% of the earth's surface, and running water covers about 0.3%. All freshwater habitats are strongly connected with terrestrial ones, with marshes and swamps constituting intermediate habitats. In addition, a large amount of organic and inorganic material continuously enters bodies of fresh water from communities growing on the land nearby. Many kinds of organisms are restricted to freshwater habitats. When organisms occur in rivers and streams, they must be able to swim against the current or attach themselves in such a way as to resist or avoid the effects of current, or risk being swept away.

Ponds and lakes, like the ocean, have three zones in which organisms occur. The *littoral zone* is the shallow area along the shore. The *limnetic zone* is the well-illuminated surface water away from the shore inhabited by plankton and other organisms that live in open water. The *profundal zone* is the area below the limits of effective light penetration.

Thermal stratification is characteristic of larger lakes in temperate regions. In summer, warmer water forms a layer at the surface known as the *epilimnion*. Cooler water, called the *hypolimnion* (about 4°C), lies below. There is an abrupt change in temperature, the *thermocline*, between these two layers. Depending on the climate of the particular area, the epilimnion may become as mush as 20 meters thick during the summer.

In autumn the temperature of the epilimnion drops until it is the same as that of the hypolimnion, 4°C When this occurs, epilimnion and hypolimnion mix—a process called fall overturn. Since water is densest at about 4°C, further cooling of the water as winter progresses results in a layer of cooler, lighter water, which freezes to form a layer of ice at the surface. Below the ice, the water remains between 0° and 4°C, and plants and animals can survive. In spring, the ice melts, and the surface water warms up. When it warms back to 4°C, it again mixes with the water below. This process is known as spring overturn. When lake waters mix, in the spring and fall, nutrients formerly held in the depths of the lake are returned to the surface and oxygen from surface waters is carried to the depths.

Lakes can be divided into two categories, based on their production of organic matter. In *eutrophic lakes* there is an abundant supply of minerals and organic matter. As the abundant organic material drift below the thermocline from the well-illuminated surface waters of the lake, it is used as a source of energy by other organisms. Most of these are oxygen-requiring organisms that can easily deplete the oxygen supply below the thermocline during the summer months. The oxygen supply of the deeper waters cannot be replenished until the waters mix in the fall. The lack of oxygen in the deeper waters of some lakes may have profound effects, such as allowing the conversion of relatively harmless materials such as sulfate and nitrate into toxic materials such as hydrogen sulfide and ammonia.

In *oligotrophic lakes*, organic matter and nutrients are relatively scarce. Such lakes are often deeper than eutrophic ones and have very clear blue water. Their hypolimnion water is always rich in oxygen. Oligotrophic lakes are highly susceptible to nutrient pollution because they naturally have such limited quantities. Excess phosphorus from sources such as fertilizer runoff, sewage, and detergents can quickly lead to harmful effects such as runaway weed growth or the promotion of 'blooms' of algae. These can rapidly deplete the lake's oxygen supply killing off the natural fish population.

BIOMES

Biomes are climatically delineated assemblages of organisms that have a characteristic appearance and that are distributed over a wide land area. Biomes are classified in several ways, but, for our purposes, we will discuss them under fourteen categories. In-depth descriptions are given for seven major categories: tropical rain forest, savanna, desert, temperate grassland, temperate deciduous forest, taiga, and tundra. The remaining seven categories are described briefly: chaparral, polar ice, mountain zone, temperate evergreen forest, warm moist evergreen forest, tropical monsoon forest, and semidesert. They vary remarkably from one another because they have evolved in regions with very different climates (Fig. 10.5).

If there were no mountains and no climatic effects caused by the irregular outlines of continents and by different sea temperatures, each biome would form an even belt around the globe. In truth, these factors greatly affect the distribution of biomes, especially elevation. Biomes that normally occur at high latitudes follow an altitudinal gradient along mountains. The summits of the Rocky Mountains are covered with a vegetation type that resembles tundra, while other forest types that resemble taiga occur farther down. We will now examine some of the distinctive features of the individual biomes.

Boreal Forest

To the south of the tundra, and below it on temperate zone mountains, is an extensive belt of forest dominated by coniferous trees. Most of these are evergreen, but large expanses of Siberian boreal forests are dominated by deciduous larches. Boreal forest climates are quite varied, but in most places the winters are bitterly cold and the summers short but often warm.

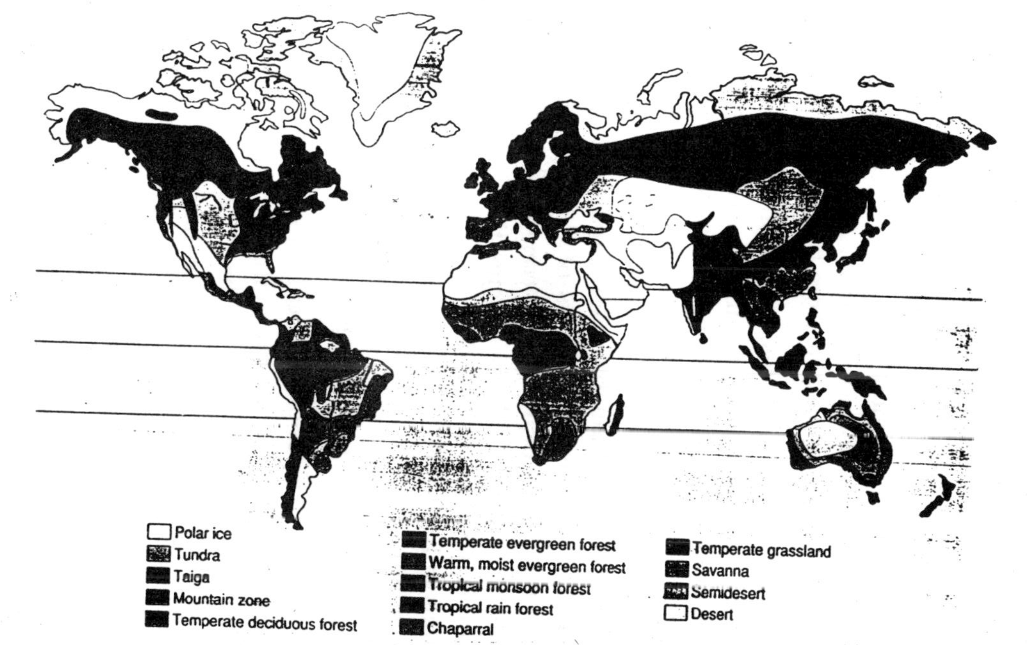

Fig. 10.5 The distribution of biomes. Each biome is similar in structure and appearance wherever it occurs on earth.

The coldest winters in the world and the greatest seasonal changes in temperatures occur in the boreal forests of Siberia. Coniferous forests occur also along the western coasts of continents at mid to high latitudes where winters are mild but very wet and summers are cool and relatively dry. These forests include the tallest trees in the world and support the highest standing biomasses of wood of all ecological communities. In the Southern Hemisphere, the southern beeches are remarkably similar in growth form to conifers. At a distance these forests are almost indistinguishable from true coniferous forests.

Whatever their climates, boreal forests are characterized by having a small number of common tree species, nearly all of which are wind-pollinated and wind-dispersed. Little of plant resources is devoted to producing structures attractive to animals, and the communities are dominated by folivores and their predators. Although conifers produce cones, they do so on an irregular basis, with years of large crops separated by intervening years of little or no cone production. Eaters of the seeds of conifers include species that store seeds, such as squirrels, jays, and nutcrackers, and species that exploit local regions of high abundance and undergo periodic irruptions as cone supplies fail, such as crossbills.

Temperate Deciduous Forest

The eastern sides of continents of mid latitudes have strong seasonal fluctuations of temperature and ample precipitation that is fairly evenly distributed throughout the year. Deciduous trees, which lose their leaves during the cold winter but produce leaves well adapted for rapid photosynthesis during the warm, moist summers, dominate these environments. Tree species richness is much greater in deciduous forests than in boreal forest, and many of the trees have animal-distursed pollon and fruits. These forests were disturbed extensively during the glacial advances of the Pleistocene. The forests richest in species occur in the southern Appalachian Mountains of the United States and in eastern China and Japan, south of the areas of glaciation.

Seasonal cycles of activity are striking in deciduous forests, the most conspicuous being the changes in the leaves of the trees themselves. Many birds migrate into this biome in the summer when insect activity is intense. Understory plants often flower and fruit early in the spring before the canopy above them has leafed out, but most trees and shrubs produce their fruits in the autumn and early winter, relying on southward migrating or overwintering birds for their dispersal.

Grassland

Grasslands occur in many climates where potential evaporation exceeds precipitation. Some grasslands are found in regions where virtually all precipitation falls in winter, whereas others occur where virtually all of it falls in summer. Some occur in regions with minor seasonal variations in temperature but others, such as the American Great Plains and the steppes of the Soviet Union, occur where seasonal changes in temperature are extreme. The key lies not in climate itself but in the resistance of grasses to fire and heavy grazing. Many grasslands revert to shrubby and woody communities if they are protected from fire and grazing.

Grasslands are structurally simple because they are dominated by nonwoody plants that die back to the ground in unfavorable periods, but they are rich in species of both grasses and *forbs:* broad-leaved species that live together with grasses. Many of these produce showy flowers, and prairies and often riots of color in late summer when forbs are in full bloom. Because all parts of the dominant plants of grasslands are eaten readily by animals, grasslands in their natural states support large populations of grazing mammals. The fraction of primary production that is eaten by herbivores is much higher than in any other terrestrial ecological community.

Cold Deserts

At mid to high latitudes, dry regions are located in the interior of large continental land masses, usually in the lee of mountain ranges. They receive light, usually fairly evenly distributed precipitation and experience great seasonal changes in temperature. They are dominated by a few species of low-growing shrubs. There is usually enough moisture to recharge the surface layers of the soil in winter, and plant growth is concentrated in spring. By early summer these deserts are often rather barren, the scant summer rains being insufficient to support much vegetation growth during the hot summers.

Cold deserts are relatively poor in species in most taxonomic groups, and because of the low rainfall, overall productivity is low. Most areas, however, have rich soils because they are young and unleached. If irrigation water is available, they are good for agriculture. Very few fleshy fruits are produced, but seed-eating birds, ants, and rodents are common.

Hot Deserts

The hot deserts of the world are concentrated in a belt around 30°N and 30°S latitude, where air from aloft is descending, warming, and picking up rather than releasing moisture. Most of them receive the bulk of their rainfall in summer when the intertropical convergence zone moves poleward but they also receive winter rains from the low-pressure cells that form over mid-latitude oceans in winter. The driest regions occur where neither summer nor winter rains penetrate regularly—such as the center of Australia and the middle of the Sahara Desert. Hot deserts support a rich and structurally diverse vegetation, except in the driest areas, and annual plants are particularly abundant. Animal-pollination and animal-dispersal of fruits are common, and great quantities of seeds are produced. Population densities of rodents and ants are often remarkably high. Lizards are common members of the diurnal fauna and snakes are common nocturnal predators.

Succulent plants that store large quantities of water in their expandable stems and carry out photosynthesis primarily with stems rather than leaves are conspicuous in many hot deserts, but they are quite rare in deserts growing on ancient and nutritionally poor soils. Australia, for example, is virtually lacking in stem-succulent plants. The reasons for this absence are not clear because the introduced stem-succulent *opuntia* cactus from the New World is a very successful invader there.

Mediterranean Biome

On the western sides of continents at moderate latitudes, where cold ocean waters circulate offshore, is found a climate that is mild and wet in winter and hot and dry in summer. It is called the Mediterranean type of climate because it was first known from that region, but it also occurs in California, central Chile, southwestern Africa, and southwestern Australia. In all areas it is dominated by low-growing shrubs with tough evergreen leaves. Annuals are conspicuous in winter and early spring. The shrubs also carry out most of their growth and photosynthesis in early spring, which is when insects are active and birds breed.

Most shrubs of Mediterranean-type vegetation produce bird-dispersed fruits, which ripen in late fall or winter when large numbers of migrants arrive from the north. One such fruit—the olive—has played a very important role in human history, providing a rich food source for people at what would otherwise be a period of low food availability. The abundant growth of annual plants in Mediterranean climates also provides a rich source of seeds that are readily stored during the hot, dry summers. Therefore, these communities support large populations of small rodents, most of which store seeds in underground burrows.

Thorn Forest

The remaining biome types are found at low latitudes where seasonal temperature fluctuations are small and the annual cycle is dominated by wet and dry seasons. In general, the length of time that a region is close to the intertropical convergence zone and, hence, receives rainfall, increases toward the equator. Thorn forests occur at high enough latitudes that the rainy season is brief, and no rain may fall for 8 or 9 months of the year. Typically, they are found equatorward of hot deserts and contain many plants similar to those found in the deserts. Animal activity is intense during the rainy season.

Africa is unusual among the tropical continents in having very large expanses that receive very low rainfall even though they are located at very low latitudes. These environments support large populations of mammals that exert a powerful impact on the community through grazing and browsing. Many of these areas revert to dense thorn forest if they are protected from fire and the activities of large mammals. Many of the herbaceous species are well adapted to heavy grazing pressures and persist only because of it.

Tropical Deciduous Forest

As the length of the rainy season increases, thorn forest is replaced by deciduous forest, which is taller in stature, has fewer succulent plants, and is much richer in species. Most of the trees, except those growing along rivers, lose their leaves during the dry season, but many of them flower then. Most trees are animal-pollinated and have animal-dispersed fruits. The community is very rich in species of both plants and animals. The long dry season is very hot and often quite windy, producing severe desiccation (drying out), even in regions that receive as much as 190 centimeters (75 inches) of rain annually.

Soils in these regions are less leached than those in winter areas. Where the underlying rocks are relatively young, these provide some of the best tropical soils for agricultural purposes. As a result, many tropical deciduous forests have been cleared for cattle ranching and crops. Where dry season water is available, irrigation is employed to yield a dry season crop as well as a rainy season one.

Tropical Evergreen Forest

Where total rainfall exceeds 250 centimeters (100 inches) of rain annually and the dry season lasts no more than a few months, evergreen forests predominate. These are the richest of all ecological communities in species of both plants and animals, and overall biological productivity is the highest of any ecological community. The soils, except where they are very young, are deeply weathered, and few minerals are released by further weathering of the soil. Therefore, these soils are usually poor for modern extractive agriculture without massive applications of fertilizers.

Tropical wet forests may support up to 300 species of trees per square kilometer, most of which are rare and nearly all of which rely on animals for transport of pollen and dispersal of fruits. Food webs are extremely complex, and many of the animal species living in these forests are yet to be described and named by scientists.

Tropical Montane Forest

Temperatures steadily drop with altitude on the slopes of tropical mountains until, if they are fall enough, they support a tundra on their summits. Biological productivity actually increases at moderate elevations because nighttime temperatures are low enough that nocturnal respiration by plants is reduced whereas daytime temperatures are still high enough that the rate of photosynthesis is not diminished. For many groups of animals, such as insects and birds, the largest numbers of species are found at approximately 1000-1500 meters of elevation rather than at sea level.

With increasing altitude on tropical mountains, the height of forests decreases, as does the importance of vines. Leaf sizes become smaller, and the growth of epiphytes becomes more luxuriant, especially at those elevations where clouds regularly form during the day and bathe the forest in moisture. At these elevations productivity is depressed because of low temperatures and the fact that the constantly saturated air reduces the rate of evaporation of moisture from leaves.

Tundra

To the north of the limits of tree growth and above the limits of tree growth in mountains is a vegetation dominated by low-growing perennial plants. These communities appear very similar in many Arctic and alpine areas, but they occur under very different climatic regimes. Arctic tundra localities have short, but often warm, summers when the sun shines 24 hours every day, whereas tropical alpine tundras have 12-hour days and 12-hour nights all year. It is never very warm at these high altitudes, and it freezes most clear nights. Some of the peculiar growth forms of alpine plants have evolved because of this regular freezing at night and thawing during the day. There is a slow, but relatively steady, rate of photosynthesis all year, and the animal communities also show relatively little seasonality.

The Arctic tundra, in contrast to alpine tundra, has a brief period of intense biological activity that is dominated by many species that exploit the tundra for only a short time and then leave. Arctic tundra is also underlaid by permanently frozen ground—*permafrost*—which restricts plant roots to superficial layers of the soil and prevents subsurface drainage. This, combined with low air temperatures and little evaporation, makes much of Arctic tundra very wet even though precipitation is low.

Taiga

Taiga is the northern forest of coniferous trees, primarily spruce, hemlock, and fir, that extends across vast areas of Eurasia and North America. Here, the winters are long and cold, and most of the limited amount of precipitation falls in the summer.

Because of the latitude where taiga occurs, the days are short in winter (as little as 6 hours) and correspondingly long in summer. During summer, plants may grow rapidly, and crops often attain a large size in a surprisingly short time: Marshes, lakes, and ponds are common here, and they are often fringed by willows or birches. Most trees in the taiga tend to occur in dense stands of one or a few species. Alders, which are common, harbonor nitrogen-fixing bacteria in nodules on their roots; partly for this reason, they are able to colonize raw, infertile soils such as those left behind by the recent retreat of glaciers or other ice.

In winter, taiga receives a deep blanket of snow. During the long, snowy season, animals must adapt to a way ot life different from the one they pursue during the brief summer, or they may migrate to escape the snow. There are changes in colour, in food habits, and in many other basic aspects of life. The snow protects the ground from freezing and allows the growth of forest; in the less snowy tundra biome, ice lies too close to the surface of the soil for forest to grow. Beneath the thick cover of snow in the taiga, there exists an active community of rodents and other animals, completely protected from most predators.

Many large mammals live in the taiga, including elk, moose, deer, and carnivores, such as wolves, bear, lynx, and wolverines. Traditionally, much fur trapping has gone on in this region, which is also an important lumber-producing region. To the south, taiga grades into forests or grasslands, depending on the amount of precipitation. Northward, it gradually gives way to open tundra.

Tundra

Farthest north in Eurasia, North America, and their associated island, between the taiga and the permanent ice, occurs the open, often boggy community known as *tundra*. This is an enormous biome, extremely uniform in appearance, that covers a fifth of the earth's land surface. Trees are small and are mostly confined to the margins of streams and lakes. In general, tundra is dominated by scattered patches of grasses and sedges (grasslike plants), heathers, and lichens, with denser stands in wet places.

Annual precipitation in the tundra is low, usually less than 25 centimeters, and the water is unavailable for most of the year because it is frozen. During the brief arctic summers, water sits on frozen ground, and the surface of the tundra often becomes extremely boggy. Permafrost, or permanent ice, usually exists within a meter of the surface.

As a taiga, herbs of the tundra are perennials that grow rapidly during the brief summers, using food stored underground. Large grazing mammals, including musk-oxen, caribou, reindeer, and carnivores such as wolves, foxes, and lynx, live in the tundra, which teems with life in the short summer. Lemmings, a kind of small rodent, are animals of the tundra. Their populations rise rapidly and then crash on a long-term cycle, with important effects on the populations of animals that prey on them.

OTHER BIOMES

Chaparral

The *Chaparral* of California and adjacent regions is historically derived from deciduous forests. Chaparral occurs in the five areas of the world with a mediterranean, or summer-dry, climate: the Mediterranean area itself, California, central Chile, the Cape region of South Africa, and southwestern Australia. Chaparral consists of evergreen, often spiny shrubs and low trees that form extensive communities in these dry-summer regions. Because of its relatively dry conditions, chaparral is frequently subjected to periodic fires. Plant species found in chaparral have adapted to these periodic fires in several ways. For example, some chaparral plant seeds can germinate only when they have been exposed to the hot temperatures generated during a fire. Those who build houses in chaparral expose themselves to the risk of being burned out, a perfectly natural phenomenon in the region. This is especially true since contemporary practices limit the occurrence of fires, which means that when they do occur, a great deal of organic material will probably have accumulated. The fires may then be very destructive, as seen, for example, in Santa Barbara, California, in the summer of 1990.

Polar Ice

Ice caps lie over the Arctic Ocean and Greenland in the north and Antarctica in the south. The poles receive almost no precipitation, so although ice is abundant, fresh water is scarce. The sun barely rises in the winter months. Life in Antarctica is largely limited to the coasts. Because the Antarctic ice cap lies over a landmass, it is not warmed by the latent heat of circulating ocean water and becomes very cold. As a result, only bacteria, algae, and some small insects inhabit the vast Antarctic interior.

Mountain Zone (Alpine)

Because increasing altitude produces many of the same changes in temperature and moisture as increasing latitude the tops of mountains have a typical windswept vegetation similar in many respects to tundra. Few if any trees are always to grow in this alpine zone, which, like polar regions, is along with life in the warm summer months. During the half winter, little grows.

Temperat Evergreen Forest

Temperate evergreen forests occur in regions where winters are cold and there is a strong, seasonal dry period. The pine forests of the western United States, the California oak woodlands, and the Australian eucalyptus forests are typical temperate evergreen forests.

Warm, Moist Evergreen Forest

Massive evergreen forests occur in temperate regions where winters are mild and moisture is plentiful. These can be seen in central China, in the pine forests covering much of the southeastern United States, and in the coastal redwood forests of northern California.

Tropical Monsoon Forest

Tropical monsoon forests occur in the tropics and semi-tropics at slightly higher latitudes than rain forests or where local climates are drier. Most trees in the forests are deciduous, losing many of their leaves during the dry season. Rainfall is typically very seasonal, measuring several inches daily in the monsoon season and approaching drought conditions in the dry season, particularly in locations far from oceans, such as central India.

Semidesert (Tropical Dry Forest)

Semidesert, or tropical dry forests, occur in tropical regions with less rain than monsoon forests but more rain than savannas. Vegetation is dominated by bushes and trees with thorns and spikes, which is why these regions are also known as thornwood forests. Plants survive on one or a few short rainy periods each year, growing intensively in response to the moisture. The brief rain is followed by a long dry season in which leaves fall from plants and there is little or no growth.

THE FATE OF THE EARTH

Now that you have completed your survey of the biomes and the corresponding ocean communities, you have gained some appreciation of the different kinds of climate under which these areas have evolved over tens or hundreds of millions of years. In just the last several hundred years, in contrast, human populations have spread over most of the land surface of the globe. Having already converted large stretches of many biomes to agriculture, urban areas, or other uses, human populations are attacking those biomes—such as tropical rain forests—that are less suitable for exploitation and about which we know much less. The human population doubled from 1950 to 1987, and most of that growth occurred in the warmer portions of the globe. If those who follow us are to find a stable ecological situation and one in which they can live out their lives in peace and relative prosperity, we must begin to manage the whole earth as a single system. In the next chapter, we will examine some of the factors that will determine our success or failure in this great enterprise.

11

Biogeochemical Cycles

The ecosystem is the most complex level of biological organization. Communities are composed of the organisms present at a particular place, while ecosystems include the nonliving factors interacting with these organisms. Individual organisms and populations of organisms in an ecosystem act as parts of an integrated whole. They adjust over time to their role in the ecosystem and relate to one another in complex ways that we only partly understand. In ecosystems there is a regulated transfer of energy—ultimately derived from the sun—and a controlled cycling of nutrients. The earth is a closed system with respect to chemicals, but an open one in terms of energy. Collectively, the organisms that occur in ecosystems regulate the capture and expenditure of energy and the cycling of chemicals. As we will see in this chapter, all organisms, including humans, depend on the ability of other organisms—plants, algae, and some bacteria—for the basic components of life.

As distinct functional units, different kinds of ecosystems have more or less clearly recognizable boundaries. Some kinds may intergrade into one another, sometimes almost imperceptibly, and boundaries then become arbitrary. Ecosystems change over time and slowly become modified into new ecosystems, whose characteristics come to differ increasingly from those that preceded them. The complex ecosystems of the tropical rain forests have changed gradually, adapting to the particular conditions of temperature seasonality, and soil that typify these places. The ecosystems of the tundra have developed in a similar way, but in relation to different environmental conditions. As the climate changes in a given place, the ecosystems present also change, as do the individual populations within those ecosystems. By this process, the overall characteristics of the populations gradually adjust to the new conditions. Not all ecosystems are natural; we may also speak of an ecosystem in an aquarium or in a cultivated field.

All of the substances that occur in organisms cycle through ecosystems in a *biogeochemical cycle*, a cyclic path involving both biological and chemical processes (Fig. 11.1). Generally, on a global scale, the bulk of these substances is not contained within the bodies of organisms, but rather exists in the atmosphere, the water, or in rocks. Carbon (in the form of carbon dioxide), nitrogen, and oxygen primarily enter the bodies of organisms from the atmosphere, while phosphorus, potassium, sulfur, magnesium, calcium, sodium, iron, and cobalt, all of which are required for plant growth, come from rocks. All organisms require carbon, hydrogen, oxygen, nitrogen, phosphorus, and sulfur in relatively large quantities; the other elements are required in smaller amounts.

The cycling of materials in ecosystems begins when they are incorporated from the atmosphere or from weathered rock into the bodies of organisms. Sometimes they pass from these organisms into the bodies of other organisms that feed on them, and ultimately through decomposition they are returned to the nonliving world. After this occurs the nutrients may begin the cycle again by being incorporated into the bodies of other organisms.

The Water Cycle

The water cycle (Fig. 11.2) is the most familiar of all biogeochemical cycles. All life depends directly on the presence of water. The bodies of most organisms consist mainly of this substance. Water is the source of hydrogen ions, whose

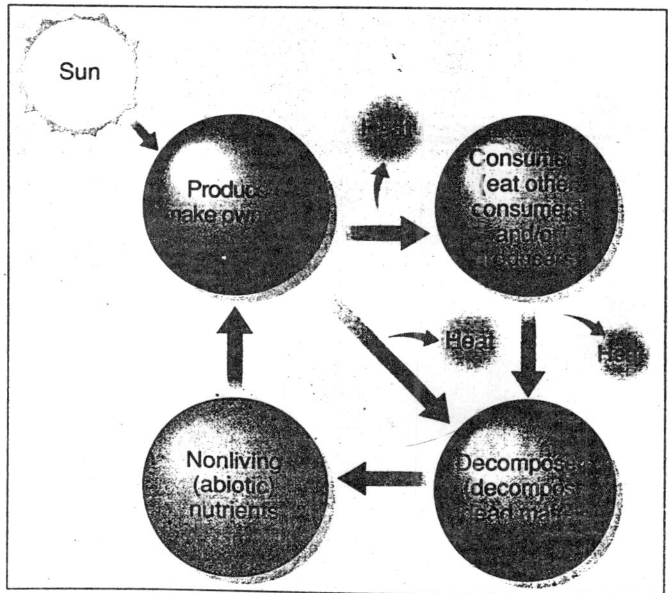

Fig. 11.1 Cycling within an ecosystem. Energy enters the ecosystem from the sun and leaves the ecosystem in the form of heat. (Energy, however, is not created or destroyed; when energy leaves an ecosystem, it is in a converted form.) Nutrients do not leave the ecosystem but, instead, cycle through it.

movements generate ATP in organisms, and for that reason alone it is indispensable to their functioning.

The oceans cover three-fourths of the earth's surface. From the oceans, water evaporates into the atmosphere, a process powered by energy from the sun. Over land areas approximately 90% of the water that reaches the atmosphere comes from plants via transpiration. Most precipitation falls directly into the oceans, but some falls on land, where it passes into surface and subsurface bodies of fresh water. Only about 2% of all the water on earth is fixed in any form—frozen, held in the soil, or incorporated into the bodies of organisms. All of the rest is free water, circulating between the atmosphere and the earth. Regardless of where this water is held temporarily, it eventually returns to the atmosphere and the oceans.

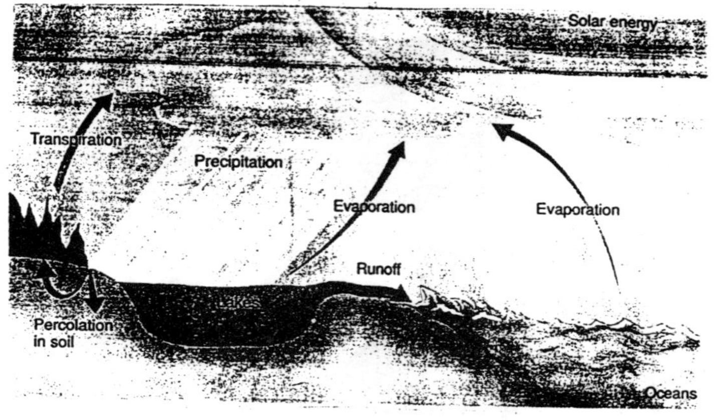

Fig. 11.2 The water cycle. Transpiration is the process by which water evaporates from the surface of plants.

Organisms live or die on the basis of their ability to capture water and incorporate it into their bodies. Plants take up water from the earth in a continuous stream. Crop plants require about 1000 kilograms of water to produce one kilogram of food, and the relationship in natural communities is similar. Animals obtain water directly or from the plants or other animals they eat. The amount of free water present at a particular place often determines the nature and abundance of the living organisms present there.

Much less obvious than surface water, which we see in streams, lakes, and ponds, is groundwater, which occurs in *aquifers*—permeable, saturated, underground layer of rock, sand, and gravel. In many areas, groundwater is the most important reservoir of water. It amounts to more than 96% of all fresh water in the United States. The upper, unconfined portion of the groundwater constitutes the *water table*, which flows into streams and is partly accessible to plants; the lower confined layers are generally out of reach, although they can be 'mined' by humans. The water table is recharged by water that percolates through the soil from precipitation as well as water that seeps downward from ponds, lakes, and streams. The deep aquifers are recharged very slowly from the water table.

Groundwater flows much more slowly than surface water, anywhere from a few millimeters to a meter or so per day. In the United States, groundwater provides about 25% of the water used for all purposes and provides about 50% of the population with drinking water. Rural areas tend to depend on wells to access groundwater almost exclusively, and its use is growing at about twice the rate of surface water use. In the Great Plains of the central United States, the extensive use of the Ogallala Aquifer as a source of water for agricultural use as well as drinking water is depleting it faster than it can be naturally recharged. This seriously threatens the agricultural production of the area and similar problems are appearing throughout the drier portions of the globe.

Because of the greater rate at which groundwater is being used, and because if flows so slowly, the increasing chemical pollution of groundwater is a very serious problem. It is estimated that about 2% of the groundwater in the United States is already polluted, and the situation is worsening. Pesticides, herbicides, and fertilizers have become a serious problem. Another key source of groundwater pollution consists of the roughly 200,000 surface pits, ponds, and lagoons that are actively used for the disposal of chemical wastes in the United States alone. Because of the large volume of water, its slow rate of turnover, and its inaccessibility, removing pollutants from aquifers is virtually impossible. Recharging the aquifers is an important strategy for conservation, but it depends on the purity of the water there initially.

Gaseous Cycles

Let us consider the gaseous cycles of two elements, carbon and nitrogen. Carbon usually is available in adequate quantities, so carbon deficiency is seldom a limiting factor in ecosystems. On the other hand, nitrogen, though it is plentiful in the atmosphere, is often a limiting factor—for example, limiting—growth—because adequate quantities of usable nitrogen compounds are not readily available for living things.

The Carbon Cycle

In the previous chapters we discussed how carbon moves through living things, but we need to say more about the *carbon cycle* on a global scale (Fig. 11.3). The global carbon cycle involves an exchange of CO_2 between the atmosphere and the main carbon reservoir, the world's oceans. At the interface between the atmosphere and aquatic systems, exchanges occur that maintain an equilibrium in the CO_2 concentration. Within

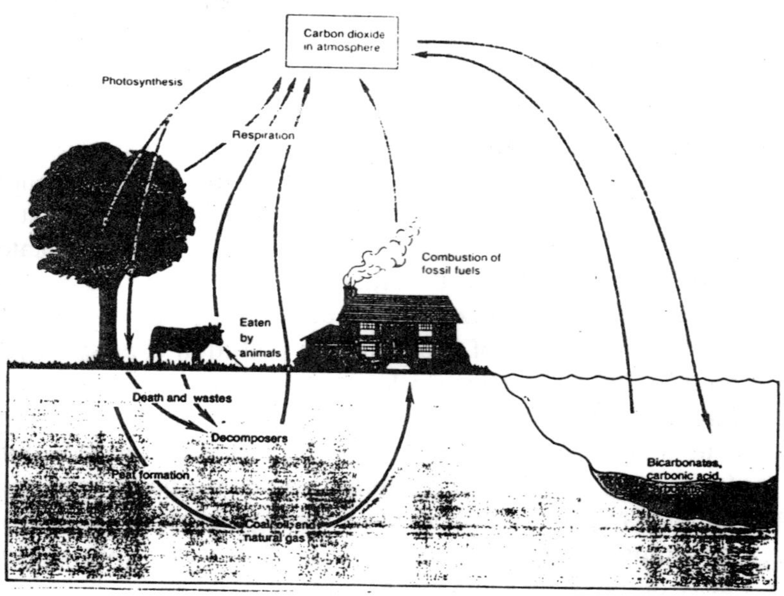

Fig. 11.3 The carbon cycle.

aquatic systems, other exchanges of CO_2 also take place. For example, excess carbon dioxide may combine with water to form carbonates and bicarbonates. Carbonates are not very soluble and they precipitate out in bottom sediments, from which they slowly return to the system. These carbonates serve to buffer the CO_2 concentration in the water because as CO_2 becomes depleted, carbonates are converted into bicarbonates and ultimately to CO_2 and water, thus increasing the CO_2 concentration in the water.

Some carbon is incorporated into forest vegetation and may remain out of circulation for hundreds of years or even longer. In very wet environments, organic matters sometimes undergoes only incomplete decomposition and accumulates as *peat*. Massive accumulations of peat during the Carboniferous period led to the production of great stores of such fossil fuels as coal, oil, and natural gas.

The world's total carbon pool is estimated to be about 55×10^{12} metric tons and is distributed among organic and inorganic forms. Carbon in fossil fuels accounts for 22 per cent of the world supply of carbon. The oceans contain 70 per cent of the inorganic carbon, mostly in the form of bicarbonate and carbonate ions, and an additional 3 per cent in dead organic matter and phytoplankton. Terrestrial ecosystems contain only 4 per cent of the total carbon pool, with forests being the main reservoirs of terrestrial organic carbon. Only about 1 per cent of the world's carbon is in the atmosphere. This is the amount potentially available for photosynthesis (Fig. 11.4).

However, because of such activities as the rapid burning of fossil fuels, the clearing of forests, and other land use, the amount of CO_2 in the atmosphere has been increasing since the Industrial Revolution. Atmospheric concentrations have risen from an estimated 260 to 300 parts per million (ppm) 100 years ago to around 330 ppm today.

Increased CO_2 concentration in the atmosphere may have a significant impact on world climate. The CO_2 in the atmosphere transmits shortwave radiation from the sun to the earth but absorbs the longer-wave radiation that passes from the earth

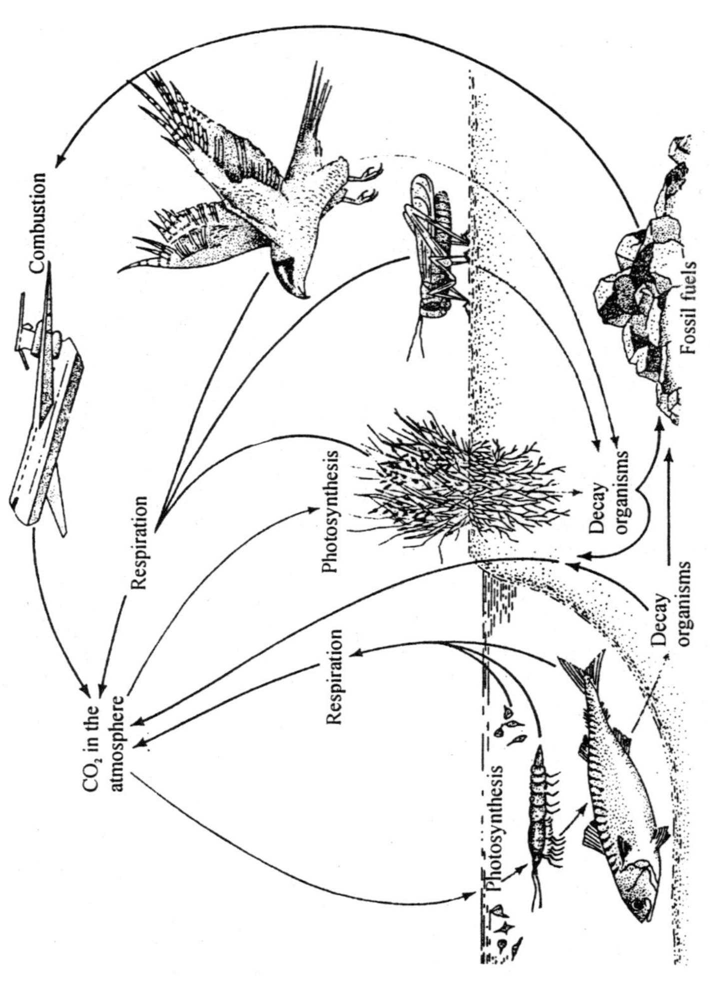

Fig. 11.4 Global carbon cycle

back toward outer space. Some of this absorbed heat energy is re-radiated toward the earth. Thus, CO_2 contributes to the *greenhouse effect* of the atmosphere, and an increase in CO_2 concentration could result in increased warming of the earth's surface. A worldwide increase in temperature could, in turn, melt the polar caps, cause the oceans to rise, increase evaporation, and decrease oceanic circulation.

However, other changes in the atmosphere that are also associated with the burning of fossil fuels could produce an opposite effect—a cooler climate. For example, an increase in the number of small particles in the atmosphere could block incoming solar radiation. Weather experts are not yet able to predict the net effect of these various changes on world climates.

The Nitrogen Cycle

Nitrogen is an abundant element. In fact, nitrogen gas (N_2) makes up about 78 per cent of the atmosphere by volume. Yet soil nitrogen deficiency is a common limiting factor of plant growth and productivity. This is because vascular plants cannot directly incorporate N_2 into organic compounds. Plants absorb nitrogen from the soil mainly as nitrate (NO_3^-) or ammonium ions (NH_4^+). Thus, these usable ions must be available in adequate quantities in the soil. Once NO_3^- or NH_4^+ is absorbed by plants, the plants—and the animals that eat them—can synthesize amino acids, proteins, and other nitrogen-containing compounds. Let us examine this *nitrogen cycle,* the process by which nitrogen atoms cycle through ecosystems.

Small quantities of NH_4^+ and NO_3^- are washed out of the atmosphere and carried down to the soil by rain. Atmospheric NH_4^+ comes from gasoline engine exhaust, industrial combustion, and similar processes, as well as from natural processes such as forest fires and volcanic eruptions Atmospheric NO_3^- production is caused when lighting or ultraviolet radiation provides energy for the oxidation of N_2 in the atmosphere by oxygen (O_2) or ozone (O_3).

However, the bulk of the nitrogen compounds absorbed by plants is not provided by these physical processes. Instead, biological *nitrogen fixation* by microorganisms—bacteria and

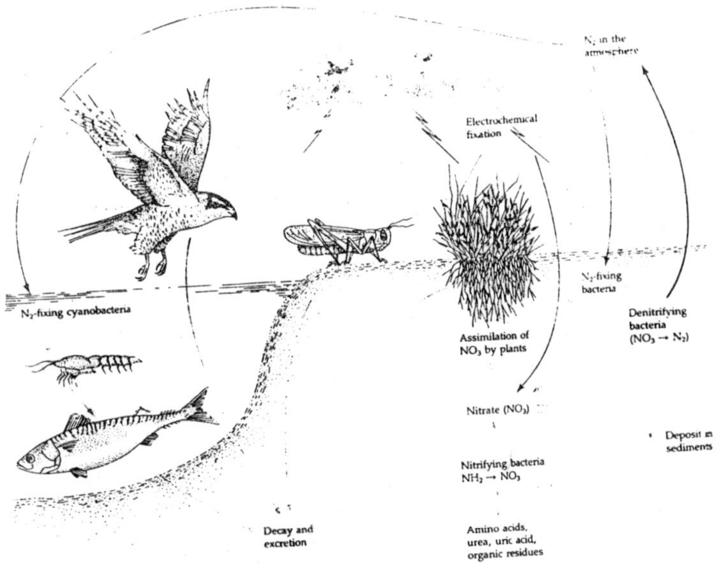

Fig.11.5 Global nitrogen cycle.

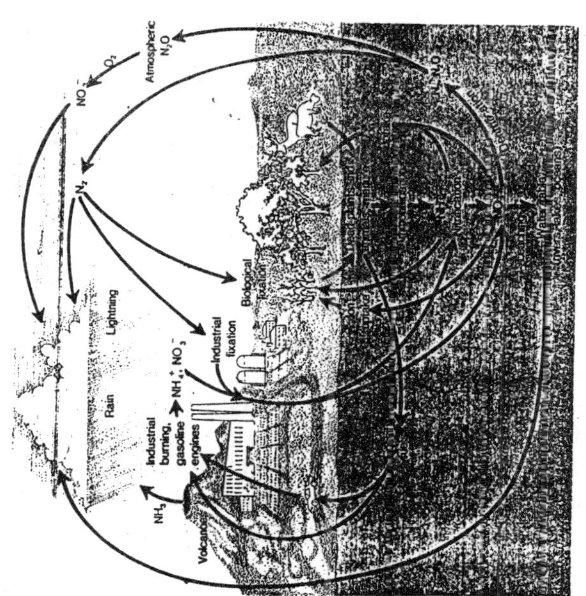

Fig.11.6 The nitrogen cycle.

cyanobacteria (blue-green algae)—provides most of the absorbable nitrogen compounds. In nitrogen fixation, ATP and reduced ferredoxin are used in the conversion of N_2 to NH_4^+. Some nitrogen-fixing microorganisms are free living; others live symbiotically with plant cells in root nodules of various plants.

Some organic nitrogen compounds quickly return to the soil in fallen leaves and animals feces. Other return only when plants or animals die. Decay processes then break down the organic compounds in several stages. In the last of these stages, called *ammonification*, soil microorganisms break down organic nitrogen to release NH_4^+. Often, other bacteria obtain energy by oxidizing NH_4^+ to produce NO_3^- in a process called *nitrification*. However, not all of the NO_3^- thus produced is absorbed by plants.

A certain proportion of the fixed nitrogen in the soil is steadily lost. Under anaerobic conditions, nitrate is often converted to nitrogen gas (N_2) and nitrous oxide (N_2O), both of which return to the atmosphere. This process, which several genera of bacteria carry out, is called *denitrification*. In its absence, all nitrogen would eventually become fixed, converted into nitrate, and washed into the oceans. Life would thus be possible only in marine and littoral habitats. Denitrification and nitrogen fixation together constitute the mechanism for returning nitrogen from the oceans to the land.

The Oxygen Cycle

Earth is the only place in the solar system where free oxygen exists in significant quantities. This free oxygen, which constitutes nearly a fifth of our atmosphere by volume, is a product of photosynthesis, carried out over more than 3 billion years of earth history. In the process of respiration, free oxygen reacts rapidly with reduced organic material. In the absence of photosynthesis, respiration would consume all organic material containing oxygen in about 50 years, but a large pool of oxygen would still remain in the atmosphere.

Sedimentary Cycles

Sedimentary Cycles differ from gaseous cycles in that the earth's crust is the major reservoir for the elements involved.

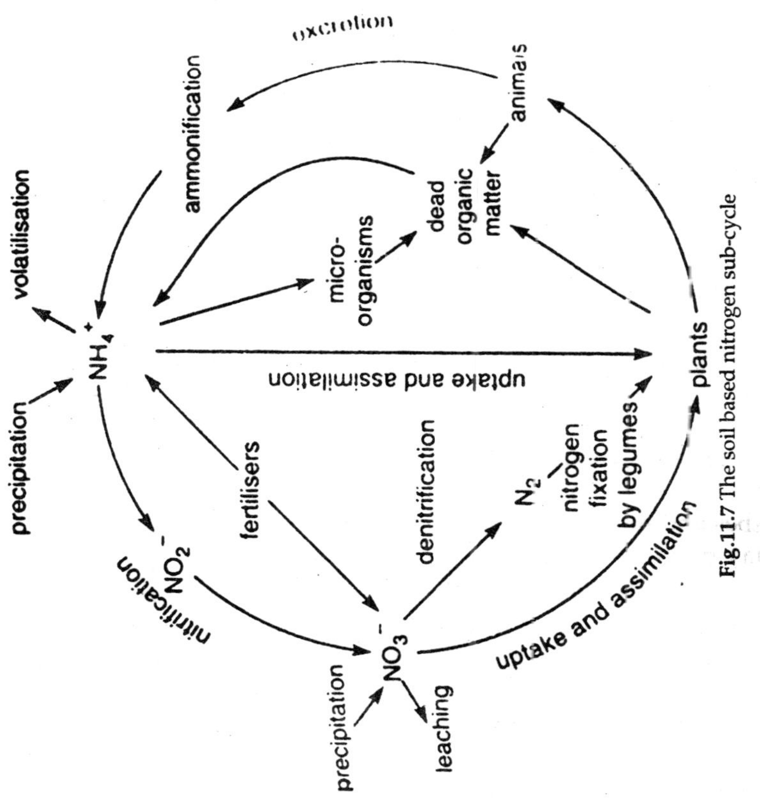

Fig.11.7 The soil based nitrogen sub-cycle

Calcium, phosphorus, and potassium all move through sedimentary cycles. Let us examine one of these cycles more closely.

The Calcium Cycle

The *Calcium cycle* is a sedimentary cycle that begins in the soil, where calcium is taken up by plants and deposited in leaves, twigs, stems, and trunks. Rain dripping from the foliage dissolves some of the calcium from the leaves and carries it back to the soil, where it is quickly taken up by the plants again. Herbivores obtain their supply of calcium from plants, and carnivores obtain their calcium by eating herbivores and other animals. Some animals have additional means of obtaining calcium. For example, many birds ingest calcium directly by picking up and eating limestone particles from the soil or along roadsides. When plants and animals die and decay, calcium is returned to the soil where it can again be taken up by plants and continue to cycle.

However, only a portion of the calcium and other nutrients taken up by plants is returned to the cycle each year. For example, in a forest ecosystem, a substantial amount of calcium is stored in woody tissues of trees, where it remains, until the trees fall to the ground and decay or until they are reduced to ashes and mineral elements by fire. Furthermore, some calcium may be lost to the ecosystem by being dissolved in rainwater that percolates through the soil. This water carries calcium and other nutrients down through the soil to the groundwater and eventually into streams, lakes and oceans. There, some of the calcium is recycled through photosynthesizing algae, small invertebrates, fish, and other aquatic and marine organisms. Another portion of this calcium is returned to land by way of sea spray as waves crash against the shore. Yet another portion becomes incorporated in the bottom sediments of lakes and oceans where it is removed from circulation for extremely long periods of time, sometimes even until geological processes uplift the lake bottom or sea floor so that it forms part of a new landmass. The calcium once again becomes available in terrestrial ecosystems.

The Phosphorus Cycle

Phosphorus is an essential element. It plays a central role in both cell metabolism and reproduction. It is contained in the energy-transferring molecules adenosine triphosphate (ATP), adenosine diphosphate (ADP) and adenosine monophosphate (AMP). It is also a key component of DNA (deoxyribonucleic acid) and RNA (ribonucleic acid), the molecules that contain and transfer the genetic code.

In essence, this is a geochemical cycle. It follows the pattern seen in the rock cycle of weathering, transport, sedimentation and uplift, followed by renewed weathering, and so on (Fig. 11.8).

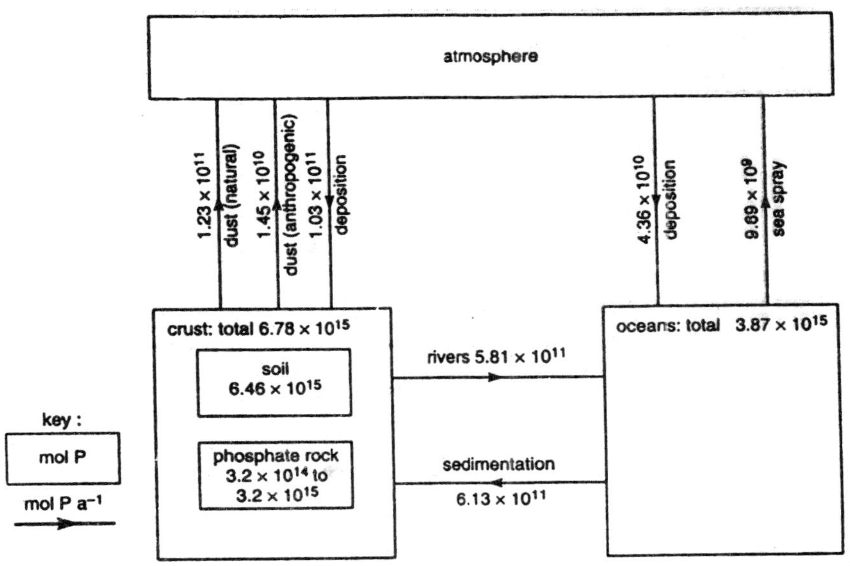

Fig. 11.8 The phosphorus cycle.

The largest reservoir within this cycle is the crust. This is because naturally occurring phosphorus containing compounds are of low solubility and extremely low volatility. Therefore, comparatively little phosphorus is present in solution in lakes, rivers or oceans and virtually none is found in the atmosphere.

The fluxes between reservoirs largely result from the movement of solid particles, borne either aerially or in suspension in water. The magnitudes of these fluxes are not accurately known because the transport processes vary with time and from place to place, as does the phosphorus content of the particulate matter in which it is transported.

The low solubility of phosphorus-containing minerals means that much of the phosphorus in soils and sediments beneath water bodies (such as lakes) is unavailable to plants and other organisms. Hence, even in environments where phosphate minerals are relativley abundant, the low concentration of phosphorus in solution may limit growth. This is particularly true for organisms that are capable of fixing nitrogen as the growth of these species is not limited by the supply of nitrogen. Therefore, the addition of phosphorus fertiliser to a soil supporting leguminous crops will often increase the rate of nitrogen fixation.

The importance of phosphorus as a crop nutrient has meant that a great deal of work has been done on the behaviour of this element in the soil environment. Within this context, three types of phosphorus are readily identified. Firstly, there is phosphate in the soil solution. Secondly, there is phosphorus that is part of the soil organic matter. Lastly, there are several forms of solid inorganic phosphate, most of which are of very low solubility. The first category is generally tiny compared with the other two. Nonetheless, it is highly important as it is probably the only form in which this nutrient is readily available for uptake by plants. The latter two forms represent the store of phosphorus within the soil.

The proportion of the total phosphorus content of a soil that is in the organic fraction varies considerably both from soil to soil and with depth within a given soil (Fig. 11.9). Typically the organic fraction accounts for 30 to 85% of the total soil phosphorus. This store is of least significance in soils that have received heavy applications of inorganic phosphate fertiliser. Conversely, it is of greatest significance both in natural soils that contain little inorganic material (e.g. peat-based soils) and in unadulterated mineral soils that are highly weathered, which occur in some tropical areas. Removal of the upper horizons from such tropical mineral soils reveals a subsoil that is so deficient in phosphorus (and frequently other nutrients) that it can only support poor plant growth.

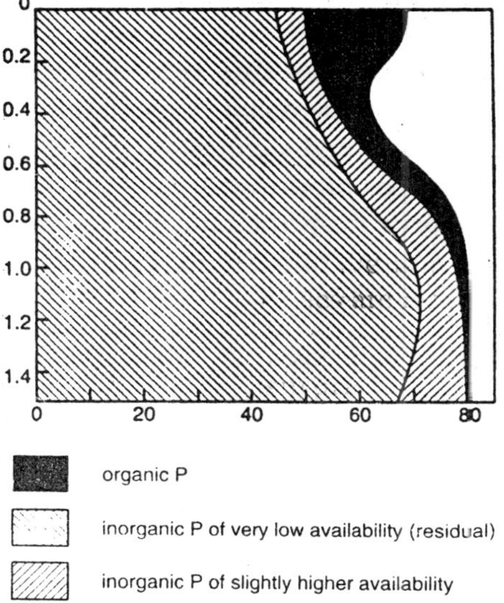

Fig. 11.9 The distribution of soil phosphorus (P) down a soil profile.

The chemical specification of phosphorus in the organic fraction of soils is not fully understood. However, it has been established that, in general, plants absorb very little of their phosphorus directly from this store.

The inorganic fraction of most soils contains both films of phosphate materials and definite phosphate compounds. The solubility of virtually all of the solid inorganic phosphates found in soils is extremely low. What is more, soluble phosphate compounds added as fertilisers are readily precipitated within the soil environment.

Human activity has significantly increased the flux of phosphorus from the crust, into both inland waters and the oceans.

In most cases the anthropogenic influx of phosphorus into inland waters has happened as a result of the addition of phosphate water softeners to detergents. These enter sewerage systems and, unless specific steps are taken to remove them, pass into the water body used to accept the sewage outfall. Such additions can lead to *algal blooms*, that is rapid growths of algae, and subsequent stagnation.

The increased phosphorus flux into the oceans has a different origin. It is mainly the result of increased soil erosion caused by modern farming and deforestation practices. It has been estimated that this has produced an increase in the rate of movement of phosphorus from the land to the oceans in the form of suspended solids from about 1.6×10^{11} mol Pa^{-1} to approximately 4.5×10^{11} mol Pa^{-1}.

The Sulfur Cycle

Sulfur is necessary for life. It has several important biochemical functions. For example, like nitrogen, sulfur has a key role in the structure and function of proteins. However, unlike nitrogen, it is not found in all amino acids (the 'building blocks' of proteins). It is present, however, in the amino acid cysteine, which contains a thiol (—S—H) group (Fig. 11.10). Proteins that contain such sulfur-containing amino acids are endowed with the ability to form *disulfide linkages* (Fig. 11.11). These can be *within* a protein molecule, that is intramolecular; or *between* protein molecules, that is intermolecular. Intramolecualr linkages of this type ensure that each protein has an appropriate shape for the function it is to perform. In contrast, intermolecular disulfide links are used to join protein molecules together. By virtue of such links, proteins can be used to from relatively rigid structures, such as hair and nails.

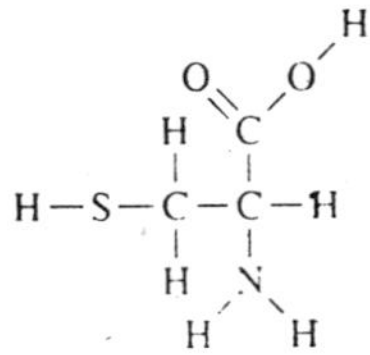

```
                H
               /
        O    O
         \\ /
      H   C
      |   |
H—S—C—C—H
      |   |
      H   N
         / \
        H   H
```

Fig. 11.10: Cysteine, a sulfur-containing amino acid.

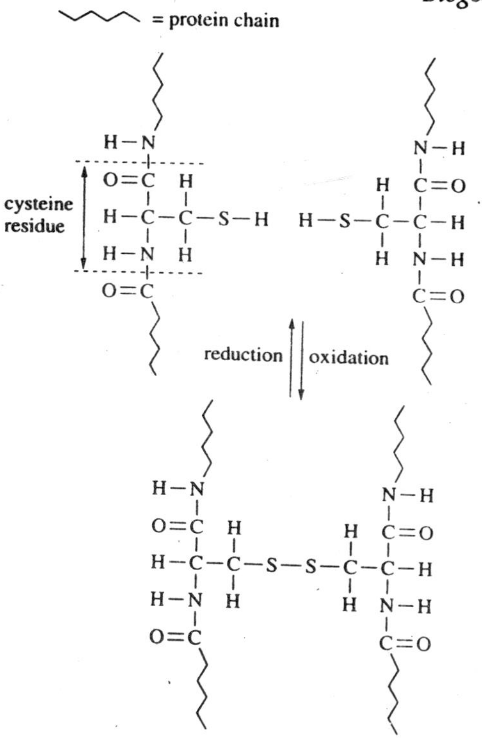

= protein chain

cysteine residue

reduction

oxidation

Fig. 11.11: Disulfide linkage formation.

Sulfur has several biochemically accessible oxidation states (Fig. 11.12). This means that microorganisms can bring about the reduction of oxidised sulfur and, under different conditions, the oxidation of reduced forms of this element. The changes in speciation brought about by such biochemical transformations are frequently responsible for the translocation of sulfur from one reservoir to another in a physically different location. Indeed, biochemically mediated redox reactions involving sulfur are of great significance in the cycling of this nutrient. It must be borne in mind, however, that purely chemical redox reactions involving sulfur also occur in nature. These are of particular significance in the atmosphere.

From Fig. 11.13 it can be seen that the major reservoirs of this element are crustal and oceanic, in comparison with which the atmospheric reservoir is tiny, though highly important.

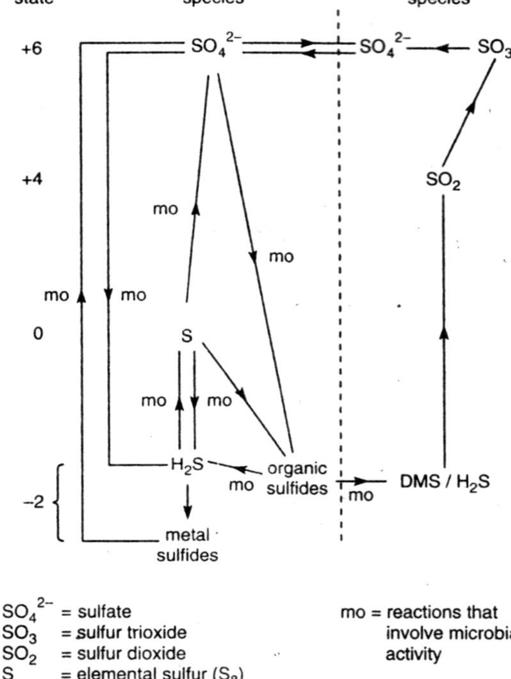

Fig. 11.12 The main environmental species of sulfur.

Let us consider the key processes that occur in the atmospheric sulfur reservoir before reviewing those that characterise its crustal and oceanic reservoirs.

At first glance, the small size of the atmospheric reservoir may seem a little surprising. After all, sulfur is found in several highly volatile chemical species that are produced in large quantities by both natural and man-made processes as shown in Fig. 11.13). The reservoir remains small because these species not only rapidly enter the atmosphere, they also rapidly leave it. In other words, the mean residence time of sulfur in the atmosphere is extremely short.

The most important sulfur-containing species entering the atmosphere are sulfur dioxide (SO_2) dimethyl sulfide ($(CH_3)_2S$, known as DMS), hydrogen sulfide (H_2S), and carbonyl sulfide (OCS).

Large amounts of hydrogen sulfide are generated under anaerobic conditions by micro-organisms, particularly in marine environments. However, in the presence of excess iron and/or similar metals it reacts to yield insoluble sulfides of extremely

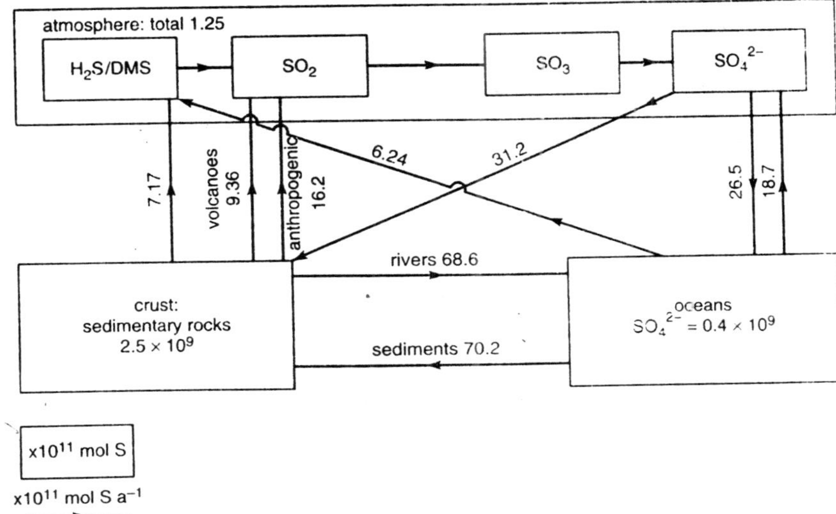

Fig.11.13 The sulfur cycle.

low volatility. Therefore, much of the sulfur present initially as hydrogen sulfide enters the crustal reservoir rather than the atmosphere. Indeed, the liberation of hydrogen sulfide from marine sediments is now considered to be of minor significance.

DMS is generated in the top 50m of the marine environment by the activity of phytoplankton. Within these organisms is found a sulfur-containing salt, dimethylsulfoniopropionate (DMSP). It is possible that this serves as an osmotic pressure regulator, stopping water from migrating out of the cells of the organisms concerned. On entering sea water DMSP degrades, forming DMS, some of which escapes into atmosphere. The concentration of DMS in sea water is low and the flux per unit area of this compound into the atmosphere is small. However, the areas involved are vast; consequently the total evolution of DMS from the oceans is enormous (upto ~5.6×10^{11} mol Sa^{-1}). This accounts for about one-tenth of the total flux of sulfur entering the atmosphere.

Carbonyl sulfide (OCS) also enters the atmosphere from the marine environment. It is produced in the upper part of the oceans by the action of sunlight on a number of organic compounds.

Sulfur dioxide enters the atmosphere as a result of essentially abiotic processes. Of particular importance are those processes associated with volcanic activity, the burning of fossil fuels and the roasting of sulfide ores.

Fig. 11.14 summarises the fate of the principal sulfur species that enter the atmosphere. As depicted in this diagram, atmospheric hydrogen sulfide and DMS are rapidly oxidised in the lower atmosphere (the troposphere:). This forms sulfur dioxide, adding to that already present from other sources. Some atmospheric sulfur dioxide is removed by the process of dry deposition, while the remainder is subjected to further oxidation to sulfate, via a complex set of reactions. The rates of deposition and oxidation of sulfur dioxide are dependent on conditions, both being more rapid in humid than in dry air. However, they are generally both in the range 1 to 10% per hour.

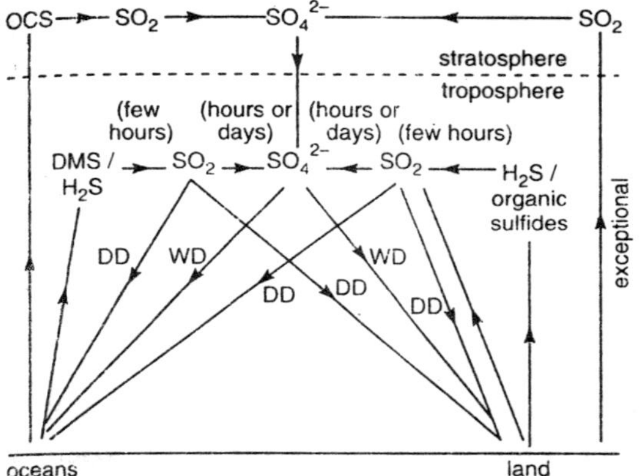

Fig. 11.14 The fate of sulfur species in the atmosphere. Horizontal arrows represent chemical reactions; the remainder represent physical translocations. Sulfur dioxide is removed from the atmosphere by dry deposition (DD) or oxidation to sulfate (SO_4^{2-}). Sulfate is mainly removed by wet deposition (WD).

In the exceptional circumstances of a major volcanic eruption sulfur dioxide may be directly injected into the upper atmosphere. Here it is oxidised to sulfate, as in the lower atmosphere.

Unlike the other sulfur species entering the atmosphere, carbonyl sulfide is resistant to oxidation within the troposphere. It survives for sufficient time to enter the stratosphere, where it is oxidised to sulfate. This happens via a photochemical reaction pathway, involving short wavelength ultraviolet radiation.

Despite the low concentration of sulfur species in the atmosphere (as low as 0.6 ppb) they are highly important. They play key roles in the nutrition of higher plants and the incidence of acid rain. Sulfur species may also play a part in the regulation of climate by providing particles around which water can condense, allowing cloud formation to occur over the remote oceans. It has also been suggested that sulfate induced stratospheric haze may accelerate the rate of ozone (O_3) depletion in the upper atmosphere.

Let us now consider the key processes that occur in the crustal and oceanic sulfur reservoirs.

Under anaerobic conditions sulfate, like nitrate, may be utilised by micro-organisms to oxidise carbohydrate during respiration. The reaction concerned may be depicted thus, using CH_2O to represent carbohydrate:

$$2CH_2O+2H^+{}_+SO_4^{2-} \rightarrow H_2S+2CO_2+2H_2O$$

Similar reactions utilising partially reduced sulfur species such as sulfite (SO_3^{2-}), thiosulfate ($S_2O_3^{2-}$) and elemental sulfur (S_8) in place of sulfate also occur.

The sulfide (S^{-11},present in H_2S) generated in these reactions will form precipitates with many metals, particularly the late transition metals and those immediately after them in the Periodic Table. The ubiquitous nature of iron means that iron sulfides are frequently formed in sediments in which sulfate reduction is occurring. The reactions form troilite (FeS) and iron pyrites (FeS_2, known as fool's gold), and may be reperesented thus:

$$3H_2S+2Fe(OH)_3 \rightarrow 2FeS+S+6H_2O$$

$$S+FeS \rightarrow FeS_2$$

The reduction of sulfate to sulfide by carbohydrate does not liberate as much energy as does the oxidation of carbohydrate with nitrate to form ammonia. Consequently, in anaerobic environments with appreciable nitrate concentrations, sulfate-reducing micro-organisms will be out-competed by those that reduce nitrate. Under these conditions ammonia, and not sulfide, will be the principal reduced species formed. Such conditions prevail in the deep waters and/or sediments of many fresh water lakes and in most waterlogged surface soils of humid regions.

The situation is different in deep marine waters and/or sediments or soils submerged beneath brackish or sea water. Here the nitirate content is relatively low and the supply of sulfate from the sea water is essentially limitless. In such sediments, the onset of anaerobic conditions generates a brief period of sulfate reduction. The bulk of the sulfide generated is precipitated as iron sulfides with a minimal evolution of gaseous hydrogen sulfide (H_2S). This process therefore results in a net flux of sulfur from the oceanic to the crustal reservoir.

Under aerobic conditions, sulfur in its reduced states (primarily sulfide, S^{-11}) can be oxidised by micro-organisms, including several bacteria of the genus *Thiobacillus*. This happens at the expense of molecular oxygen (O_2) in aerobic soils, sediments and water bodies. The ultimate product of these reactions is sulfate (SO_4^{2-}), in which the sulfur is in the plus six oxidation state. These processes are thermodynamically favourable and result in the liberation of energy that can be utilised by the organisms concerned. The following equation represents an example of this type of reaction:

$$H_2S+2O_2 \rightarrow H_2SO_4$$

Note that the sulfate is produced in the form of sulfuric acid (H_2SO_4). This means that exposure to the air of material with a significant sulfide content results in its rapid acidification. When soils that have been flooded for appreciable lengths of time with brackish or sea waters are drained, their pH values drop to as low as 1 or 2. Soils of this type are known as *cat-clays*. They cover significant areas of land in south-east Asia and in tidal areas of other parts of the world. Unless these soils are kept submerged, they soon become too acidic to support higher plant growth.

Sulfuric acid acidification can also be a significant problem in mining area, where reduced sulfur species are oxidised in spoil heaps.

The redox reactions considered in the foregoing discussions have considerable implications in plant nutrition. This is because the availability of soil sulfur for uptake by higher plants is highly dependent on its chemical speciation. In general, plants exclusively absorb sulfur from the soil as sulfate. In most surface soils of humid regions this is a small fraction of the total sufur present, even under aerobic conditions. The major soil reservoir of this bioelement is the organic fraction. Sulfur in this store becomes available when it is mineralised by soil organisms.

Overall, most plants are supplied with adequate amounts of sulfur, particulalry in industrialised regions. In many cases this is due, in part at least, to the ability of plants to absorb sulfur from the atmosphere. This can occur to a significant extent even in soils with sufficient sulfate. Plants grown in such soils can obtain 25-35% of their sulfur directly from the air.

The most significant human intervention into the sulfur cycle occurs as a product of the combustion of fossil fuels. These fuels contain a small but significant amount of sulfur, which is liberated as sulfur dioxide when they are burnt. This rapidly oxidises in the atmosphere and is deposited as acid rain. The implementation of pollution control and abatement measures designed to control the acid rain problem will reduce atmospheric concentrations of this bioelement. This may well necessitate an increase in the use of sulfur-based fertilisers in the future.

Sodium, Potassium, Calcium and Magnesium Cycles

Potassium, magnesium and calcium are needed by organisms in relatively large amounts: they are macro-nutrient elements. In comparison, sodium is needed in lesser amounts: it is a micro-nutrient element. These ions have several biological functions.

Sodium and potassium are members of group 1 of the Periodic Table, whereas calcium and magnesium are members of group 2. These elements have no redox chemistry in the environment. Sodium and potassium are found in the plus one oxidation state only (i.e. as Na^+ and K^+), while calcium and magnesium are exclusively in the plus two oxidation state (i.e. as Ca^{2+} and Mg^{2+}).

The geochemical cycles entered into by sodium, potassium, calcium and magnesium are depicted in Fig. 11.15. From this it can be seen that there are many similarities in the environmental behaviour of these elements. They all enter into an essentially sedimentary cycle of weathering, transport, sedimentation and uplift followed by renewed weathering, and so on. In this respect, their environmental behaviour is similar to that of phosphorus.

Table 11.1 The main biological functions of sodium, potassium, calcium and magnesium in human beings and plants.

Element	*Main functions*	
	Plants	*Human beings*
Sodium (Na)		Control of the balance of electrical potentials across cell membra-nes. Involved in nerve impulses.
Potassium (K)	Main cation within cells. Key role in guard cell operation. Acts as a cofactor for some enzymes.	Control of the balance of ele-ctrical potentials across cell membranes. Involved in nerve impulses.
Calcium(Ca)	Acts as a cofactor for some enzymes. For-mation of middle lam-ella (the 'gum' betwe-en adjacent cell walls)	Bone formation. Cell membr-rane stabilization. Involved in blood clotting. Involved in muscle contraction.Acts as a cofactor for some enzymes.
Magnesium (Mg)	Key component of chlorophyll.Acts as a cofactor for some enzymes.	Involved in nerve impulses. Acts as a cofactor for some enzymes.

While the cycles of sodium, potassium, calcium and magnesium are essentially similar, there are important differences in detail. These are largely attributable to the influences of biological activity, differences in solubility of some of their salts and the differing ability of these ions to be entrapped within clays.

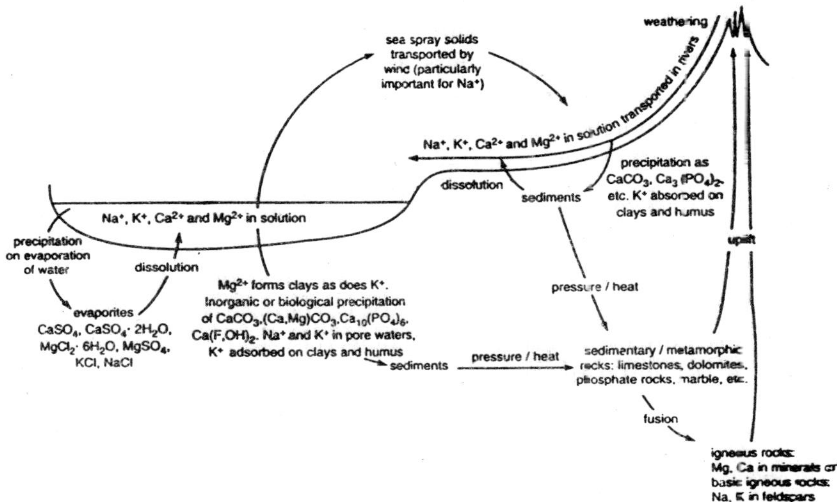

Fig. 11.15 The geochemical cycles of sodium (Na), potassium (K), calcium (Ca) and magnesium (Mg).

With this in mind, it is interesting to contrast the relative concentrations of these ions in typical sea and river waters (Table 11.2). The dominant cation in river water is calcium, while that in sea water is sodium.

Table 11.2 The dissolved sodium (Na), potassium (K), calcium (Ca and magnesium (Mg) content of typical sea and river waters.*

Nature of E	*River water*		*Sea water*	
	[E]/mM	$\frac{[E]}{\Sigma[E]}$	[E]/mM	$\frac{[E]}{\Sigma[E]}$
Na^-	0.274	0.31	470	0.88
K^-	0.059	0.067	1.02	1.9×10^{-3}
Ca^{2-}	0.375	0.43	10.2	0.019
Mg^{2-}	0.171	0.19	53.6	0.10

$\Sigma[E]/mM(river) = 0.274 + 0.059 + 0.375 + 0.171 = 0.879$

$\Sigma[E]/mM(sea) = 470 + 10.2 + 1.02 + 53.6 = 534.82$

* Note that river waters are more variable in their dissolved solids contents than are sea waters.

The dominance of calcium over sodium in river water is essentially a reflection of the fact that more calcium than sodium is dissolved by percolating waters prior to their entry to rivers. This occurs for two reasons. Firstly, calcium minerals are more susceptible to weathering than are sodium silicates. Secondly, the concentration of calcium in the crust is higher than that of sodium (Ca is 3.6% w/w of the crust, while Na is 2.8% w/w).

In contrast to river water, sea water contains a much higher content of dissolved solids (typical total dissolved solids values being 34.4g dm^{-3} for sea water and 0.118 g dm^{-3} for river water). Significantly, the concentrations of carbonate and calcium species within sea water are sufficient to exceed the solubility product of calcium carbonate. Consequently this may precipitate from solution, ultimately forming limestones (this is usually a biomediated process). There is no similar mechanism for sodium precipitation in open waters, hence the dominance of sodium over calcium in the oceans.

During limestone formation magnesium carbonate is frequently precipitated along with the calcium carbonate. However, the solubility of magnesium carbonate is greater than that of calcium carbonate. Hence, much less magnesium than calcium is precipitated in this way and limestones are predominantly calcium carbonate. Consequently, sea water contains more magnesium than calcium, even though the rivers flowing into the oceans supply more calcium than magnesium.

The dominance of magnesium over calcium in sea water is maintained under conditions of evaporite formation. Evaporite deposits contain both calcium salts (e.g. anhydrite, $CaSO_4$ and gypsum, $CaSO_4.2H_2O$) and magnesium salts (e.g. $MgCl_2.6H_2O$ and $MgSO_4$). However, of these the calcium sulfate salts are of lowest solubility and hence precipitate first, thus ensuring the dominance of magnesium over calcium in the remaining water.

Of the four ions discussed in this section, potassium has the lowest concentration in both river and sea waters. This is, in part at least, due to the relatively high concentration of this ion in biological material, typically being 15 times that of sodium.

In addition, outside biological systems, potassium is strongly absorbed onto negatively charged clays and humic materials. It is also relatively easily entrained into new silicate minerals, a trait it shares with magnesium.

Human interventions in the cycles discussed here are primarily deliberate. These are intended to increase agricultural productivity.

Potassium is commonly a limiting plant nutrient. This is a consequence of the relatively high demand that plants have for this element, coupled with its propensity to be bound to, and within, clays. Potassium deficiency is particularly common on heavily cropped land, where it is removed in large amounts with the harvest. Under these conditions additions of potassium fertilisers are necessary for the maintenance of soil fertility.

Calcium and magnesium are also deliberately added to agricultural land. These are added in the form of oxides, hydroxides or carbonates, materials known as *lime*. Clearly such additions increase the concentrations of the nutrients calcium and magnesium. Perhaps more importantly, they also have indirect beneficial effects associated with the decrease in acidity that liming brings about. For example, raising the pH of highly acidic soil to about 6.5 by the addition of lime will enhance the availability of phosphorus and molybdenum while decreasing the concentrations of iron, aluminium and manganese to sub-toxic levels. Without liming, the fertility of many soils in humid regions could not be maintained.

12

Pollution

Pollution is an undesireable change in the physical, chemical or biological characteristic of our air, land and water that may or will harmfully affect human life, or that of desirable species, our industrial processes, living conditions, and cultural assets, or that may or will waste or deteriorate our raw material resources. Pollutants are residues of the things we make, use and throw away. Pollution increases not only because as people multiply the space available to each person becomes smaller, but also because the demands per person are continually increasing, so that each throws away more year by year. As the earth becomes more crowded, there is no longer an 'away'. One person's trash basket is another's living space.

It has already been proved that pollution is now the most important limiting factor for man. The effort that must now be put into pollution abatement and prevention may well provide the negative feedback that will prevent man from completely raping the earth's resources, and thereby destroying himself. The problem is different only in aspect in the sharply divided world of man: in the undeveloped nations (70 per cent of the world's people) shortage of available food and resources is associated with chronic pollution and disease caused by human and animal wastes, while in the affluent or developed nations (30 per cent of the world's people) agro-industrial chemical pollution is now more serious than organic pollution. In addition, global pollution of air and water mostly emanating from the developed countries threatens everyone.

THE COST OF POLLUTION

The cost of pollution is measured in three ways, all of which add up to a terrible and increasingly intolerable burden to human society: (1) The loss of resources through unnecessary wasteful exploitation, since, as the National Academy's report puts it "pollution is often a resource out of place." (2) The cost of pollution abatement and control (Fig. 12.1a). While the clean up of sewage and solid wastes (refuse) is now the most expensive, the cost of abatement of the much more poisonous wastes from motor vehicles and power generation is protected to increase 100 times in the next 30 years. (3) The cost in human health. Recognition of this aspect of pollution cost will probably do more to alert egotistical and self-centered man to the rising danger than the other kind of costs which can be too well hidden by short-term "cost-benefit" manipulations at the local level (Fig. 12.1b). As human mortality from infectious deseases shows a precipitous decline, mortality and sickness from environmentally related respiratory diseases and cancer has shown an equally precipitous rise. In a recent review of the human health cost of air pollution Lave and Seskin (1970) estimate that a 50 per cent reduction in air pollution in urban areas alone could save two billion dollars annually in the aggregate cost of medical care and work hours lost in sickness, and this does not include the "cost" of human misery or death and disability caused by automobile and industrial accidents. As environmental stresses on the human body increase, many medical scientists fear a "backlash" in infectious diseases not only because of lowered body resistance but because viruses and other disease organisms will increasingly slip through water treatment and food processing plants as the quality of water and food at the intake deteriorates. Both water treatment and waste treatment (up to now considered as separate problems) must now be linked into a 'recycle' system. The behavioural consequences of crowding and breakdown in social structure that accompanies any decline in the quality of the

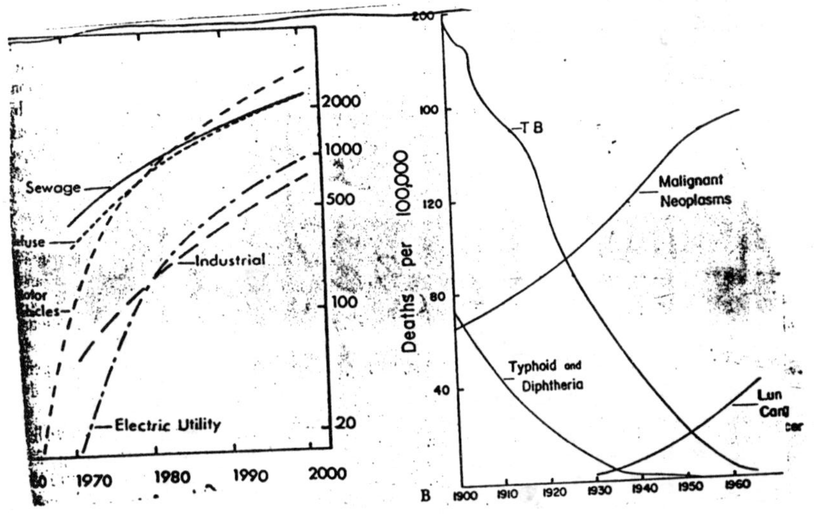

Fig. 12.1(ab) The cost of pollution. **A.** Projection of the cost to the consumer, in millions of dollars, of the cntrol of primary pollutants in the Delaware River Basin. Note that sewage and refuse (solid wastes) provided the drain on the tax dolla: in 1970, but that the cost of controlling pollution from industrial development, motor vehicles, and power generation will increase 10– to 100–fold by the year 2000 unless future expansion is planned and controlled. (Graph prepared from data in "Waste Management and Control," National Academy of Sciences Spilhaus), 1966), **B.** The dramatic decline in mortality from infectious diseases (tuberculosis, typhoid, and diphtheria) accompanied by the equally dramatic rise in malignancies believed to be related to pollution. (Data in B.D. Grove and A.M. Hetzel: "Vital Statistics rates in the United States, 1940-1960." National Center for Health Statics, PHS Publ. No. 1677).

environment has already been noted. Maurice Visscher (1967) in another National Academy report states that mental health is probably now the major cause of human morbidity and disability.

THE KINDS OF POLLUTION

Classifying pollution can be as difficult and confusing as classifying lakes or other natural phenomena. Classifications according to environment (air, water, land etc.) and pollutant (lead, carbon dioxide, solid wastes, etc.) are, of course, widely used approaches.

AIR POLLUTION

Types of air pollution

Air pollution is not a new problem. In 1273 King Edward I banned use of sea-coal in London, and Parliament established other controls in 1306. A 1661 report, Fumiffugium, attributed one-half of all deaths in London to pulmonary conditions brought on by polluted air. Great Britain's Public Health Act of 1848 included provisions for smoke abatement.

With the use of coal for power in industry and locomotives, smoke became a nuisance in many cities. After conditions became so bad you could barely discern the sun at midday. Pittsburgh pioneered in attacking the smoke problem, mainly by restricting the burning of soft coal.

Near the beginning of last century, acid fumes from a smelter at Copper Hill, Tennessee, completely denuded the land in the area, creating a "devil's playground" of barren eroded hills. Other forms of air pollution have since been found to affect all kinds of vegetation.

In the 1940s another type of air pollution was recognized in the Los Angeles area. This pollution was most severe on sunny days, when there were also temperature inversions. Vegetation damage was observed in 1944, and as the problem became more severe, eye irritation caused tearing. In the 1950s it was demonstrated that auto exhaust gases, when irradiated by sunlight, formed compounds that had oxidizing and irritating qualities. This chemical soup was called photochemical smog.

Effects of Air Pollutants

Human and Animal Health: In December 1930, a mysterious irritating fog formed in the Meuse Valley in Belgium. People became ill on the second day and men donned gas masks to search vainly for some unknown poison gas cache, perhaps left over from World War I. Not until 1948, in Donora, Pennsylvania, did people recognize that air pollution confined by a temperature inversion could be deadly. Four thousand deaths during a 1952 air pollution episode in London attracted the world's attention.

Air pollution exposes humans to poisoning by toxic materials in the environment. Toxic substances enter the human body by ingestion, by absorption through the skin or eyes, by means of a puncture or injection, or by inhaling a dust or gas. Air pollutants enter the body through the respiratory system. The cleansing mechanisms for the lungs bring some particles up to where they are swallowed or expelled.

In general, four factors influence how a toxic substance will affect an individual: concentration, duration of exposure, toxicity, and individual susceptibility.

Concentration and Duration of Exposure: Major air pollution episodes resulting in human deaths have involved a complex interaction of high pollution levels, stagnant air, a concentration of pollutants within valleys or other topographic basins, and a condition of atmospheric stratification referred to as an inversion, which places a 'lid' over the polluted area and prevents dispersion of the pollutants.

Under normal conditions the temperature of air declines with height, but under conditions of zero or low wind an inversion might occur—a condition where a band of warmer air overlays cooler air. Such inversions may be caused by any of several conditions. When the sun sets, the ground cools more rapidly than the air. Nocturnal cooling of air next to the ground frequently leaves a layer of warmer air immediately above. A nocturnal inversion breaks up when the early morning sun warms the earth, heating the air next to it. On a larger scale, the overrunning of cooler air by a warmer air mass or the subsidence of a warmer air mass over a layer of cooler air causes an inversion that will not break up until the weather system changes. Also, a wedge of cooler air may move in underneath warmer air.

When no inversion exists, a discharge of warm gases from a smokestack will be buoyed upward and dispersed by the wind. However, when there is an inversion, there is little if any horizontal air motion. The stack discharge rises until it encounters the inversion layer, whereupon it ceases rising. This condition may be observed in a mountain valley after the sun goes down. Smoke from a cabin will rise to the inversion layer and then spread outward.

Inversions lead to a concentration of pollutants discharged into the atmosphere. The pollutants accumulate at the inversion level, gradually spreading out horizontally in all directions and diffusing downward. Since the concentration of pollutants will be a maximum at the inversion layer, this bodes ill for persons who live on upper floors in high-rise buildings if that is the level of the inversion layer. The concentrations at ground level can also exceed safe limits during an inversion with deadly results.

Where mountain valleys or other physical features confine the air, inversions cause more acute concentrations of pollutants than a flat terrain, where relatively unlimited horizontal diffusion occurs. When nocturnal inversions break up vertical mixing takes place that brings the pollutants accumulated near the inversion elevation to the ground level. Thus, the highhest concentration of pollutants at ground level may be at the time of the breakup of the inversion.

Toxicity : Two different substances can cause different effects even though the concentrations and duration of exposure are identical and the test animals exposed are as nearly alike as possible. The difference in effect is caused by toxicity. Toxicity is commonly measured by how much of a substance kills 50 per cent of exposed animals, a quantity called LD_{50}. The National Institute of Occupational Safety and Health publishes a list of toxic substances and their known toxic effects.

In some cases people do not seem to react to a toxic substance until some level of exposure, significantly above zero, is reached. The level where physiological reactions of humans or test animals begin to be observed is called the *threshold level*. The concept of a threshold level is also used to examine the effects of radiation. Some substances do not appear to have a threshold. In other words, any exposure—no matter how small—causes some reaction. Substances that have no threshold are considered most hazardous because even one molecule at a vital place could cause trouble. Those that emit ionizing radiation are considered to be in this category, and asbestos, and perhaps some toxins, may be.

Individual Susceptibility: Individual susceptibility depends on the person's health history. People with lung and heart ailments are most affected by air pollutants, as are the very young and the aged. Individuals who are allergic to certain substances are sensitive to lower exposures and have worse reactions when exposed to those substances.

We are also aware that greater reaction is produced when a person is exposed to two or more of certain substances simultaneously than to either substance alone. This effect is called *synergism*. Such combined effects have been noticed in community air pollution episodes where the concentration of a pollutant in the air appears to cause adverse reactions at levels below those observed in laboratory experiments.

Gases are readily carried to the depths of the lungs, but there are natural defenses to keep out particles. Some particles impinge on mucus in the nasopharynx area (Fig. 12.2) and are washed out. The nasal and bronchial passages are lined with hairlike *cilia* that tend to sweep particles out. As a result, particles of dust, carbon and pollen larger than 10 microns are

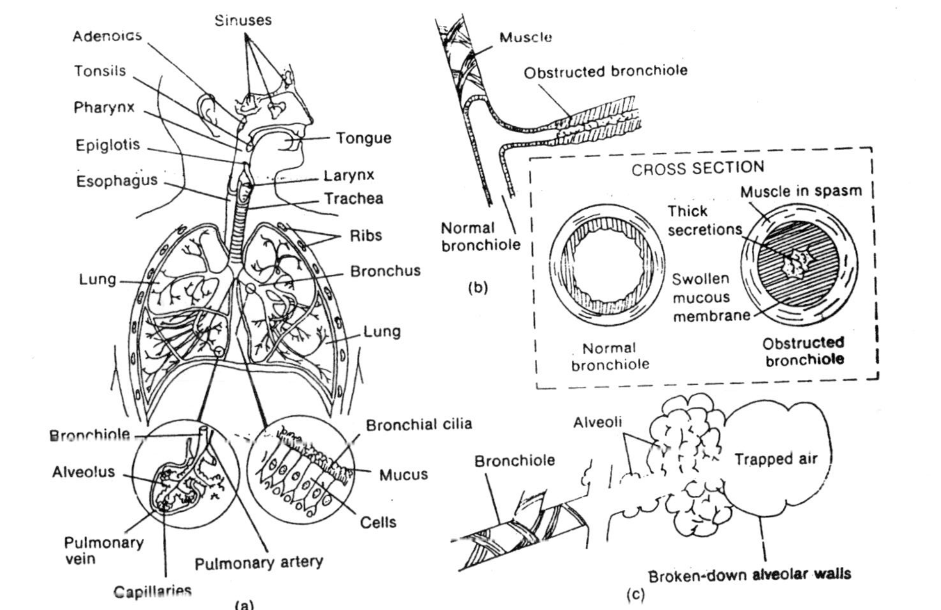

Fig. 12.2 *(a)* The respiratory system. *(b)* Obstruction of a bronchiole during an asthma attack. *(c)* In pulmonary emphysema, loss of elasticity and deterioration of alveoli walls deter exhalation of carbon dioxide gas.

usually kept out of the lungs (Fig. 12.3). Particles less than 1 micron readily pass to the lungs and become lodged. Coughing dislodges and expels some, but not all, such particles. Particles less than 0.01 micron will act like chemical molecules and can be absorbed through cell walls in the lungs.

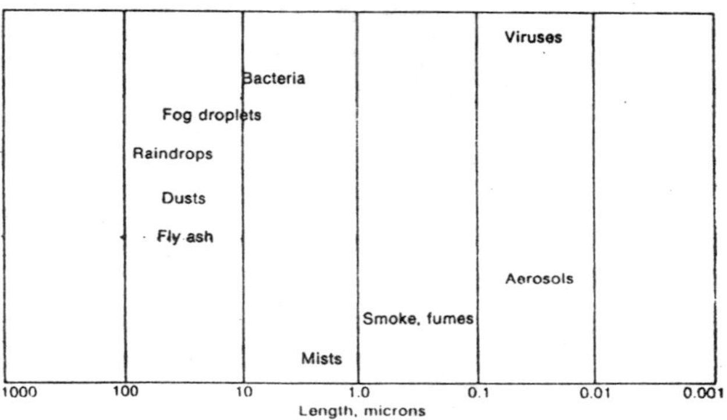

Fig. 12.3 Sizes of selected particulates.

Pollutants entering the body affect specific organs (Fig. 12.4). Gases such as chlorine, ammonia, and ozone are pulmonary irritants. Silica and asbestos are notorious among irritating dusts for causing fibrosis in the lungs. Some substances (beryllium, for example) are granuloma-producing agents, and some metals produce fever. Carbon monoxide (CO) and hydrogen sulfide (H_2S) are asphyxiating pollutants. Lead other metals, and some pesticides cause systemic poisoning.

Perhaps the people most aware of distress from air pollution are the *bay fever* sufferers. They are sensitive to natural pollens of flowers, trees, and grasses. These people and others can also become allergic to dusts and other pollutants or contract diseases through some airborne fungi.

As scientists link some 60 to 90 per cent of cancer to environmental stimuli, air pollutants become suspect. More lung cancer occurs among persons living in urban and industrial areas than among those in rural areas and among those in areas of low pollution. However, cancer cannot definitely be attributed to air pollution on this basis alone.

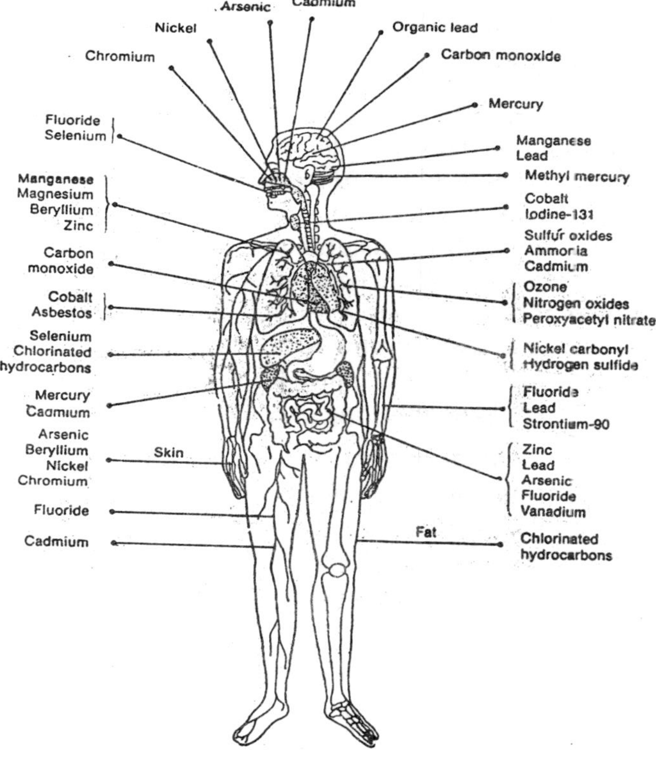

Fig. 12.4 Main targets of major air pollutants in the human body.

Years of research have preceded the identification of certain carcinogenic air pollutants. Most lung cancer deaths have been attributed to smoking cigarettes. Exposure to asbestos, especially when accompanied by smoking, has been associated with *mesothelioma*, a relatively rare tumor. *Chimney sweep's disease* is now recognized as a form of cancer (cancer of the scrotum). Organic compounds found in diesel exhaust and other combustion emissions are also suspected carcinogens. In addition to possibly causing cancer, some pollutants are suspected to be *mutagenic* (causing mutations in genes) or *teratogenic* (causing birth abnormalities).

Plants and Material Goods

Air pollution effects on plants can best be seen near the source of pollution. For example, tree foliage along turnpikes is damaged in a band where fumes from diesel truck exhaust touch the leaves. Cement dust deposited on leaves, when moistened, will form incrustations; other dusts plug the leaf openings, or *stomata*. Where ozone levels are high, pine needles turn brown and die (*necrosis*). Sulfur oxides can cause acute injury resulting in tissue drying to an ivory colour or darkening to a reddish brown. Chronic injury leads to pigmentation of leaf tissues and a gradual yellowing called *chlorosis* (Table 12.1).

Another control measure involves removing SO_2 by scrubbing stack gases. Flue gas desulfurization is 95 per cent efficient. In 1967 Japan began a programme to reduce sulfur emissions. Between 1970 and 1975, its SO_2 levels fell 50 per cent while energy consumption rose 120 per cent. As a result, Japan set new goals to limit SO_2 emissions even more.

Other studies indicate that acid deposition is also a serious problem in the western United States, although it is not yet well understood. The causes of acid deposition in the west are different from those in the east. Nitrogen oxides from transportation sources contribute more significantly to acid deposition in the west. Smelters and power plants presently account for the sulfur oxide emissions. Western power plants are likely to become a major source of SO_2.

Table 12.1 Examples of Pollutants Toxic to Plants

Pollutant	*Toxic level (parts per million)*	*Indicator Plants*	
		Sensitive	*Resistant*
Sulfur dioxide	0.1-3.0	Pumpkins	Potatoes
		Barley	Onions
		Squash	Corn
		Alfalfa	Maple
		Cotton	Most trees
		Wheat	
		Apples	
Fluoride	0.0001	Gladioli	Alfalfa
		Tulips	Roses
		Prunes	Tobacco
		Apricots	Tomatoes
		Pine	Cotton
Ozone	0.15	Tobacco	Mint
		Tomatoes	Geraniums
		Muskmelons	Carrots
		BeansGladioli	
		Spinach	Peppers
		Potatoes	Beans
Oxidant smog	0.2	Petunias	Cabbages
		Lettuce	Corn
		Oats	Cotton
		Pinto beans	Wheat
		Bluegrass	Pansies

In 1991 the Environmental Protection Agency ordered the coal-burning Navajo Generating Station to reduce sulfur dioxide emissions by 90 percent between 1995 and 1999. This action was taken to protect the once relatively clean air of Grand Canyon National Park, which is only 16 miles away from the power plant, and now likely to be polluted by the station.

It is important to aggressively control acid deposition in the West before it causes the serious damage to ecosystems that has been identified in the East. Acid deposition could threaten eleven national parks in the West, and human populations in urban areas could also be threatened—highly acidic fog has been reported around Los Angeles and San Francisco.

Some progress is being made in understanding and controlling acid deposition. A comprehensive, multiagency monitoring programme called the National Acid Precipitation

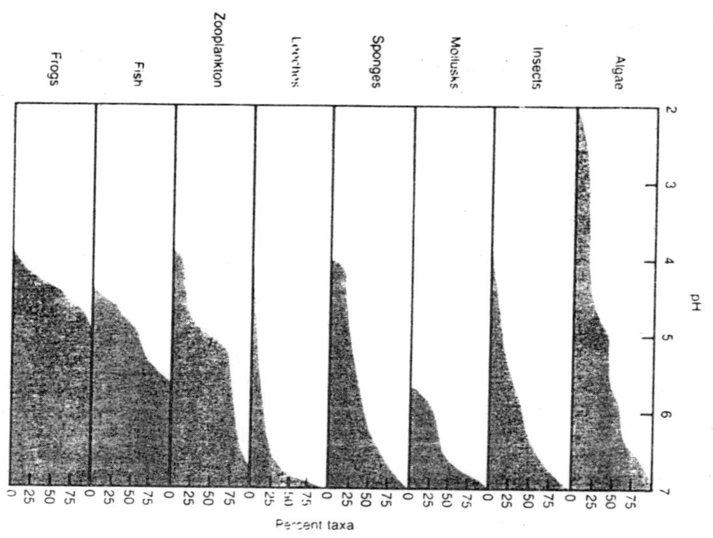

Fig. 12.5 As the pH of a lake goes down, the diversity of aquatic life is reduced until very few species survive.

Assessment Programme (NAPAP) was created in the late 1970s. NAPAP established a nationwide system of regional monitoring networks and a long-term research programme. The network has proven very helpful to researchers and decision-makers who are trying to understand and control acid deposition.

NOISE POLLUTION

Noise may be considered a form of air pollution. Since sound pressure waves move in every direction, their energy is spread over a larger and larger surface area as the sound moves away from the source. Consequently, the amount of sound energy spread over a square meter decreases in relation to the square of the distance from the source. Distance is therefore an important factor in reducing the effect and perception of noise (Fig. 12.6). While the effects of noise usually decrease with distance from the source, nearby surfaces such as building walls reflect noise and increase the perceived intensity. If noise is

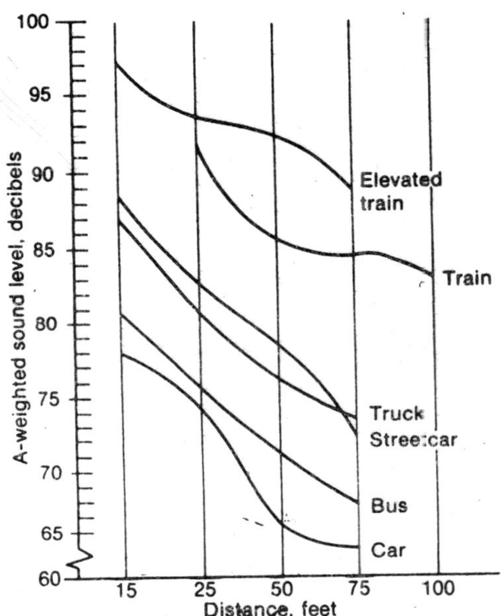

Fig. 12.6 A survey of specific noise sources shows that distance is a factor in noise intensity.

channelled between buildings or between other surfaces such as hot air ducts, the intensity is also increased. Some surfaces—curtains, rugs, acoustical tile, and others absorb noise and decrease its effects.

Industrial Noise

A number of industrial operations generate noise. Stamping metal into auto fenders, punching holes into metal plates, riveting plates together, and crushing different materials all produce impact noise, and grinding and drilling metal produce continuous noise. Rapid air motion caused by jets of air, blowers, and fans, and vibration of equipment also cause noise. Although industrial noise mainly affects workers in the industry, some of this noise also reaches nearby homes.

Community Sources of Noise

Most community noise originates from transportation. Transportation noises are generated by a vehicle's power unit,

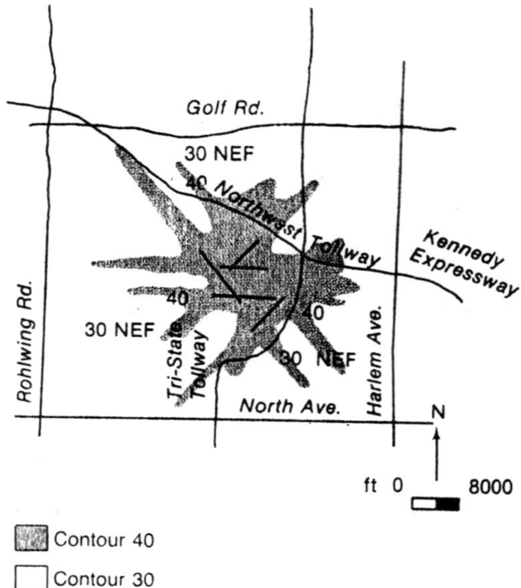

Fig. 12.7 Noise intensity for Chicago's O'Hare International Airport. The Noise Exposure Forecasts (NEF) contours for airports are determined by the number of daytime and nighttime jet aircraft operations, major runway locations and flight paths, and whether supersonic aircraft will use the airport. Outside contour 30, land is normally acceptable for residential housing, but hospitals, schools, and churches may need special construction to shield aircraft noise.

such as a jet's engine or a truck's motor, and from the contact between the tyres or wheels and the road or rails. The greatest amount of aircraft noise is produced upon landing and takeoff because more power is used at these times. Homes and businesses under and near the landing and takeoff path receive most of this noise (Fig. 12.7).

Greater vertical distances, like greater horizontal distances, reduce sound intensity. For example, the higher an airplane flies upon leaving or approaching an airport, the less the noise is perceived on the ground. However, because street noises are channelled up between buildings on each side of the street, taller structures up to three stories high receive more intense noise from street sources at their upper floors (Fig. 12.8).

Aside from the effect vehicle noise has on the surrounding community, the noise also has an impact on the vehicle's operator and occupants. The operators of large earth-moving machines frequently suffer hearing loss. Dangerous situations arise when vehicle noise masks the sounds of sirens and the driver is unaware of the warning sound of an emergency vehicle.

A residential community abounds with its own sources of noise that add to those from industry and transportation. The variety is almost endless: lawn mowers, loud radios and televisions, motorbikes, banging metal garbage cans, power tools, construction projects, amplified rock music, loud conversations, children's screams, barking dogs, roosters, and so on. The familiarity of a sound as well as its characteristics determine how annoying an individual will find the noise.

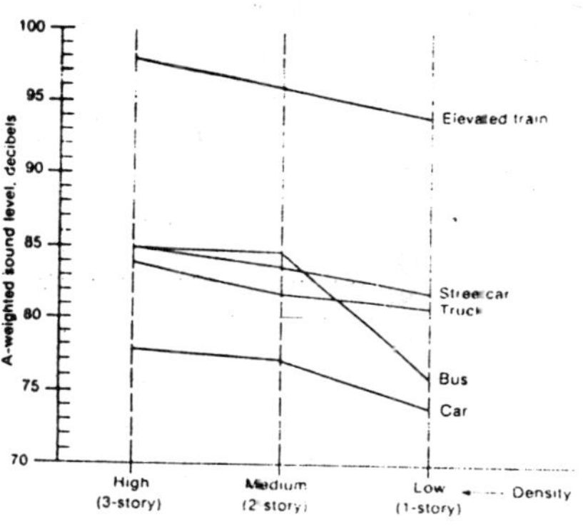

Fig. 12.8 Building height as a factor in noise propagation. Data suggest that ground-level noise is higher on streets with taller buildings.

People are more annoyed by noise in the evening and at night than at any other time. People on daytime work schedules use this time for rest and relaxation.

Residents of multiple-family dwelling units frequently complain of noise transmission between units. It is not just a matter of annoyance but affects the privacy of conversation and other activities. Mechanical services and equipment such as elevators, air-conditioners, and refuse handling also make noise that disturbs tenants.

Noise Control Methods

Noise is amenable to the same principles of environmental control advanced for other environmental problems. Substitution of a quieter machine design, process, or material may be an easy and effective means of eliminating or reducing a noise problems. This principle can be or has been applied in several ways. For example, a low speed propeller fan makes less noise than a high speed one, and a squirrel-cage blower makes less noise than a propeller-type fan. Because of improved design of the complete system, the new subway trains in San Francisco come into the station in a *swoosh* rather than a roar. Installing devices that resist motion and thereby damp vibration, can reduce noise from vibrating surfaces. Forces that cause vibration or using stiffer materials that resist vibration can sometimes be isolated by mounting the equipment on springs or resilient materials such as rubber and by using flexible connections. Jet exhausts can be modified to produce less air turbulence, and mufflers reduce noise from engine and air exhausts.

If noise cannot be reduced to acceptable levels in industry, personal protection or special design considerations may be necessary. Fitted earplugs sometimes protect the ear more effectively, but the visibility of earmuffs makes their use easier to supervise.

LAND POLLUTION

The land pollution has been mainly caused by solid wastes and chemicals. One of the major pollution problems of large cities has been the disposal of solid waste material including farm and animal manure, crop residues (agricultural wastes), industrial wastes like chemicals, fly ash and cinders which have been residues of combustion of solid fuels, garbage, paper, cardboard, plastics, rubber, cloth, leather, construction rubbish, brick, sand metal and glass resulting from demolition of buildings, dead animals like cattles, dogs, cats, birds, containers discarded manufactured products like old refrigerators, washing machines and autos. Huge quantities of unwanted material bring about serious disposal problems.

Main Land Pollutants

1. The major amount of solid rubbish has been provided by our households in the form of domestic wastes. Some common examples include groceries, food scraps, vegetable remains, packing material, paper, remnants of used coal, ash, wood, metals, plastics, ceramics, glass, etc.

Many of these have been non-reusable. All these form heaps of municipal refuse. It is not properly disposed of, this can prove perilous. Such places often become a dwelling place for rats, flies, bacteria, mosquitoes and a large number of other vectors, having the potential of causing many human diseases.

2. Modern agriculture is mainly responsible for polluting soil through the non-judicious use of chemical fertilizers, herbicides, insecticides and fumigants. Most of these have been stable chemicals and remain in the soil for long periods without degradation, and exert cumulative effect. Pollution caused by pesticides has been among the greatest causes of concern in the field of agricultural pollution. Pesticides residues occur in soil, air and water as well as in living organisms. Besides killing the living organisms present on the surface of the soil, they reach even the deeper layers through tilling and irrigation on the land, killing still more living forms. With the continuous use the soil microorganisms lose their capacity of nitrogen fixation.

3. Areas surrounding smelting and mining complexes are generally soiled by metals like cadmium, zinc, lead, copper, arsenic and nickle. These act as phytotoxic even in small quantities. They also make plants unsafe for human and animal consumption. Zinc along with cadmium, has been released into the environment during the use or breakdown of lubricating oils, vehicles tyres, galvanized metals and fertilizers.

4. The main sources of land pollution have been the industries like pulp and paper mills, oil refineries, power and heating plants, chemicals and fertilizer manufactures, iron and steel plants, plastic and rubber producing complexes and so on. Thousands and thousands tons of solid wastes have been either dumped or burnt or emptied into rivers. Most industrial furnaces give rise to a grey, power residue of unburnt material called fly ash, an important pollutant, besides huge mounds of solid wastes.

TRANSPORTATION OF SOLID WASTE

Of the total, 90 per cent solid waste is produced from agriculture and mining. The solid wastes could be usually disposed of with little difficulty in the rural areas where they are generated. Solid-waste collection, processing and disposal have been far more of a complex management programme in built-up urban communities. Urban-solid-wastes sources have been of two principal categories: industrial and municipal. Municipal waste is a collective term referring to residential, commercial, institutional, and demolition solid wastes. These subcategories will be described individually.

Residential Wastes

In urban areas, the volume and composition of residential waste may vary considerably in different communities and climates. In general per-capita waste production has increased with per-capita income, although in some localities solid-waste production has actually decreased while presumably income has not. They quantity of residential solid waste in a city also varies with the season of the year. Wastes like garden trimmings autumn leaves, clean-up campaigns, summer produce, Christmas residues, etc., are all seasonal in their generation.

The type of housing also influences the waste within a city. High-density areas produce less solid waste per capita; apartments produce solid waste with little or no grass and leaves. Areas having large numbers of garbage-disposal units produce much less food wastes. Typical residential solid waste by weight has been found to range 35 to 45 per cent paper, 10 to 20 per cent food waste, 5 to 10 per cent glass and ceramics, 5 to 10 per cent metals, 5 to 15 per cent grass and leaves, 2 to 6 per cent textiles, and usually less than 4 per cent wood, plastics, rubber, leather, and inert materials.

Commercial Wastes

The aggregate composition of commercial solid waste has been found to be similar to that of residential waste, but individual generation points have tendency to produce a more homogeneous waste stream. That is, restaurants obviously produce more food and related packaging wastes, markets produce some food and mostly paper, of carton wastes, whereas hotels produce considerable textile wastes, as well as more normal trash. The amount of solid waste generated per collection unit has been found to be much higher for a commercial solid-waste source than for a residential solid-waste source; because of this, collection vehicles in commercial areas fill up rapidly.

In dense urban areas, commercial solid-waste generation has been more than in less-populous land-use zones. Collection routes in which commercial and residential wastes may be picked up by the same organization may be combined so that vehicles could efficiently make both residential and some commercial stops. In many cities, commercial wastes are collected by private contractors with some municipal regulation. It is desirable that local administrators must be able to control the standards for all collections within a community to insure uniform, fair, and sanitary service conditions.

Industrial Wastes

These include food-processing residues, boiler ash, wood, plastic and metal scraps, chemical sludges, packaging waste, debris, dirt, etc. Increased industrial application of air and water-pollution controls will also cause an increase in the solid wastes which have been destined for land disposal. Industrial waste collection and disposal are usually undertaken by contractors, and less often by the producer. Industrial waste-generation quantities in a given area depends upon the types and sizes of the industries present rather than area population.

Demolition Wastes

Solid wastes due to demolition and construction activities have been including mixed lumber, roofing and sheeting scraps, stone, rubble, broken concrete, asphalt, brick, plaster, wallboard, pipe, wire, and other residual building materials.

Bulky Wastes

These are including trees, telephone poles, bushes, abandoned autos and auto parts, tyres, stoves, refrigerators or other large appliances, furniture, crates, mattresses, etc. Conventional collection vehicles fail to transport such material. Special open-truck vehicles have to be used for collection of bulky wastes which otherwise may be dumped and abandoned on public or private property. Tow trucks with dollies could be used to tow away abandoned vehicles. Vehicles collecting shrub and tree branches have been generally more efficient if equipped with volume-reduction equipment such as a chipper of leaf mulcher to reduce the cellulose wastes.

Hazardous Wastes

Wastes having pathological, explosive, radioactive, or toxic materials are regarded hazardous. Dangerous liquids and solid wastes sometimes get containerized and included in the solid-waste stream. Collection disposal crews must have the training to identify all hazardous wastes. Once hazardous wastes are identified, they must be handled separately, with appropriate safety procedures. Fire-fighting equipment and protective clothing must be present on special and regular collection vehicles for handling hazardous materials and coping with emergencies when accidents requiring emergency action take place.

Storage

Storing solid wastes prior to collection disallows offending aesthetic tastes, attraction of vectors, and excessive odours. Storage devices must be convenient for the user and facilitate, safe, efficient collection, processing, and disposal. Types of storage devices must be described as follows:

Containers

Rigid containers must be made from durable metal or heavy plastic, and covers should fit tightly. Pick-up handles should be strong because if a handle may break it could cause serious injury to the user or collector. Damaged containers having sharp edges should be immediately replaced to avoid cuts and bruises. Containers should be stored on a flat concrete base or elevated above the ground for reducing metal container corrosion and spilling. Some vinylplastic containers may get oxidized and crack with time or cold-exposure; polyethylene and certain durable plastics may have a superior useful life.

Bags

Plastic and paper storage bags and holders are having certain advantages and disadvantages. The various advantages have been as follows:

1. one-way disposal without retrieval;
2. a universal and standard-size, cleaner container;
3. faster collection because they do not have to be uncovered and returned after being emptied;
4. reduced spillage and resultant litter;
5. easier storage and handling as liners or in suspended bag holders;
6. reduced back strains and similar collector and user injuries because bags are far lighter, will not hold heavy loads, and are easier to handle than rigid cans; and
7. more sanitary and cause less noise, dust, odour, and microorganism release; and adaptable to mechanized collection by truck equipment.

In summary, plastic bags have been simpler to store, more expandable, less expensive, and are easier to tie and handle than comparable paper bags. However, both paper and plastic bags one more easily punctured by sharp objects, and dogs, raccoons, rats and the like can tear these open. Bags have been a continuing cost and contribute to solid waste in themselves. Thirty-two to 40-gallon bags and rigid containers are generally the standard for solid-waste collection. Plastic storage bags (polyethylene) have been commonly 1.5 mil thick while paper bags have been commonly two-ply and have plastic additives for wet strength. Soild-waste generation points that produce large amounts of solid waste between collections should use large bins (1.5-15 cu yd) rather than numerous smaller bags or containers. The larger bins also may use stationary compactors to reduce storage volumes and collection and disposal costs.

Twice-weekly collection has been desirable since fly propagation may be reduced, over once-per-week collection by 75 per cent. The bins should be seamless, watertight, sturdy, and maintained frequently to prevent vermin infestation and odours. Bin collection has been useful for handling large volumes of industrial, commercial, agricultural and high-density residential wastes.

Drop Boxes

Drop boxes have been larger storage bins and are useful for servicing rural, industrial, and demolition sites. Large bins with or without stationary compactors cost less per unit of solid waste than small bags, cans or bins. Centrally located boxes of 20- to 40-cu-yd capacity can service a large rural area at less cost than a small sanitary landfill. Vehicles capable of hoisting these boxes onto their chassis can be used for collection and transport. An emply box is normally dropped off when a full box is collected to avoid leaving the area without a solid-waste-storage device. This method needs one vehicle pickup trip for each collection, but could be an economical procedure if properly planned and managed.

Other Storage Devices

Backyard solid-waste shelters that need emptying by a collection crew using hand tools are inefficient, cause odours, support rats, flies, and other vermin, and create a fire hazard. Underground pit storage of containers is also undesirable because spilled garbage is difficult to reach and clean out, resulting in odours and rat and fly propagation. Hard-to-reach containers in pits may also need collectors to use dangerous lifting positions.

Compactors

Storage containers equipped with heavy-duty compaction devices may be able to reduce space requirements by increasing solid-waste densities severalfold. In communities which have outlawed private on-site solid-waste incinerators, compactors can get installed in the former incinerator locations. Household units, a faily new innovation, are currently being used only in more affluent residences. Compactors, shredders, and other volume-reduction devices obviously do not decrease the total weight of generated solid waste.

Source Segregation and Reclamation

Producer-provided segregation of reclaimable materials at the source has been a simple method of solid-waste separation. Segregated materials have minimum contamination, but need special equipment or extra trips for handling and collection.

Resource conservation has been presently the major motivation for waste segregation. For communities incinerating solid waste, the separation of combustibles and non-combustibles may be desirable. The dominant use of sanitary landfills for solid-waste disposal reduces the need for source segregation. Class-I disposal sites having little potential for groundwater contamination require no waste segregation since they have no limitation on acceptable wastes. Class-II disposal sites (located above the groundwater table and protected from surface runoff, with controllable surface drainage) can be able to accept ordinary household and commercial solid waste, including putrescible organics and most common scrap metal.

Source segregation of metal cans, paper and other valuable components may be becoming a public responsibility in the future. Many communities have in the past been practicity separate collection of garbage, cans, glass and paper products, but the trend has been toward combined collection to reduce collection and disposal costs. Unlimited collection of all generated solid waste at the source should cause a cleaner community; reduced service pickup or restricted costly collection systems result in open dumping and litter.

WATER POLLUTION

Oil

In marine systems, a common contaminant is crude oil. Oil can kill directly through coating and asphyxiation, especially acute for intertidal life, or by poisoning by contact or ingestion, as in plants and preening birds, respectively. Water-soluble fractions can be lethal to fish and invertebrates and may disrupt the body insulation of birds, resulting in their death from hypothermia. The jackass penguin, endemic to South Africa, has suffered severe population declines from oil lost by tankers rounding the Cape of Good Hope.

Though natural seepage of oil into the oceans, mainly at junctions of tectonic plates, occurs at the rate of 600,000 tons per year, one large tanker can carry 200,000 tons of oil. The *Amoco Cadiz*, wrecked off France in 1978, was a 230,000 ton vessel. Oyster beds along the Brittany coast were totally ruined, fishing and resort industries were affected, and in tidal flats and salt marshes, effects lasted for seven years. Estimates of the total damages at the time were of the order of $30 million. It is important to remember that clean-up operations with detergents can prove more devastating to marine life than the oil itself, which is what happened in Britain after the *Torrey Canyon* disaster. Sometimes, it is best merely to encourage the natural breakdown of oil. Bioremediation following recent oil spills suggest that fertilizer applications significantly increase the rate of biodegradation of oil by existing microorganisms.

A particularly severe oil spill resulted from the wreck of the *Exxon Valdez* in Alaska at 12.40 A.M. on Good Friday, March 24,1989. The rocks at Bligh Island tore five huge gashes in the hull of the ship, one 6 ft high by 200 ft long. The result was one of the worst oil spills in U.S. waters; 10.7 million gal spilled, most of it in the first 12 hours. The accident took place in fine weather, clear visibility, and no traffic and was clearly the result of human error. A week after the spill, the resultant slick covered nearly 900 sq mi. Hundreds of miles of shoreline were covered with oil, in places as much as 6 in. deep. Officially, 27,000 birds, 872 sea otters, and untold numbers of fishes died, although the true numbers are probably higher because many dead birds and otters probably sank and were not recovered. These deaths resulted because birds and otters depend on the insulation provided by their feathers and fur to help them maintain proper body temperature. A coating of oil destroys that insulating property, and the animals literally die of exposure.

The effects also carried over into the terrestrial ecosystem when bears, otters, and bald eagles feasted on the oily carrion washed upon the beach. Sitka black-tailed deer ate kelp on the beaches. Few of these animals were expected to be found dead on the beaches because they generally return to their normal habitats before the effects become apparent. Still, the Fish and Wildlife Service found more than 100 dead eagles, and most pairs in the area failed to produce young that year. Much money was spent in the rescue and rehabilitation of oiled wildlife, including sea otters and birds. It is valuable to examine the success of these types of procedures, particularly as it seems that money might be better spent in other protective measures.

In total, 357 sea otters were captured and delivered to rehabilitation facilities. Of these, 123 died in captivity. Thirty-seven of the 234 survivors were judged unsuitable for return to the wild and were transferred to aquaria and other permanent holding facilities; 25 of these animals were still alive 10 months later. The remaining 197 survivors were released by August 1989, 45 of them with surgically implanted radios. Twenty-two of the instrumented animals were dead (11) or missing (11) the following spring, thus indicating relatively low post-release survival of the captured and treated animals. At best, 222 sea otters (the 197 released and 25 in captivity) were captured and rehabilitated. A total of 878 dead sea otters were found, and this probably represents only 20 per cent of the total killed by oil.

Capture and rehabilitation costs for sea otters alone were $18.2 million. Assuming that 222 otters were saved (maximum possible), costs exceeded $80,000 per animal. Post-oil spill capture and rehabilitation probably cannot be used to substantially reduce the otter losses. Perhaps a more effective strategy to protect populations of sea otters is to spend the equivalent amount of money to reduce the risks and to enhance populations of threatened species in anticipation of potential catastrophic loss. Money could perhaps be better spent in properly documenting population densities and studying their population biologies. Such documentation is necessary if we are to know the impact of any impending catastrophe.

Eutrophication

Eutrophication is the enrichment of waters with nutrients primarily phosphorus and nitrogen. This usually leads to enhanced plant growth. These changes may occur as a result of anthropogenic changes (cultural eutrophication) or as a result of succession, the natural aging of a lake (natural eutrophication). Bodies of water that are not rich in nutrients are called *oligotrophic,* and those rich in nutrients are called *eutrophic.* The most common attributes of a eutrophic lake are blooms of algae that make the water more turbid, more unattractive to swimmers and less suitable to certain kinds of fish.

What is the state of eutrophication of lakes around the world? The data indicate that there is great disparity between countries and regions. Of the three-quarters of a million lakes in Canada, the great majority, 75 per cent, are still oligotrophic, although the lakes of the southern, more densely populated regions are predominantly eutrophic, which reflects the increase in eutrophication associated with anthropogenic effects. Among the smaller U.S. lakes, there are many that are surrounded by farmland. Their condition, therefore, reflects eutrophication from agriculture and animal husbandry. The result is that up to 70 per cent of U.S. lakes may be eutrophic. In Europe, too, many lakes seem to be eutrophic.

However, the great majority of lakes and rivers in the eutrophic category are relatively small in both surface area and volume, averaging only 2.2 km^3 in volume compared to 67.6 km^3 for oligotrophic lakes. Smaller lakes are perhaps more susceptible to cultural eutrophication than bigger lakes. Because there are many small lakes and few big ones, the percentage of lakes that are eutrophic is high. However, the volume of freshwater in a eutrophic condition is small, only 12 per cent of the volume compared to 52 per cent classified as oligotrophic. The other 36 per cent of lake water is in intermediate condition. It is clearly difficult to reach a consensus on the state of eutrophication of U.S.lakes. On the other hand, many of the smaller water bodies are important either for drinking-water supply or for recreational purposes and should be cleaned up. There is clearly room for improvement.

The data for rivers are less definitive, but eutrophication effects are generally less acute than for standing waters because nutrient inputs are often quickly washed away. On the other hand, the data for human-constructed reservoirs show much higher rates of eutrophication than the natural lakes, although the volume of these waters is small in comparison to the area and volume of natural lakes.

How can we control cultural eutrophication? First, we have to know what the anthropogenic causes are. The degree of eutrophication for the Great Lakes shows a striking resemblance to the maps of human population density along the shores. Similarly, areas mostly affected by eutrophication in the Mediterranean coincide with densely populated lands, most of which are either areas of intensive agriculture or high industrial development. High population densities lead to cultural eutrophication via three pathways:

1. A strong tendency for urban waste to increase and be discharged directly into waterways. An added factor to this since the 1940s is the use of detergents containing polyphosphates.
2. Rapid industrialization with a corresponding increase in industrial wastes of all kinds.
3. Intensification of agriculture and the increased use of chemical fertilizers, especially those containing phosphorus, concentration of livestock breeding, and direct discharge of agricultural wastes, rich in nitrogen, into waterways.

The measures to control eutrophication fall into two main headings: preventive measures and corrective measures. Preventive measures include:

1. Treatment of waste waters (removal of phosphorous and nitrogen);
2. Diversion of waste waters from lakes or rivers.

3. Primary sedimentation basins in waste-water streams to let phosphorus and nitrogen-rich material settle to the bottom where it can be removed;
4. Watershed protection (reforestation, restriction of livestock, controlled fertilization/irrigation);
5. Substitution of phosphate detergents by other detergents not rich in phosphorus.

Corrective measures to try to bring eutrophic lakes back to an oligotrophic condition include:

1. Physical manipulation (withdrawal of water; aeration to increase levels of oxygen);
2. Chemical manipulation (application of herbicides to kill blooms of algae or large plants [macrophytes]);
3. Biological manipulation (mechanical harvesting of algae, macrophytes; direct manipulation of the food chain by adding exotic fish).

Treatment of waste waters is already underway in many areas. The reduction of polyphosphates in detergent has been imposed by law in Canada and some U.S. states bordering on the Great Lakes. In Sweden, 80 per cent of treatment plants include a third stage for the elimination of phosphorus, and only 20 per cent of waters discharged into the waterways receive no treatment. Strategies for control of nitrogen outputs, especially from agricultural activities, are in less well-developed stages; but there are some dramatic success stories in lake restoration.

It would be unfair to create the impression that water pollution is a totally insoluble problem or that levels of pollution must inevitably rise. The complete case study of the eutrophication of Lake Washington and its reversal to an unpolluted condition has been well documented by Lehman (1986). Lake Washington at Seattle is a moderately deep (65m) warm basin that discharges into Puget Sound via a system of locks and canals built in 1916. Seattle began discharging raw sewage into Lake Washington at the beginning of the twentieth century, but in 1926 this trend began to change as sewage was diverted into treatment plants and thence into Puget Sound. By 1941, the last sewer outfall into the lake was removed, but thereafter the suburbs of Seattle began to expand, and by 1953, ten new treatment plants had sprung up around the lake and were discharging 80 million litres into it daily. Alternative options were not as readily available to the suburbs as they had been to Seattle itself. By this time, both the scientists at the University of Washington and the lay public were aware that algae had increased in the lake, and a species indicative of classical lake deterioration, *Oscillatoria rubescens,* was found in the lake for the first time. This alga had been connected with the early stages of decline in water quality in the classical examples from Europe, particularly Lake Zurich, and in Madison, Wisconsin. Scientists monitoring the Lake Washington situation were able to predict with great certainty the demise of the lake and the newspapers were quick to pick up these predictions. Such attention by the media paid off because by 1958, local politicians voted money to clean up the lake. Ground-breaking ceremonies for the clean-up campaign were held in 1961, not a moment too soon as this body of water had already been christened "Lake Stinko" by the local press. Visibility in lake water had declined from 4 m in 1950 to less than 1 m in 1962. One by one, the waste-treatment plants around the lake had their effluent diverted. The trend of lake deterioration stopped in 1964, and by 1965 algal abundance was decreasing and water transparency increasing. Surprisingly, *Oscillatoria* persisted into the 1970s, but by 1975, it was gone and phosphorus concentrations had leveled off. The lake was clean again; visibility was as high as 12 m at times! The whole process was later repeated in Canada, this time in a controlled-environmental setup, by limnologists in a large-scale series of experimental lakes in Ontario. The results showed beyond a shadow of doubt that phosphorus was the master controlling agent of eutrophication.

RADIOACTIVE WASTES

The two major concerns about atomic energy is that it generates radioactive waste and that the possibility of an accident in a nuclear-power plant could spew radiation over a wide area.

From radioactive wastes comes ionizing radiation—radiation with sufficient energy for its interactions with matter to produce an ejected electron and a positively charged ion. The biological problem is that such interactions in the cells of living organisms can cause genetic and physiological damage and even death. Most phases of the fuel cycle of a nuclear reactors produce radioactive substances. Some are very short-lived and can be released in carefully controlled quantities to the atmosphere or bodies of water, but others can have long half-lives and must be kept away from the environment and from people for a very long time. Strontium-90 has a half-life of 28 years, which means that it would take 500 years to drop to one-millionth of its original activity and the recommended level of safety. Plutonium has a half life 24,400 years. Fortunately, these wastes are not produced in large quantities: one ton of spent nuclear fuel when reprocessed gives 500 liters of waste. The disposal of these wastes, however, causes many problems because they must be isolated from the biosphere for 250,000 years, a high societal commitment. Suggestions for disposal have included sinking them into the Antarctic ice cap, firing them into the sun, or placing them on interplate subductions to be carried into the Earth's core. The most popular, however, is geological disposal in rock formations without any groundwater circulation; salt deposits (especially abandoned salt mines) are favoured.

Studies of the effects of ionizing radiation on ecosystems have been carried out in oak-pine forests, in tropical rain forests, in hardwood forests in the northern and southern United States, and in a shortgrass prairie. One useful generalization to emerge is that the sensitivity of plants depends on the ratio of photosynthetic tissue to total tissue. Thus, the most sensitive plants are trees, which have a relatively low ratio of photosynthetic mass (leaves) to nonphotosynthetic mass (stem and root), then herbs, grasses, and finally algae, which are usually quite resistant. Among trees, pines, which are evergreen, are more sensitive than deciduous hardwoods, in which the investment in leaves is smaller. A generalization that applies to both plants and animals is that sensitivity to radiation is correlated with size, the largest species being the most sensitive, as they are to stress in general.

Problems resulting from nuclear-power plants are not simply going to go away; 1 kg of uranium 235 produces as much commercial energy as 2,500 tons of coal, and nuclear power is thus a very attractive energy source. In addition, use of nuclear power would lessen the production of SO_2 and CO_2 from the burning of coal. Acid rain and the threat of global warming would both diminish.

Nuclear Accidents

The greatest risk of nuclear power may lie not in waste production but in accidents at the plant. No full *meltdown* has yet occurred in the United States, but there have been a large number of minor accidents, the most famous of which was the Three Mile Island incident on March 28, 1979, in Harrisburg, Pennsylvania. A faulty water pump caused overheating, which was followed by a whole series of human errors, which in turn allowed a small amount of radioactive water to be released into a nearby river. No deaths or serious injuries were reported, but the clean-up cost at least $1.5 billion and lasted more than 10 years. Other locations have not been so lucky; the Chernobyl disaster in Russia occurred on April 26, 1986. Two explosions inside one of the four water-cooled reactors blew the roof off the reactor building and set the graphite core on fire. As usual, human error was to blame; engineers had turned the automatic safety systems off for an experimental test. At least 31 plant workers died from exposure to radiation, and 200 others suffered acute radiation sickness. Winds carried radioactive material over much of Europe (Fig. 12.9). In the nearest village, trees in an area about 6 km^2 were killed within a few days of the accident. By 1991, the area of dying trees had reached 38 km^2. In this area, rodents showed strong population declines, and even insect populations, normally highly radio-resistant, were depressed.

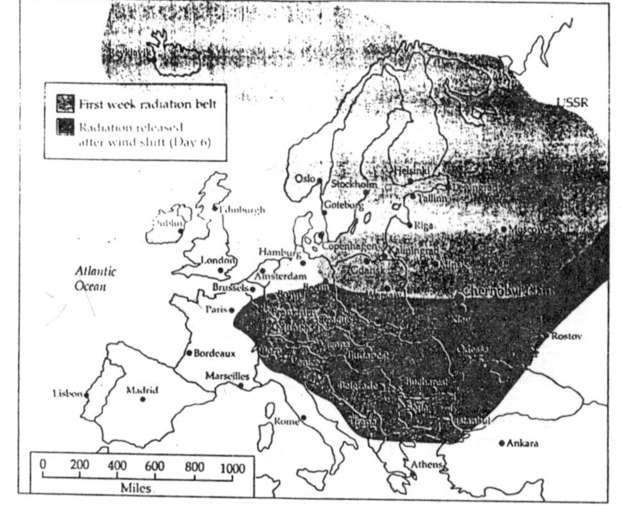

Fig. 12.9 Spread of radioactive fallout over parts of the former Soviet Union and much of eastern and western Europe after the Chernobyl accident.

OUR IMPACT ON THE ENVIRONMENT

The simplest way to gain a feeling for the dimensions of the global environmental problem that we face is simply to scan the front pages of any newspaper or news magazine or to watch television. Although they are only a sampling, features selected by these media teach us a great deal about the scale and complexity of the challenge we face. We will discuss a few of the most important here.

Nuclear Power

At 1:24 A.M. on April 26, 1986, one of the four reactors of the Chernobyl nuclear power plant blew up. Located in Ukraine, 100 kilometers north of Kiev, Chernobyl was one of the largest nuclear power plants in Europe, producing a thousand megawatts of electricity, enough to light a medium-sized city. Before dawn on April 26, workers at the plant hurried to complete a series of tests of how the generator Reactor Number 4 performed during a power reduction and took a foolish short-cut: they shut off all the safety systems. Reactors at Chernobyl were graphite reactors designed with a series of emergency systems that shut the reactors down at low power, because the core is unstable then—and the workers turned these emergency systems off. A power surge occurred during the test, and there was nothing to dampen it. Power zoomed to hundreds of times maximum, and a white-hot blast with the force of a ton of dynamite partially melted the fuel rods and heated a vast head of steam that blew the reactor apart.

The explosion and heat sent up a plume 5 kilometers high, carrying several tons of uranium dioxide fuel and fission products. Over 100 megacuries of radioactivity were released, making it the largest nuclear accident ever reported. By way of comparison, the Three Mile Island accident in Pennsylvania in 1979 released 17 curies, millions of times less. This cloud traveled first northwest, then south-east, spreading the radioactivity in a band across central Europe from Scandinavia to Greece. Within a 30-kilometer radius of the reactor, at least one-fifth of the population, some 24,000 people, received serious radiation doses (greater than 45 rem). Thirty-one individuals died as a direct result of radiation poisoning, most of them firefighters who succeeded in preventing the fire from spreading to nearby reactors.

For the rest of Europe, the radiation dose was much lower but still significant. Data indicate that radiation outside of the immediate Chernobyl area will be expected to be responsible for from 5000 to 75,000 cancer deaths because of the large numbers of people exposed.

The Promise of Nuclear Power: Our industrial society has grown for over 200 years on a diet of cheap energy. Until recently, much of this energy has been derived from burning wood and fossil fuels: coal, gas, and oil. However, as these sources of fuel become increasingly scarce and the cost of locating and extracting new deposits more expensive, modern society is being forced to look elsewhere for energy. The great promise of nuclear power is that it provides an alternative source of plentiful energy. Although nuclear power is not cheap—power plants are expensive to build and operate—using current technology and its raw material, uranium ore, is so common in the earth's crust that it is unlikely we will ever run out of it.

The burning of coal and oil to obtain energy produces two undesirable chemical by-products, sulfur and carbon dioxide. The sulfur emitted from burning coal is a principal cause of acid rain, while the CO_2 produced from the burning of all fossil fuels is a major greenhouse gas (see discussion of global warming on next page). For these reasons, we need to find replacements for fossil fuels, apart from the fact that in time they will eventually run out.

For all of its promise of plentiful energy, nuclear power presents several new problems that must be mastered before its full potential can be realized. You have met one serious challenge already in this chapter: the need to ensure safe operation of the world's approximately 390 nuclear reactors. A second challenge is the need to safely dispose of radioactive wastes produced by the plants and to safely decommission plants that have reached the end of their useful lives (about 25 years). In 1990 approximately 53 plants were more than 25 years old, and not one has been safely decommissioned. The third challenge is the need to guard against terrorism and sabotage, because the technology of nuclear power generation is closely linked to that of nuclear weapons.

For these reasons, it is important to continue to investigate and develop other alternatives to fossil fuels, such as solar energy and wind energy, which hold great promise when properly developed. The generation of electricity by burning fossil fuels accounts for up to 15% of global warming gas emissions in the United States. As much as 75% of the electricity produced in the United States and Canada currently is wasted through use of inefficient appliances, according to scientists at Lawrence Berkeley Laboratory. The use of highly efficient motors, lights, heaters, air conditioners, refrigerators, and other technologies that are currently available could lead to saving large amounts of energy and greatly alleviate the problem of global warming gas emission. For example, a new, compact fluorescent light bulb uses only 20% of the amount of electricity of conventional lighting, provides equal or better lighting, lasts up to 13 times longer, and provides substantial cost savings.

Carbon Dioxide and Global Warming

By studying earth history and making comparisons with other plantes, scientists have determined that concentrations of gases, particularly carbon dioxide, in the atmosphere maintain the average temperature on earth about 25^{0}C higher than it would be if these gases were not present. Carbon dioxide and other gases trap the longer wavelengths of infrared light, or heat, radiating from the surface of the earth. By doing so, they create what is known as the *greenhouse effect*. The atmosphere acts like the glass of a gigantic greenhouse surrounding the earth.

Roughly seven times as much carbon dioxide is locked up in fossil fuels, approximately 5 trillion metric tons, as exists in the atmosphere today. Before widespread industrialization, the concentration of carbon dioxide in the atmosphere was approximately 260 to 280 parts per million (ppm). Since the extensive use of fossil fuels, the amount of carbon dioxide in the atmosphere has been increasing rapidly. A major proportion of carbon dioxide added to the atmosphere also comes from the destruction of forests, which releases large amounts of carbon dioxide when the trees and other vegetation is burned. During the 25-year period starting in 1958, the concentration of carbon dioxide increased from 315 ppm to more than 340 ppm. Climatologists have calculated that the actual mean global temperature has increased about 1^0C since 1900, a phenomenon known as *global warming*.

In a recent study, the U.S.National Research Council estimated that the concentration of carbon dioxide in the atmosphere would pass 600 ppm (roughly double the current level) by the third quarter of the next century, and that level might be expected as soon as 2035. Such concentrations of carbon dioxide, if actually reached, would lead to a global surface-air-warming of between 1.5^0 and 4.5^0C. The actual increase might be considerably greater, however, because a number of trace gases, such as nitrous oxide, methane, ozone, and chlorofluorocarbons, are also increasing rapidly in the atmosphere as a result of human activities. These gases have warming or "greenhouse" effects similar to those of carbon dioxide. They all absorb infrared wavelengths more efficiently than carbon dioxide, and their combined effect might be more significant. For example, one of these gases, methane, has increased from 1.14 ppm in the atmosphere in 1951 to 1.68 ppm in 1986—nearly a 50% increase.

Major problems associated with climatic warming include rising sea levels. Sea levels may have already risen 2 to 5 centimeters due to this reason. If the climate becomes so warm that the polar ice caps melt, sea levels would rise by more than 150 meters, flooding the entire Atlantic coast of North America for an average distance of several hundred kilometers inland. Increased levels of carbon dioxide would cause some plants to grow more vigorously than they do at present and would greatly change the relationships between plants and other organisms in natural communities—giving some an advantage and suppressing others. Since the effects of the atmospheric changes described will vary locally, they are perhaps even better described as leading to *global climate change* than simply to global warming. Distribution of precipitation is difficult to model and of great importance to local productivity. Certainly, changing climatic patterns are likely to make some of the best farmlands much drier than they are at present. If the climate warms as rapidly as many scientists project, the next 50 years may be marked by greatly altered weather patterns, a rising sea level, and major shifts of deserts and fertile regions. Our effects on the world's atmosphere must be better understood and considered seriously for our common welfare in the future.

The Threat of Pollution

The pollution of the Rhine is a story that can be told countless times in different places in the industrial world, from Love Canal in New York to the James River in Virginia to the town of Times Beach in Missouri. Nor are all pollutants that threaten the sustainability of life immediately toxic. Many forms of pollution arise as by-products of industry. For example, the polymers known as plastics, which we produce in abundance, break down slowly, if at all, in nature. Scientists are attempting to develop strains of bacteria that can decompose plastics, but their efforts have been largely unsuccessful. Consequently, virtually all of the plastic items that have ever been produced are still with us. Collectively, they constitute a new form of pollution for which there is, as yet, no solution.

Water pollution is another serious problem that exists on a global scale. There is simply not enough water available to dispose of the diverse substances that today's enormous human population produces continuously. Despite the implementation of ever improved methods of sewage treatment throughout the world, our lakes, streams, and groundwater are becoming increasingly polluted. Household detergents containing phosphates that flow into oligotrophic lakes can lead to their eutrophication. This leads to an overgrowth of algae and a rapid deterioration of water quality.

Widespread agriculture, carried out increasingly by modern methods, causes large amounts of many new kinds of chemicals to be introduced into the global ecosystem. These include pesticides, herbicides, and fertilizers. Industrialized countries like the United States now attempt to carefully monitor side effects of these chemicals. Unfortunately, large quantities of many toxic chemicals that were manufactured in the past still circulate in the ecosystem.

For example,the chlorinated hydrocarbons, a class of compounds that includes DDT, Chlordane, lindane, and dieldrin, have all been banned for normal use in the United States, where they were once widely used. They are still manufactured in the United States for export and used in other countries, often finding their way back to the United States as contaminants on fruits and vegetables. Chlorinated hydrocarbon molecules break down slowly and accumulate in animal fat. Furthermore, as they pass through a food chain, they become increasingly concentrated. DDT caused serious problems by leading to the production of thin, fragile eggshells in many predatory bird species in the United States and elsewhere until the late 1960s, when it was banned in time to save the birds from extinction. Chlorinated compounds have other undesirable side effects and exhibit hormone like activities in the bodies of animals. These effects, and the appropriate action to take concerning the chemicals, are under active investigation.

Obviously, a "back to nature" approach, one that ignores the important contributions made to our standard of living by the intelligent use of chemicals, will not allow us to care adequately for the needs of the current world population. Nor would it allow us to feed the additional billions of people who will join us during the next few decades. On the other hand, it is essential that we use our technology as intelligently as possible and with due regard for the protection of the productive capacity of all parts of the earth.

Acid Precipitation

The four smokestacks are part of the Four Corners power plant in New Mexico. This facility burns coal, sending smoke up high into the atmosphere with these stacks, each over 65 meters tall. The smoke that the stacks belch out contains high concentrations of sulfur dioxide and other sulfates, which produce acid when they combine with water vapour in air. The intent of those who designed the plant was to release the sulfur-rich smoke high up in the atmosphere, where winds would disperse and dilute it. This sort of solution to the problem posed by burning high sulfur coal was first introduced in Britain in the mid-1950s and rapidly became popular in the United States. Basically, the stacks carry the acids produced by the fuels away from the areas where they are produced: London no longer suffers from acid fogs, but the forests and lakes of Sweden are being destroyed.

Environmental effects of this acidity are serious. Sulfur introduced into the upper atmosphere combines with water vapour to produce sulfuric acid, and when the water later falls as rain or snow, the precipitation is acid. Natural rainwater rarely has a pH lower than 5.6; in the northeastern United States, rain and snow now have a pH of about 3.8, about a hundred times as acid as the usual limit. Because the prevailing winds in the temperate latitudes (where most industries are concentrated) are westerlies, sulfur emissions released by plants in the midwestern United States primarily return to earth in rain and snow that falls in the eastern United States and Canada. Similar patterns occur in Europe.

Acid precipitation destroys life. Thousands of lakes in northern Sweden and Norway no longer support fish; these lakes are now eerily clear. In the northeastern United States and eastern Canada, tens of thousands of lakes are dying biologically as a result of acid precipitation. At pH levels below 5.0, many fish species and other aquatic animals die, unable to reproduce under these conditions. In southern Sweden and elsewhere, groundwater is now found to have a pH between 4.0 and 6.0, its acidity resulting from acid precipitation that is slowly filtering down into the underground reservoirs, thus threatening water supplies of future generations.

There has been enormous forest damage in the Black Forest in Germany and in the forests of the eastern United States and

Canada. It has been estimated that at least 3.5 million hectares of forest in the northern hemisphere are being affected by acid precipitation, and the problem is clearly growing.

Its solution at first seems obvious: capture and remove the emissions instead of releasing them into the atmosphere. However, there are serious difficulties in executing this solution. First, it is expensive. Reliable estimates of the cost of installing and maintaining the necessary "scrubbers" in the United States are of the order of 4 to 5 billion dollars per year. Although this is less than 1% of the amount that will ultimately be spent to "bail out" failed savings and loan associations, our national priorities are not yet clearly focused on a healthy environment. An additional difficulty is that the polluter and the recipient of the pollution are far from each other, and neither wants to pay for what they view as someone else's problem. The Clean Air Act revisions of 1990 addressed this problem in the United States significantly for the first time, and substantial progress has been made in industrialized policies worldwide in implementing a solution.

The Ozone Hole

The swirling colours of fig. 12.10 are a view of the South Pole in 1989 as viewed from a satellite. This is a computer reconstruction in which colours represent different concentrations of *ozone* (O_3), a different form of oxygen gas than O_2. As you can see, over Antarctica there is an "ozone hole" that is about the size of the United States, within which the ozone concentration is much less than elsewhere. This ozone hole was first reported in 1985 by British environmental scientists. Reviewing available satellite data, we now know the zone of ozone thinning appeared for the first time in 1975. The hole is not a permanent feature, but rather one that becomes evident each year for a few months at the onset of the Antarctic spring. Every September from 1975 onward, the ozone "hole" has reappeared. Each year the layer of ozone is thinner and the hole is larger, sometimes reaching southern New Zealand, Australia, and southern South America. In 1985 the minimum ozone concentration in the hole was 30% lower than five years earlier.

The major cause of the ozone depletion was suggested in the early 1970s, but general acceptance was slow. Chlorofluorocarbons (CFCs) are chemicals that have been manufactured in large amounts since they were invented in the 1920s. They are largely used in cooling systems, fire extinguishers, and styrofoam containers. CFCs were percolating up through the atmosphere and reducing O_3 molecules to O_2. Although other factors have also been implicated in ozone depletion, the role of CFCs is so predominant that worldwide agreements to phase out their production by the year 2000 have been signed. Production of CFCs and other ozone destroying chemicals was banned in the United States after 1995. Nonetheless, large amounts of CFCs that were manufactured earlier are moving slowly upward through the atmosphere. The ozone layer will be further depleted before it begins to reform.

Thinning of the ozone layer in the stratosphere, 25 to 40 kilometers above the surface of the earth, is a matter of serious concern. This layer protects key biological molecules, especially proteins and nucleic acids, from the harmful ultraviolet rays that bombard the earth continuously from the sun. Life on land may have become possible only when the oxygen layer was sufficiently thick that it generated enough ozone, in chemical equilibrium with the oxygen, that the surface of the earth was sufficiently shielded from these destructive rays. This factor may account for the billions of years in which all life was aquatic. There may have been too much opportunity for damage to biologically significant molecules for organisms to exist on dry land.

Ultraviolet radiation is a serious human health concern. Every 1% drop in the atmospheric ozone content is estimated to lead to a 6% increase in the incidence of skin cancers. Skin cancers are one of the more lethal diseases afflicting humans. Humans are relatively resistant to increased ultraviolet radiation, but other organisms, such as photosynthetic plankton species that are so important to global productivity, are apparently such more susceptible.

Destruction in the Tropics

More than half of the world's human population lives in the tropics, and this percentage is increasing rapidly. For global stability, and for the sustainable management of the world ecosystem, it will be necessary to solve the problems of food production and regional stability in these areas. World trade, political and economic stability, and the future of most species of plants, animals, fungi, and microorganisms depend on our addressing these problems.

Many people in the tropics have traditionally engaged in *shifting agriculture*—clearing a patch of forest, growing crops for a few years, and then moving on. The fertility of the soil has by then returned to its original, very low level, and the temporary enrichment that resulted from cutting and burning the forest has been exhausted. Such agricultural systems work well where human populations are relatively low. But as these numbers grow, there is little opportunity even for the cultivation of traditional crops such as manioc (tapioca, cassava). Firewood gathering is also hastening the demise of many tropical forests. About 1.5 billion people worldwide, a third of the global population, depend on firewood as their major source of fuel. They are cutting the local supplies faster than the trees can regenerate themselves.

Shifting agriculture is a major factor in the destruction of tropical forests, but there are a number of other significant factors. We can illustrate them by reference to the tropical rain forest, biologically the richest of the world's biomes. Most other kinds of tropical forest, such as seasonally dry forests and savanna forests, have already been largely destroyed. These forests tend to grow on more fertile soils and were exploited by humans a long time ago. In the mid-1990s, it is estimated that about 5.5 million square kilometers of tropical rain forest still exist in a relatively undisturbed form. This area, about two-thirds of the size of the United States (excluding Alaska), represents about half of the original extent of the rain forest. From it, about 160,000 square kilometers are being clear cut per year, with perhaps an equivalent amount severely disturbed by shifting cultivation, firewood gathering, and allied practices.

The total area of tropical rain forest destroyed—and therefore permanently removed from the world total—amounts to an area greater than the size of Indiana each year. At such a rate, all of the tropical rain forest in the world will be gone in about 30 years, but in many regions, the rate of destruction is much more rapid. As a result of such overexploitation, experts predict there will be little undisturbed tropical forest left anywhere in the world by early in the next century. Many areas now occupied by forest will still be tree covered, but those trees will represent only a small percentage of those that now grow in these areas.

Not only does the loss of tropical forests represent a tragic loss of largely unknown biodiversity, as we will discuss below, the loss of the forest themselves is ecologically a serious matter. Tropical forests are complex, productive ecosystems that function well in the areas where they have evolved. When we cut a forest or open a prairie in the North Temperate Zone, we provide the basis for a farm that we know can be worked for generations. In most areas of the tropics, we simply do not know how to engage in continuous agriculture. When we clear a tropical forest, we engage in a one-time consumption of natural resources that will never be available again. The complex ecosystems that have been built up over billions of years are now being dismantled, in almost complete ignorance, by the human species.

What biologists must do is to learn more about the construction of sustainable agricultural ecosystems that will meet human needs in tropical and subtropical regions. The undisturbed tropical rain forest has one of the highest rates of net primary productivity of any plant community on earth, and it is therefore logical to assume that it can be harvested for human purposes in a sustainable, intelligent way.

The Loss of Biodiversity

The most serious and rapidly accelearating of all global environmental problems, is the loss of *biodiversity*. Based on the loss of species of well-known groups of organisms, including mammals, birds, and plants, over the past 300 years, and taking into account the rapid loss of habitat that is occurring at present, especially in the tropics, scientists such as E.O. Wilson of Harvard University have calculated that as much as 20% of the world's biodiversity may be lost during the next 30 years. Since we have named no more than 15% of the world's eukaryotic organisms, and a much smaller proportion of those in the tropics, it is obvious that we will not even know of the existence of many of the organisms that we are driving to extinction.

These losses will affect not only poorly known groups. As many as 50,000 species of the world total of 250,000 species of plants; 4000 of the world's 20,000 species of butterflies; and nearly 2000 of the world's 9000 species of birds could be lost during this short period of time. Considering that our species has been in existence for only 500,000 years of the world's 4.5 billion year history, and that our ancestors developed agriculture only about 10,000 years ago, this is an astonishing rate of loss.

This loss is important for several reasons. First, many feel that on moral, ethical, and aesthetic grounds, we do not have the right to drive to extinction such a high proportion of what are, as far as we know, our only living companions in the universe. We should take positive, international Biodiversity Convention, introduced at the Earth Summit in Rio de Janeiro in June 1993.

Second, organisms are our only means of sustainability. If we want to solve the problem of how to occupy the world on a continuing basis, it will be the properties of organisms that make it possible. As sources of food, medicine, clothing, biomass (for energy and other purposes), and shelter, organisms are the only sustainable source. We have examined only a minute proportion of the existing kinds of organisms to see whether their properties are of interest to us. Only during the past few decades have we discovered the techniques that make possible the transfer of genes from one kind of organism to another. We are jest beginning to be able to enumerate genetic differences between organisms and to use their genes for our advantage. We are killing species off at a rate that has not been approached for the past 65 million years. Far more species are in danger of extinction within our lifetimes than the total number that became extinct at the end of the Cretaceous period. One can scarcely think of anything that we could passively allow to happen that would be more detrimental to the human future.

Third, organisms occurring in communities function to preserve soils, regulate water and nutrient cycles that are essential to plants, modulate characteristics of the atmosphere, and absorb pollution. By destroying them in the relentless drive toward 'development,' we are creating conditions of instability and unproductivity, including desertification, waterlogging, mineralization, and many other undesirable outcomes throughout the world. If we know so little about the properties of the organisms themselves, we know much less about the ways in which ecosystems and communities function. We are destroying them at our own peril.

What biologists can do in the face of this crisis is to help design intelligent plans for locating those organisms most likely to be of use and saving them from extinction. They must also participate in sound, globally based schemes to preserve as much as possible the biological diversity of life on earth. With the loss of tropical forest and other biological communities throughout the world, we are permanently losing many opportunities not only for knowledge, but for increased prosperity. It is biologists who must understand this message and inform their fellow citizens of its importance to them.

ENVIRONMENTAL SCIENCE

Environmental scientists attempt to find solutions to environmental problems, considering them in a broad context. Unlike biology or ecology, which are sciences that seek to learn general principles about how life functions, environmental science is an applied science that is dedicated to solving problems. Its basic tools are derived from ecology, geology, meteorology, social sciences, and the many other areas of knowledge that bear on the functioning of the environment and our management of it. Environmental science addresses the problems created by rapid human population growth, an increasing need for energy, a depletion of resources, and a growing level of pollution. These problems are both unavoidable and obvious.

The problems faced by our severely stressed planet are not insurmountable. A combination of scientific investigation and public action, when brought to bear effectively, can solve environmental problems that seem intractable. How is success to be achieved? Viewed simply, there are five components to solve any environmental problem:

1. *Assessment*: The first stage in addressing any environmental problem is scientific analysis, the gathering of information. Data must be collected and experiments performed to construct a 'model' that describes the situation. Such a model can be used to make predictions about the future course of events;

2. *Risk Analysis*: Using the results of scientific analysis as a tool, it is possible to analyze what could be expected to happen if a particular course of action were followed. It is necessary to evaluate not only the potential for solving the environmental problem, but also any adverse effects that a plan of action might create. Often an environmental impact statement is prepared at this point;

3. *Public Education:* When a clear choice can be made among alternative courses of action, the public must be informed. This involves explaining the problem in terms the public can understand, presenting the alternative actions available, and explaining the probable costs and results of the different choices;

4. *Political Action:* The public, through its elected officials, makes a choice, selecting a course of action and implementing it. Choices are particularly difficult to implement when environmental problems transcend national boundaries.

5. *Follow-Through*: The result of any action taken should be carefully monitored to see whether the environmental problem is being solved and, more basically, to evaluate and improve the initial evaluation and modeling of the problem. Every environmental intervention is an

experiment, and we need the knowledge gained from each one for its future applications.

WHAT BIOLOGISTS HAVE TO CONTRIBUTE

The development of appropriate solutions to the world's environmental problems must rest partly on the shoulders of politicians, economists, bankers, engineers—many different kinds of people. However, the application of basic biological principles is vital to finding permanent solutions to these problems and achieving a stable, productive world.

It is clear that a scientific education has become necessary for everyone, so that we may understand the basis for our continued existence on earth and the steps that we will need to take to improve the quality of our lives. Biology should play a major part in that education. It is of critical importance in improving the standard of living for our fellow humans and in helping to guide us towards sustainability. Biological literacy is no longer a luxury for intelligent people who want to play a constructive role in improving the world; it has become a necessity for everyone.